Regulated Proteolysis in Microorganisms

SUBCELLULAR BIOCHEMISTRY

Recent Volumes in this Series

Volume 52	**A Handbook of Transcription Factors** Edited by Tim Hughes
Volume 53	**Endotoxins: Stricture, Function and Recognition** Edited by Xiaoyuan Wang and Peter J. Quinn
Volume 54	**Conjugation and Deconjugation of Ubiquitin Family Modifiers** Edited by Marcus Groettrup
Volume 55	**Purinergic Regulation of Respiratory Diseases** Edited by Maryse Picher and Richard C. Boucher
Volume 56	**Water Soluble Vitamins** Edited by Olaf Stanger
Volume 57	**Aging Research in Yeast** Edited by Michael Breitenbach, Michal S. Jazwinski and Peter Laun
Volume 58	**Phosphoinositides I: Enzymes of Synthesis and Degradation** Edited by Tamas Balla, Matthias Wymann and John D. York
Volume 59	**Phosphoinositides II: The Diverse Biological Functions** Edited by Tamas Balla, Matthias Wymann and John D. York
Volume 60	**Adherens Junctions: From Molecular Mechanisms to Tissue Development and Disease** Edited by Tony Harris
Volume 61	**Epigenetics: Development and Disease** Edited by Kundu and Tapas Kumar
Volume 62	**The Eukaryotic Replisome: a Guide to Protein Structure and Function** Edited by MacNeill and Stuart
Volume 63	**vGPCR Signalling Complexes – Synthesis, Assembly, Trafficking and Specificity** Edited by Dupré J. Denis, Hébert E. Terence and Jockers Ralf
Volume 64	**Reprogramming Microbial Metabolic Pathways** Edited by Wang Xiaoyuan, Chen Jian and Quinn Peter
Volume 65	**Protein Aggregation and Fibrillogenesis in Cerebral and Systemic Amyloid Disease** Edited by Harris J. Robin
Volume 66	**Regulated Proteolysis in Microorganisms** Edited by David A. Dougan

For further volumes:
http://www.springer.com/series/6515

David A. Dougan
Editor

Regulated Proteolysis in Microorganisms

Springer

Editor
David A. Dougan
Department of Biochemistry
La Trobe Institute for Molecular Science
La Trobe University, Melbourne,
VIC, Australia

ISSN 0306-0225
ISBN 978-94-007-5939-8 ISBN 978-94-007-5940-4 (eBook)
DOI 10.1007/978-94-007-5940-4
Springer Dordrecht Heidelberg New York London

Library of Congress Control Number: 2013934137

Printed on acid-free paper

Springer is part of Springer Science+Business Media (www.springer.com)

For Kaye and Matthew

Preface

All cells are composed of thousands of different proteins, each with a specific function. Collectively these proteins contribute to the proper function and maintenance of cells. As such it is not surprising, that regulating the integrity and concentration of each protein in the cell, not only under normal conditions but also under conditions of stress, is a fundamentally important biological process. For many years, it was believed that gene expression through regulated transcription and translation was primarily responsible for altering the abundance of individual proteins. Protein degradation was thought of only as a mechanism to recycle amino acids in a slow and somewhat non-selective manner. However, in the past 30 years, it has become evident that regulated protein degradation plays an important role in the cell's response to changing environmental conditions. Indeed in 2004 the world's attention was focussed on regulated proteolysis, when Aaron Ciechanover, Avram Hershko and Irwin Rose were awarded the Nobel Prize in Chemistry, for their fundamental discovery of Ubiquitin-mediated protein degradation. Although this research centred largely on regulated proteolysis in eukaryotes, it stimulated much research on related proteolytic systems in bacteria and other microorganisms. Indeed, during the past 10 years there have been numerous significant advances in this field.

The aim of this book is to highlight and compare the different proteolytic systems found in a selection of model and medically relevant microorganisms; from Gram-negative and Gram-positive bacteria (i.e. *Escherichia coli* and *Bacillus subtilis*, respectively), Archaea and *Saccharomyces cerevisiae*, to important pathogenic bacteria (i.e. *Mycobacterium tuberculosis*). The first chapter provides a general overview of the different proteolytic machines in *Escherichia coli*, focussing primarily on the mechanism of action of ClpAP and ClpXP (the two most extensively characterised AAA+ proteases) and the adaptor proteins that regulate substrate delivery to these machines. Chap. 2 takes an historical look at the first characterised, and most broadly conserved, ATP-dependent protease – Lon – and finishes with an elegant model for the allosteric-activation of protein degradation by this protease. Chap. 3 continues with a mechanistic analysis of the membrane bound ATP-dependent protease, FtsH. This chapter, also briefly examines the many physiological roles this protease plays, primarily focussing on its role in the regulation of lipid synthesis. Many of these proteolytic machines also play important physiological roles during conditions of environmental or proteotoxic stress. The next four chapters focus on the physiological role of these machines in controlling a variety of stress response pathways in model and pathogenic strains of bacteria. The many and varied roles of regulatory proteolysis in the model Gram-positive bacterium, *B. subtilis*, are discussed in Chap. 4, while the two subsequent chapters (Chaps. 5 and 6) examine the importance of regulatory proteolysis in controlling distinct stress response pathways in *E. coli*. Chap. 5 describes the role these machines play in regulating the heat-shock response and the general stress response, while Chap. 6 centres on the role of proteolysis in controlling of the envelope stress response. Chap. 7 continues with the theme of regulatory proteolysis, focussing on its contribution to virulence in a number of pathogenic strains of bacteria. The next part (Chaps. 8 and 9) highlight the role of regulated protein degradation in *Saccharomyces cerevisiae*. Chap. 8 focuses on a single AAA+ protein, Cdc48 – as a key regulator of intracellular protein degradation in yeast. Cdc48 is not only an important regulator of a number of proteasome-mediated degradation pathways, including endoplasmic reticulum associated degradation (ERAD), but also plays a crucial role in autophagy and endolysosomal protein degradation. Chap. 9 highlights the contribution of the different AAA+ proteases to protein homeostasis in mitochondria, focussing primarily on the role of Lon, *i*-AAA and *m*-AAA in yeast but also touches on the role of ClpXP in the mitochondrion of higher eukaryotes. Finally, the novel "ubiquitin-like" protein modifications that were recently discovered in *Mycobacterium* sp. and Archaea are covered in the last two chapters (Chaps. 10 and 11, respectively). Both chapters discuss the current understanding of these types of protein modification and their possible link to proteasome-mediated degradation. In *Mycobacterium* sp., the process of protein modification has been termed pupylation as it involves the attachment of a novel prokaryotic ubiquitin-like protein (PUP) to a protein substrate. Chap. 10 provides a comprehensive biochemical description of pupylation, and includes a detailed structural analysis of several diverse components involved in this pathway, including the proteasome. Like *Mycobacterium* sp., Archaea also contain a functional proteasome and an "ubiquitin-like" protein modification system. However in contrast to bacteria (i.e. *Mycobacterium* tuberculosis) and Eukaryota, protein modification in Archaea involves the attachment of a novel protein known as small archaeal modifying protein (SAMP). The final chapter (Chap. 11) describes our current understanding of this modification process in Archaea, by SAMP (termed sampylation) and although the physiological role of this process is currently unclear, this chapter reflects on the possibility that sampylation is linked to regulatory proteolysis. Collectively, the book provides a comprehensive guide to regulatory proteolysis in distinct organisms. It illustrates the diverse mechanisms that AAA+ protease machines have evolved to selectivity recognise proteins for degradation in a spatial and temporal manner, while avoiding the unregulated degradation of the vast and concentrated pool of proteins in the cell.

As a final note, I would like to thank each of the authors, firstly for the quality of the chapters they have contributed, but also for their patience during the production of this book. I would also like to sincerely thank the anonymous reviewers for their time, effort and invaluable expertise. I would also like to extend my thanks to Thijs van Vlijmen and Springer SBM for the opportunity to edit this book, it's been an incredible learning experience. My thanks also extend to all the members of my laboratory for their patience during the production of this book – undoubtedly, you will soon be wishing I was editing another one.

David A. Dougan

Contents

Contributors

Sarah E. Ades Department of Biochemistry and Molecular Biology, The Pennsylvania State University, University Park, PA, USA

Sarah E. Barchinger Department of Biochemistry and Molecular Biology, The Pennsylvania State University, University Park, PA, USA

Graduate program in BMMB, The Pennsylvania State University, University Park, PA, USA

Lone Brøndsted Department of Veterinary Disease Biology, Faculty of Life Sciences, University of Copenhagen, Frederiksberg C, Denmark

Alexander Buchberger Department of Biochemistry, Biocenter, University of Würzburg, Würzburg, Germany

K. Heran Darwin Department of Microbiology, New York University School of Medicine, New York, NY, USA

David A. Dougan Department of Biochemistry, La Trobe Institute for Molecular Science (LIMS), La Trobe University, Melbourne, VIC, Australia

Dorte Frees Department of Veterinary Disease Biology, Faculty of Life Sciences, University of Copenhagen, Frederiksberg C, Denmark

Eyal Gur Department of Life Sciences, Ben-Gurion University of the Negev

Department of Life Sciences, The National Institute for Biotechnology in the Negev, Beer-Sheva, Israel

Hanne Ingmer Department of Veterinary Disease Biology, Faculty of Life Sciences, University of Copenhagen, Frederiksberg C, Denmark

Huilin Li Department of Biochemistry and Cell Biology, Stony Brook University, Stony Brook, NY, USA

Brookhaven National Laboratory Biology Department, Brookhaven National Laboratory, Upton, Brookhaven, NY, USA

Julie A. Maupin-Furlow Department of Microbiology and Cell Science, University of Florida, Gainesville, FL, USA

Dimce Micevski Department of Biochemistry, La Trobe Institute for Molecular Science (LIMS), La Trobe University, Melbourne, Australia

Noël Molière Institute for Microbiology, Leibniz Universität Hannover, Hannover, Germany

Teru Ogura Department of Molecular Cell Biology, Institute of Molecular Embryology and Genetics, Kumamoto University, Kumamoto, Japan

Takashi Okuno Department of Material and Biological Chemistry, Faculty of Science, Yamagata University, Yamagata, Japan

Ralf Ottofueling Department of Biochemistry, La Trobe Institute for Molecular Science (LIMS), La Trobe University, Melbourne, Australia

Marie I. Samanovic Department of Microbiology, New York University School of Medicine, New York, NY, USA

Kaye N. Truscott Department of Biochemistry, La Trobe Institute for Molecular Science (LIMS), La Trobe University, Melbourne, VIC, Australia

Kürşad Turgay Institute for Microbiology, Leibniz Universität Hannover, Hannover, Germany

Wolfgang Voos Institut für Biochemie und Molekularbiologie (IBMB), Universität Bonn, Bonn, Germany

Linda A. Ward Department of Biochemistry, La Trobe Institute for Molecular Science (LIMS), La Trobe University, Melbourne, VIC, Australia

Part I
AAA+ Proteolytic Machines

Chapter 1
Machines of Destruction – AAA+ Proteases and the Adaptors That Control Them

Eyal Gur, Ralf Ottofueling, and David A. Dougan

Abstract Bacteria are frequently exposed to changes in environmental conditions, such as fluctuations in temperature, pH or the availability of nutrients. These assaults can be detrimental to cell as they often result in a proteotoxic stress, which can cause the accumulation of unfolded proteins. In order to restore a productive folding environment in the cell, bacteria have evolved a network of proteins, known as the protein quality control (PQC) network, which is composed of both chaperones and AAA+ proteases. These AAA+ proteases form a major part of this PQC network, as they are responsible for the removal of unwanted and damaged proteins. They also play an important role in the turnover of specific regulatory or tagged proteins. In this review, we describe the general features of an AAA+ protease, and using two of the best-characterised AAA+ proteases in *Escherichia coli* (ClpAP and ClpXP) as a model for all AAA+ proteases, we provide a detailed mechanistic description of how these machines work. Specifically, the review examines the physiological role of these machines, as well as the substrates and the adaptor proteins that modulate their substrate specificity.

E. Gur (✉)
Life Sciences Department, Ben-Gurion University of the Negev,
Beer-Sheva 84105, Israel

The National Institute for Biotechnology in the Negev,
Beer-Sheva 84105, Israel
e-mail: gure@bgu.ac.il

R. Ottofueling • D.A. Dougan (✉)
Department for Biochemistry, La Trobe Institute for Molecular Science,
La Trobe University, Melbourne 3086, Australia
e-mail: d.dougan@latrobe.edu.au

D.A. Dougan (ed.), *Regulated Proteolysis in Microorganisms*, Subcellular Biochemistry 66,
DOI 10.1007/978-94-007-5940-4_1,

General Introduction

The bacterial cytosol is a complex mixture of macromolecules (proteins, DNA and RNA), which perform a variety of different functions. Given that proteins play a central role in many of these important cellular tasks, their correct maintenance within the cell is critical for cellular viability, not only under normal cellular conditions but also under conditions of stress. As such, a bacterial cell contains a network of molecular chaperones and proteases (often referred to as the protein quality control (PQC) network) dedicated to maintaining homeostasis of protein folding. Chaperones function to protect functional proteins against unfolding and to refold misfolded and aggregated species. The role of proteases is to remove unwanted and hopelessly damaged proteins.

In the bacterial cytosol, protein degradation is performed mainly by a number of different ATP-dependent proteolytic machines. In general these machines are composed of two components, a peptidase and an unfoldase. Invariably, the unfoldase is a member of the AAA+(ATPase associated with diverse cellular activities) superfamily and as such these molecular machines are commonly referred to as AAA+proteases [1]. In Gram-negative bacteria, such as *Escherichia coli* there are generally five different AAA+proteases (ClpAP, ClpXP, HslUV, Lon (also refereed to as LonA) and FtsH). In contrast most Gram-positive bacteria, such as *Bacillus subtilis*, contain up to seven different AAA+protease (ClpCP, ClpEP, ClpXP, HslUV (CodXW), LonA, LonB and FtsH). Interestingly, in bacteria belonging to the Actinobacteria and Nitrospira phyla (e.g. in *Mycobacterium tuberculosis* (*Mtb*)) one or more of these AAA+proteolytic machines is replaced by the proteasome (for a detailed review of this AAA+machine, and its physiological role in *Mtb* please refer to Darwin and colleagues [2]). Regardless of their origin, these machines can be divided into two broad groups; those that contain the unfoldase and peptidase components on separate polypeptides (e.g. ClpAP, ClpCP, ClpEP, ClpXP and HslUV (CodXW)), and those that contain both components on a single polypeptide (e.g. LonA, LonB and FtsH) (see Fig. 1.1).

This review will focus on the "two-component" proteolytic machines, primarily those from *E. coli* (e.g. ClpAP and ClpXP), with a brief comparison to the equivalent machines (e.g. ClpCP and ClpXP) in the model Gram-positive bacterium, *B. subtilis*. However, for an extensive review on regulatory proteolysis in *B. subtilis* please refer to [3]. Likewise, for a detailed review on the "single polypeptide" proteases, i.e. Lon and FtsH please refer to [4] and [5], respectively.

Structure and Function of the "ClpP Containing" Proteases (ClpAP, ClpXP and ClpCP)

As mentioned above, bacteria contain a wide variety of different proteolytic machines, of which ClpXP is certainly the best-studied AAA+protease [1]. ClpXP is known to play a number of critical roles in a wide variety of bacterial species, from the control of different stress response pathways in Gram-positive and Gram-negative

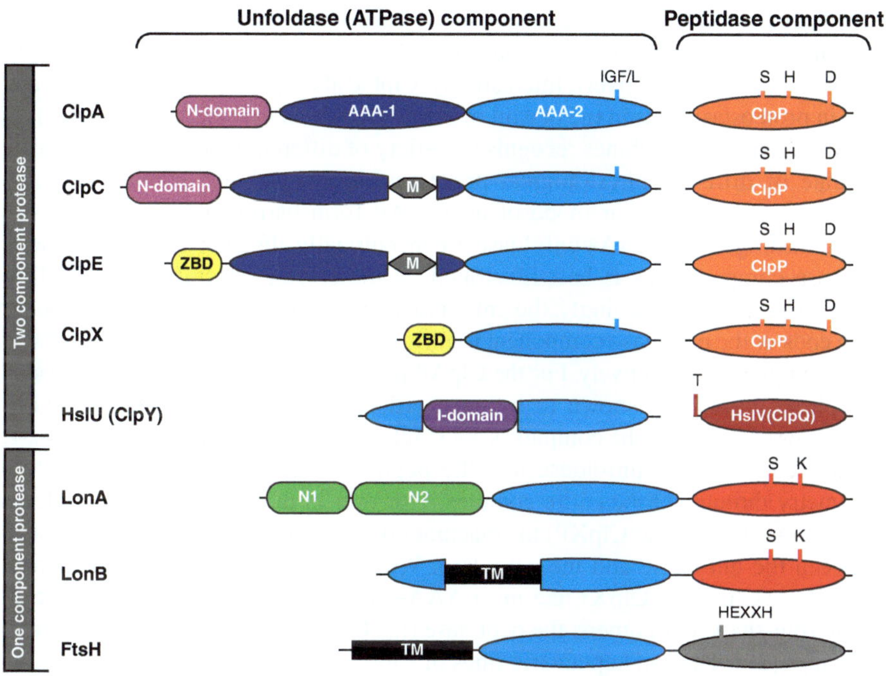

Fig. 1.1 Cartoon representation of the various different AAA+ proteases in bacteria. AAA+ proteases can be separated into two different groups. Two component proteases (e.g. ClpAP, ClpCP, ClpXP, ClpEP and HslUV) contain the unfoldase and peptidase components on separate polypeptides. One component proteases, contain the peptidase and the unfoldase on a single polypeptide (e.g. LonA, LonB and FtsH). The unfoldase component contains one or more AAA+ domains, responsible for ATP-dependent unfolding of the substrate. All unfoldase components also contain at least one accessory domains (e.g. ClpA and ClpC contain a conserved N-terminal domain (N-domain, *pink*), ClpC and ClpE contain a middle domain (M, *grey*), ClpE and ClpX contain a Zinc binding domain (*ZBD, yellow*), HslU contains an accessory domain inserted into the AAA+ domain (I-domain, *purple*), LonA contains two N-terminal domains unrelated to the N-domain of ClpA and ClpC (N1 and N2, *green*), while LonB and FtsH both contain a single transmembrane (TM) region), which serve various different functions (see main text for details). In the case of the ClpP-binding unfoldase components, the AAA-2 domain contains an IGF/L loop for interaction with ClpP. The protease components are responsible for cleavage of the unfolded substrate. In the case of ClpP, hydrolysis of the polypeptide is catalysed by the catalytic triad (S, H and D), while FtsH and HslV contain either a conserved HExxH motif or an N-terminal threonine (T) respectively, for peptide bond cleavage

bacteria (see [6, 7]) to the regulation of virulence through the degradation of key factors that control virulence (see [8]). ClpXP has also been shown to play an important role in regulating mitochondrial protein homeostasis (proteostasis) in eukaryotes such as worms [9, 10] and plants [11]. Surprisingly however, this proteolytic machine is absent from most fungi including, *Saccharomyces cereviseae* [12, 13]. For a detailed review of about the role of these AAA+ proteases in regulating mitochondrial function please refer to [14]. Although the AAA+ proteases ClpAP and

ClpCP are not as widely conserved as ClpXP, these proteases do, nevertheless, control a number of key proteolytic/regulatory pathways in Gram-negative and Gram-positive bacteria, respectively. Interestingly, ClpCP also appears to play an important role in proteostasis within the plastid of plants (for a recent review see [15, 16]).

Although these machines recognise a variety of different substrates and regulate a range of different physiological processes, each machine shares a common architecture and a similar mode of action. All form barrel-shaped complexes in which the oligomeric AAA+ unfoldase is concentrically aligned with the oligomeric protease component as is best illustrated by the crystal structure of the HslUV complex [17, 18]. Interestingly, the unfoldase component may be located at either or both ends of the peptidase component to form single-headed (1:1) or double-headed (2:1) complexes, respectively. For the ClpAP protease, the symmetric double-headed complexes have been shown to be most efficient at processing substrates [19]. Regardless of whether the complexes are single- or double-headed, both oligomeric components (i.e. the unfoldase and the peptidase) generally exhibit a six-fold symmetry throughout the entire complex. However in the Clp protease complexes (e.g. ClpAP, ClpCP and ClpXP) the machines display a unique symmetry mismatch between the unfoldase and the peptidase. While the AAA+ unfoldase component (i.e. ClpA, ClpC and ClpX) like most AAA+ proteins studied to date, form hexameric ring-shaped oligomers, the peptidase (i.e. ClpP) is composed of two heptameric rings [20]. The two heptameric rings of ClpP stack back-to-back, encapsulating the catalytic (active site) residues of ClpP within a barrel shaped tetradecamer. This symmetry mismatch poses some interesting questions. How do these two rings (the hexameric unfoldase and the heptameric peptidase) interact to form a functional complex, and how many subunits are required for a functional interaction. Regardless of whether the protease complex is symmetric or asymmetric, all AAA+ proteases undergo three basic steps in order to degrade a substrate protein (see Fig. 1.2). In the first step, the substrate is recognised by the unfoldase, although in some cases substrate recognition may be facilitated by an adaptor protein (see later). In bacteria, substrates are usually recognised via short sequence specific motifs (termed degrons), which are often located at the N- or C-terminus of the substrate protein. Following recognition, the substrate is then unfolded in an ATP-dependent fashion (Fig. 1.2, step 2). The unfolded substrate is then translocated into the associated peptidase, where the polypeptide chain is hydrolysed into small peptide fragments (~3–8 amino acids long), which have been proposed to egress through the holes in the sidewall of the peptidase, although this method of egress remains somewhat controversial (Fig. 1.2, step 3) [21, 22].

The Peptidase ClpP

The ClpP peptidase is synthesized as a zymogen, containing a N-terminal propeptide [23], which is autocatalytically cleaved upon oligomerization, resulting in the formation of a proteolytically active oligomer. ClpP is a serine protease, composed of a

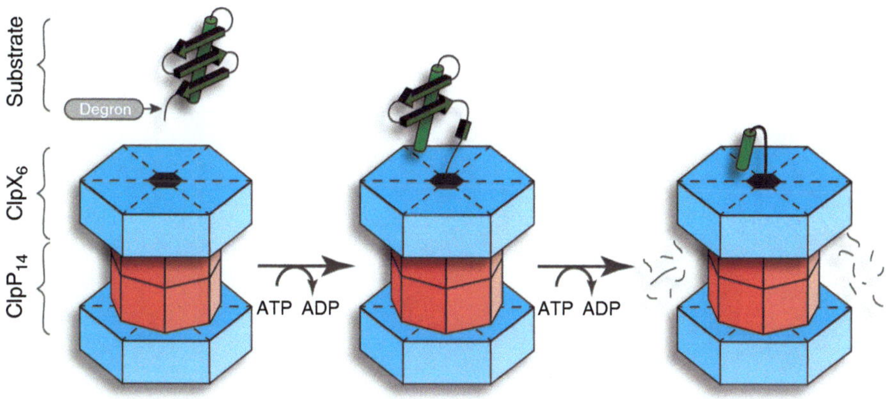

Fig. 1.2 Cartoon illustrating the main steps involved in substrate recognition and degradation by AAA+ proteases. The unfoldase (e.g. ClpX) forms a hexameric ring-shaped structure (*blue*) at one or both ends of the peptidase (e.g. ClpP), which forms two heptameric rings stacked back-to-back (*red*). The substrate (*green*) contains a degradation signal (*degron*) often located at the N- or C-terminus of the protein. The degron is recognised by the unfoldase and the substrate protein unfolded, in an ATP-dependent fashion, then translocated into the peptidase where the protein in cleaved into small peptide fragments, which diffuse through holes in the side-wall of the peptidase

Ser-His-Asp catalytic triad (Fig. 1.1), which exhibits chymotrypsin-like activity, that is, it cleaves peptide bonds mostly after non-polar residues [24, 25]. The active peptidase is a barrel-shaped oligomer composed of two heptameric rings, stacked back-to-back [20], that forms a degradation chamber in which the proteolytic active sites are sequestered away from cytosolic proteins (Fig. 1.3a). Each monomer of ClpP resembles a hatchet and consists of three subdomains: a handle, a globular head and a N-terminal loop. The heptameric ring is formed by the interaction of seven subunits through the head subdomain, and the tetradecamer is formed by the interaction of two heptameric rings through the handle subdomain (Fig. 1.3a). Entry into the catalytic chamber of this serine peptidase is restricted to a narrow entry portal (~10 Å) at both ends of the barrel-shaped complex. The N-terminal peptides of ClpP flank the axial pore and are proposed to act as a gate for entry into the proteolytic chamber. As a result of this narrow axial entry portal, folded proteins are excluded from entering the catalytic chamber, although small peptides and unfolded proteins can be degraded in an ATPase independent fashion, albeit unfolded proteins are degraded very slowly in the absence of the ATPase [26]. Importantly, the degradation of unfolded substrates can be accelerated by the addition of a cognate unfoldase (i.e. ClpX, ClpA or ClpC), which implies that entry into ClpP is gated and that this gated-entry can be activated by the unfoldase. Indeed, recent cryo-EM reconstructions have shown that binding of ClpA triggers a change in the N-terminal loops of ClpP, from a "down" conformation where they block entry to the catalytic chamber, to an "up" conformation which permits access to the chamber [27]. Consistent with a "gating" role for the N-terminal loops of ClpP, deletion of these loops was

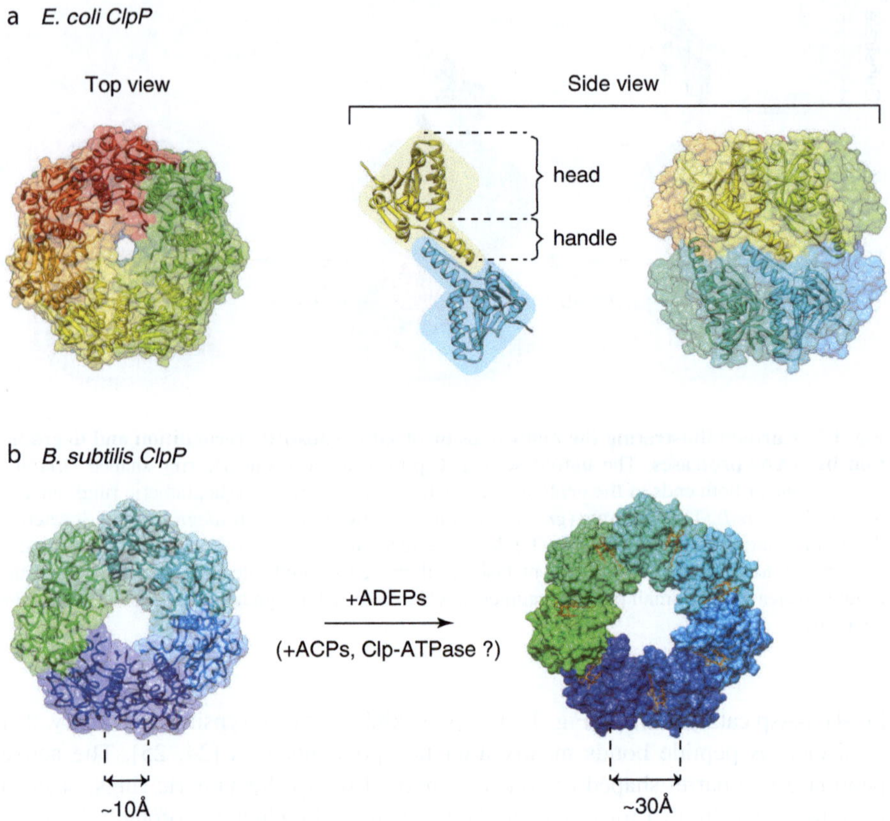

Fig. 1.3 Oligomeric structure of ClpP. (**a**) ClpP (PDB: 1TYF) forms two heptameric ring-shaped oligomers (*Top view*) stacked back-to-back (*Side view*) to create a barrel-shaped oligomer. Interactions between adjacent head subdomains drive oligomerisation of the seven-membered ring, while interactions between the handle subdomain of two heptamers are responsible for formation of the tetradecamer. (**b**) In the absence of the unfoldase, the entry portal into the catalytic chamber of ClpP (PDB: 3KTH) is narrow (~10 Å), in the presence of chemical activators of ClpP (i.e. ADEPs, ACPs and potentially the unfoldase), the entry portal into the catalytic chamber of ClpP (PDB: 3KTI) is opened (~30 Å)

shown to accelerate the degradation of short peptides [28]. The cognate AAA+ unfoldase also mediates the degradation of folded substrate proteins by actively unfolding and translocating the substrates through the axial pore and into the proteolytic chamber of ClpP. Indeed, it appears that the oligomeric structure of ClpP has been carefully designed to prevent widespread and indiscriminate degradation of cellular proteins by regulating substrate access to its proteolytic chamber. Consistent with this idea, several recent studies have identified a series of novel antibiotics (e.g. acyldepsipeptides (ADEPs) and ACPs) that activate ClpP (in the absence of its cognate unfoldase) for unregulated protein degradation [29–34]. This activation of ClpP results in the unregulated degradation of nascent polypeptides and unfolded proteins in the cell [34], and in a recent study ADEP was shown to inhibit cell division of Gram positive bacteria, through the ClpP-mediated degradation of FtsZ, a key protein required for septum formation [35].

Based on a series of biochemical and structural studies, these chemical activators of ClpP dock into a hydrophobic pocket located on the surface of ClpP (Fig. 1.3b). Firstly, and most importantly, ADEP binding to this hydrophobic pocket results in opening of the ClpP pore (from ~10 Å in the absence of ADEP to ~21–27 Å in the presence of different forms of ADEP). This "gated-opening" of the ClpP pore, is proposed to be sufficient to allow entry of unfolded proteins into the proteolytic chamber of ClpP (where the catalytic residues are located) and possibly the primary reason for degradation of unfolded substrates. Interestingly, in the case of *B. subtilis* ClpP, ADEP not only triggers opening of the pore, but also triggers oligomerisation of ClpP from free "inactive" monomers to "active" tetradecamers [32], a step that is normally controlled by the cognate unfoldase, ClpC [36]. Similarly, ADEP activation of human ClpP for unregulated degradation is also likely to result from assembly of the ClpP tetradecamer [37] a process that normally requires the assistance of ClpX [38]. As a consequence, ADEP also appears to be a competitive inhibitor of unfoldase binding to ClpP, preventing the regulated degradation of substrates that would normally be delivered to ClpP by the unfoldase component [32]. As such, the ADEP-bound conformation of ClpP has been proposed to mimic the unfoldase-bound conformation of ClpP. Surprisingly, binding of ClpA to ClpP, as measured from sections of the ClpAP cryo-EM structure, appears to have little effect on the size of the ClpP pore (diameter ~12 Å) [27] and hence it has been suggested that the size of the pore may vary with translocation of different substrates [39]. Nevertheless, it remains to be seen, if an ordered arrangement of the N-terminal loops on ClpP (as observed in the *B. subtilis* ClpP-ADEP structure) or a disorder arrangement of the N-terminal loops of ClpP (as observed in the *E. coli* ClpP-ADEP complex) resembles the unfoldase bound complex.

The Unfoldase Components (ClpX/ClpA/ClpC)

In *E. coli*, ClpP forms proteolytic complexes with both ClpA and ClpX, while in *B. subtilis*, ClpP associates with three different unfoldases, ClpC, ClpX and ClpE [3]. Although the overall architecture of the different unfoldase components is similar, each unfoldase contains a unique organisation. While ClpA, ClpC and ClpE each contain two AAA+ domains, ClpX only contains a single AAA+ domain (Fig. 1.1). Regardless of the number of AAA+ domains present, each unfoldase contains one or more accessory domains. In the cases of ClpA and ClpX, a single accessory domain is located at the N-terminus of the protein, while both ClpC and ClpE contain two accessory domains, one at the N-terminus of the protein and the other located between the two AAA+ domains, termed the middle or M-domain (Fig. 1.1). In general, these accessory domains are required for the binding of substrates and/

or adaptor proteins. In the case of ClpA, the N-terminal domain is essential for docking of the adaptor protein ClpS [40–42] but also required for the recognition, and hence degradation of some substrates [43, 44]. Similarly, the N-terminal domain of *B. subtilis* ClpC is essential for the ClpP-mediated degradation of most substrates [45, 46]. However in this case, the N-domain is thought not to be directly involved in substrate recognition but rather plays a crucial role in binding adaptor proteins (i.e. MecA and McsB), which are required for ClpC oligomerisation and/or substrate delivery [36, 47, 48]. Interestingly in the case of *B. subtilis* ClpC, the second accessory domain (the M-domain) located between the two AAA+domains, also plays an important role in the recognition of adaptor protein, however the details of substrate delivery by these adaptor proteins is currently unknown [36, 46–48]. For further details regarding the mechanism of action of ClpCP please refer to [3].

In the case of ClpX (and ClpE from Gram-positive bacteria) the N-terminal accessory domain (residues 1–60 in *E. coli* ClpX) is a C4-type Zinc binding domain (ZBD), which contains four Cysteine residues that coordinate a single Zn atom. In *E. coli* ClpX, this domain forms a very stable dimer [49], and is responsible for the recognition of several substrates (such as λO and MuA) but not SsrA-tagged proteins [50–52]. This domain is also essential for the recognition of the adaptor proteins, SspB [50, 52, 53] and UmuD [54], discussed in more detail later.

Given that *E. coli* ClpX is, by far the most extensively characterised Clp-ATPase, this section will focus primarily on the structure and function of ClpX. However, many of the features described here for the AAA+domain of ClpX are likely to be generally applicable to most AAA+proteases. At a structural level, the AAA+domain (~200–250 a.a.) is composed of two subdomains – a large N-terminal subdomain, which forms an α/β wedge-shaped Rossman fold and a small C-terminal subdomain, which forms a α-helical lid across the nucleotide-binding site [55, 56]. ATP is bound in a cleft between the large and small subdomain of a single subunit and the large subdomain of the adjacent subunit. As such, these interactions provide much of the driving force for formation of the hexamer. To date, several highly conserved sequence motifs have been identified within the AAA+domain, each of which is responsible for a specific function [57]. The Walker A motif (GXXXXGK [T/S], where X=any amino acid) is required for ATP binding and facilitates oligomerization of the protein into ring-shaped hexamers. The Walker B motif (hhhhDE, where h=any hydrophobic amino acid) is required for hydrolysis of bound ATP and hence drives conformational changes in the protein, mediating substrate binding and translocation. The central pore of the hexamer is comprised of several important motifs and loops (e.g. the pore-1 loop) involved in substrate binding [58–61]. The Sensor 1 and 2 motifs, together with the arginine fingers, are proposed to couple the nucleotide-bound state of the oligomer with conformational changes in the subdomains, which through movement of the substrate-binding loops, results in substrate unfolding and translocation [55, 58]. Despite the broad sequence conservation of AAA+domains, individual AAA+domains appear to serve different functions in proteins that contain two or more AAA+domains (i.e. ClpA or ClpC) [62]. For example, the first AAA+domain (D1) in ClpA is crucial for oligomerisation while the second AAA+domain (D2) is primarily responsible for ATP hydrolysis [63]. Interestingly, variants of ClpA lacking ATPase activity in either D1 or D2, are only able to process substrates with "intermediate" or "low" local stability respectively, suggesting that each domain can function independently, at least to a limited extent [64]. However, the ATPase activity of both domains is required for the efficient processing of substrates with "high" local stability [64] indicating that both domains work together to unfold and translocate substrates into ClpP.

As viewed from the top (or ClpP distal face) of the unfoldase, the ClpX hexamer can be divided into six units, each of which was composed of a small AAA+ subdomain from one subunit with a large AAA+ subdomain of the adjacent subunit [55, 56]. Recently, it was shown that the structures of all six of these units were highly superimposable [55] and hence it was proposed that each unit forms a functional rigid body (Fig. 1.4a, b). Despite the high degree of structural similarity between each rigid body unit, the overall shape of the ClpX hexamer is asymmetric, which suggests that the angle of the hinge between the rigid body units (i.e. the angle between the large and the small subdomains within a single subunit of ClpX) varies. This difference in the angle between the rigid body units results in a different ability of each subunit to bind nucleotide. Based on this description, each subunit within the ClpX hexamer can be classified into one of two groups; type 1 subunits, which are able to bind nucleotide (referred to as L, for "loadable"), and type 2 subunits, which are unable to bind nucleotide (referred to as U, for "unloadable"). In the crystal structure of ClpX, the hexamer is composed of four L (or type 1) subunits and two U (or type 2) subunits arranged in the following manner, L-L-U-L-L-U (Fig. 1.4c). Therefore, given that ATP binding and hydrolysis is expected to stabilise the L conformation, while the release of ADP is predicted to result in an transition from the L to the U conformation, it is proposed that the ATPase activity of ClpX will promote domain rotations within a subunit that will propagate around the hexamer and drive transition of the other subunits, in a chain reaction. These ATPase-induced conformational changes are proposed to form an integral part of the mechanism for substrate translocation by ClpX into ClpP (see later).

The Unfoldase-Peptidase Complex

Given that the AAA+ unfoldase component (i.e. ClpX, ClpA or ClpC) is hexameric and the associated peptidase (e.g. ClpP) is formed by two heptameric rings, the resulting proteolytic machines, ClpXP (ClpAP and ClpCP), exhibit an asymmetry between the two components. This asymmetry, although not unique in biology, poses several interesting questions. How do the two components interact with one-another? How many of these features per hexamer (i.e. how many subunits) are required for formation of a functional complex? Not surprisingly, the formation of the complex is transient, and efficient interaction of the two components is dependent on nucleotide-bound state of the unfoldase. Specifically, formation of the ClpXP complex is only supported by ATP, ATPγS (a slowly hydrolysable analogue of ATP) or a ClpX mutant that is defective in ATP hydrolysis [65]. In contrast, the complex

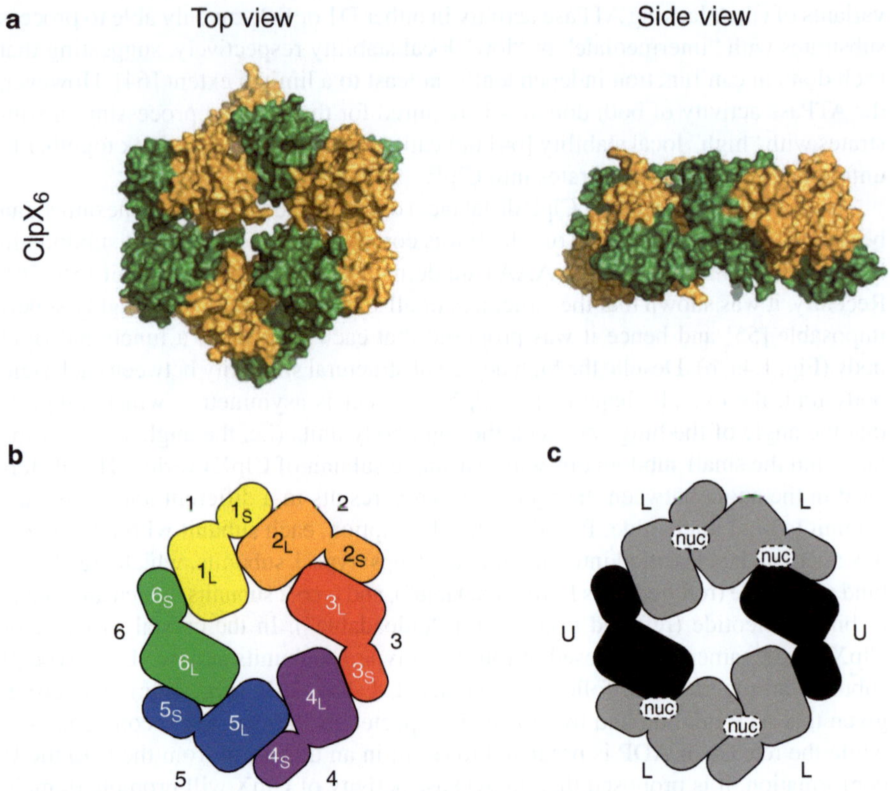

Fig. 1.4 Oligomeric structure of ClpX. In the presence of nucleotide, ClpX forms a hexameric *ring-shaped* oligomer. (**a**) Surface representation of the ClpX hexamer (PDB: 3HWS). (**b**) Cartoon, illustrating the asymmetric organisation of the ClpX hexamer. (**c**) The asymmetric organisation of the ClpX hexamer results from a differential binding of nucleotide (nuc) within the hexamer. Nucleotides are bound in a cleft formed by the large and small domain of one subunit and the large domain of the adjacent subunit. Depending on the orientation of the small and large domain within a subunit, a subunit can be classified into two types; loadable (L) which are able to bind nucleotide and unloadable (U) which are unable to bind nucleotide. The arrangement of these different subunit types, within the ring gives rise to an asymmetric appearance of the hexamer

dissociates in the presence of ADP or in the absence of nucleotide [66, 67]. This interaction, (i.e. between the two components), is mediated by two sets of contacts; one at the periphery of the interface and the other near the central pore. The peripheral contact occurs between a flexible loop on ClpX and a hydrophobic pocket on the surface of ClpP, and is important for a strong, nucleotide-independent interaction with ClpP. The flexible loop contains a conserved tripeptide motif ([L/I/V]-G-[F/L]) and as such is often referred to as the IGF/L-loop (Fig. 1.5a). This motif is unique to ClpP-binding unfoldases (i.e. ClpA, ClpC, ClpE and ClpX) and is essential for interaction with ClpP [68, 69]. Consistently, mutation of this motif dramatically reduces the affinity of ClpX to ClpP [67, 68]. The second contact

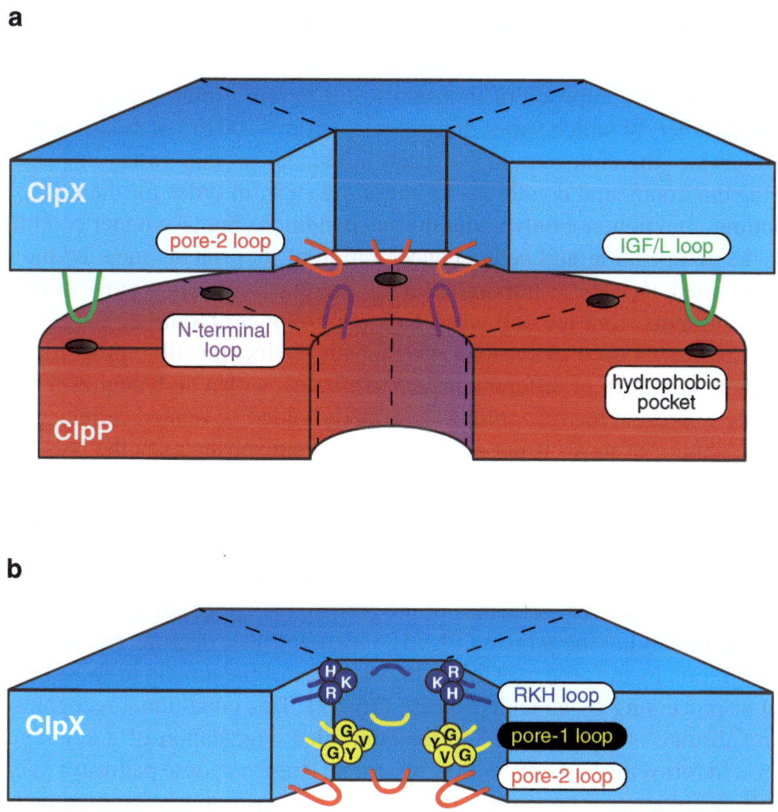

Fig. 1.5 ClpP-binding and substrate interaction is mediated by several loops and pockets. (**a**) Cut-away view of ClpX (*blue*), highlighting the important interactions that contribute to complex formation with ClpP (*red*). The IGF/L loops (*green*) on ClpX form a static interaction with the hydrophobic pocket on ClpP (*black*). ClpXP complex formation is modulated by the nucleotide state of ClpX, through a set of dynamic interactions, between pore-2 loops of ClpX (*red*) and the N-terminal loop of ClpP (*purple*). (**b**) The substrate is recognised and translocated through the pore via a set of conserved pore loops; RKH (*blue*), pore-1 (*yellow*) and pore-2 (*red*). These loops move up and down the pore of ClpX in a nucleotide-dependent fashion, thereby translocating the substrate into ClpP

occurs between two loops; one loop (termed the pore-2 loop) protrudes from the axial pore of ClpX, and interacts with the N-terminal loop of ClpP [21, 70, 71]. This interaction, between the two axial loops, appears to be highly dynamic and is dependent on the nucleotide-state of individual subunits of ClpX [71]. Although the ClpXP complex is asymmetric, both sets of loops (the IGF/L-loop, for docking into the hydrophobic pocket on ClpP and the two axial pore loops) appear to be flexible enough that contacts from each subunit of ClpX contribute to the interaction. Indeed loss of a single IGF-loop, within the ClpX hexamer, is sufficient to reduce ClpP binding and activity, while loss of more than one contact per hexamer completely abolishes ClpP binding [71].

Degradation Recognition Motifs (Degrons)

A bacterial cell is composed of thousands of different proteins, the concentration (or copy number) of which varies dramatically (from ~100 to 10^5 molecules per cell) [72]. Likewise, the concentration of each individual protein varies in response to changing environmental conditions or stress. As such, in order for the cell to maintain optimal function, not only under normal conditions but also under conditions of stress, the composition and active concentration of its proteins must be monitored and maintained. Hence it is important for the cell to specifically remove unwanted or damaged proteins from the cell when they are no longer required. To achieve this, bacterial proteases need to combine two seemingly incompatible properties, broad recognition of a range of different protein substrates, with a high degree of substrate specificity to prevent the recognition of properly folded or wanted cellular proteins.

A key feature of most, if not all, bacterial protein substrates is the presence of a specific amino acid motif, often referred to as a degradation tag or degron [73]. These degrons are generally located at the N- or C-terminus of the protein, although in some cases they are located internally. Although most degrons are intrinsic to the target protein, a handful of degrons (e.g. the SsrA tag and some N-end rule substrates) are not defined by the primary sequence of the protein, but rather are added (either co- or post-translationaly) to the protein [74, 75]. Often, intrinsic degrons are only revealed (for recognition by the protease) following exposure of the protein to stress (e.g. heat-shock) or processing by an endoprotease [76–79]. This conditional recognition of a protein substrate is ideally suited to the controlled degradation of a key regulatory protein, and forms the basis of controlling several stress response pathways in bacteria (see [6]). In some cases however, a degron may be constitutively exposed under normal conditions, in order to maintain low levels of the protein (e.g. SigmaS) [80].

Trans-translation and the SsrA-Tag: A Specific Protein Tagging System in Bacteria

Messenger RNA molecules normally contain a stop codon at the 3′ end of the transcript, which serves not only to signal the end of translation, but also triggers ribosome dissociation. In some cases however, as a result of truncation of the mRNA or errors during its transcription, the lack of a stop codon in the mRNA sequence caused "stalling" of protein synthesis [81–83]. To overcome this problem, bacteria possess a conserved mechanism, to restart translation and allow ribosome dissociation. This mechanism (illustrated in Fig. 1.6), often referred to as trans-translation, is sensed by an empty A-site and signalled by stalling of the translating ribosome [84]. This signal results in the recruitment of a specialised RNA molecule into the empty A-site of the ribosome. This RNA, encoded by *ssrA* (small stable RNA gene A) [85] has been termed a tmRNA as it functions both as a tRNA and as an mRNA [84, 86, 87]. The tRNA-like structure can be charged with alanine at its 3′ end, while an extended

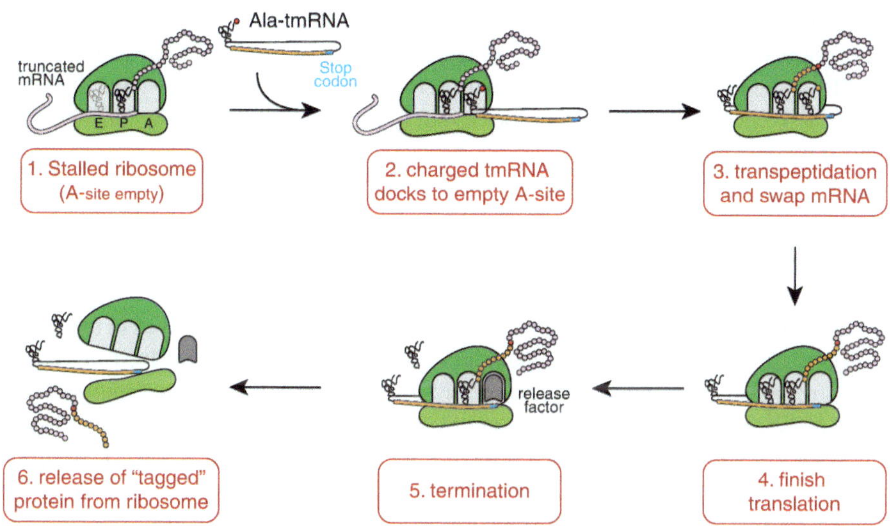

Fig. 1.6 Cartoon, illustrating the process of trans-translation. *1.* Truncated mRNA (lacking a stop codon) cause "stalling" of the ribosome. *2.* This "stalling" triggers binding of a tmRNA into the empty A-site of the ribosome. *3.* Following a transpeptidation reaction, the truncated mRNA is replace with the mRNA from the tmRNA and *4.* translation proceeds, resulting in *5.* correct termination of protein synthesis *6.* rescuing the ribosome and releasing the "tagged" protein for targeted degradation by ClpXP

loop within the same RNA molecule encodes a short open reading frame (ten amino acids in *E. coli*) that ends in a stop codon. Following docking of the charged tmRNA into the empty A-site, the alanine is transferred to the nascent polypeptide and the open reading frame (encoded by the mRNA portion of the tmRNA) is translated. Noteworthy, trans-translation results in the attachment of a short C-terminal extension (termed the SsrA tag) to the incompletely synthesised protein.

Importantly, given that SsrA-tagged proteins are produced from aberrant or incomplete mRNA, it is unlikely that they will be able to fold. For this reason, interaction of SsrA-tagged proteins with chaperones is wasteful, as attempts to refold trans-translation products would be futile. Rather, SsrA-tagged proteins are rapidly degraded by proteases. In *E. coli*, the SsrA tag is 11 amino acids long (AANDENYALAA) and substrates tagged with the sequence are recognised by ClpXP, ClpAP and FtsH [81, 88–90]. Despite the fact that the SsrA tag is recognised by several different proteases *in vitro*, the *in vivo* degradation of these substrates is almost exclusively performed by ClpXP [81, 91].

Nevertheless, this tag has been used extensively as a model degron to study the function of both ClpXP and ClpAP. As such, it has proved to be a powerful research tool to study the mechanism of protein recognition and degradation by AAA+ proteases. A major advantage of the SsrA tag, as a research tool to study protein degradation, is that any protein can be converted into a ClpXP (or ClpAP) substrate, simply through the attachment of the SsrA tag to its C-terminus. This has permitted a detailed mechanistic analysis of protein degradation using a range of different substrates with a variety of unique or desired features (i.e. green fluorescent protein (GFP) or the I27 domain of the human titin) to examine unfolding [92–95]. Likewise, it has also served as an excellent tool to study the mechanism of adaptor-mediated substrate delivery (see below).

Other ClpX Recognition Motifs

Apart from the specific recognition of the SsrA-tag, ClpX is also involved in the recognition of several other proteins, including a number of proteins involved in various stress response pathways. In order to determine the complete substrate-binding repertoire of *E. coli* ClpX, a mutant version of ClpP was used to capture the physiological substrates of ClpXP *in vivo* [96]. Using this approach, ~100 putative ClpXP substrate proteins were identified [96, 97]. Following verification of several of these proteins (either by *in vitro* or *in vivo* degradation assays) five different ClpX "recognition" motifs were proposed [96]. Of the five different "recognition" motifs, two were located near the C-terminus of the protein and three near the N-terminus of the protein (Fig. 1.7). While both classes of C-terminal motifs (C-motif 1 and 2, Fig. 1.7) shared homology with known ClpXP substrates (i.e. the SsrA-tag and MuA, respectively), only a single N-terminal motif (N-motif 1, Fig. 1.7) had been observed previously (i.e. λO) [98].

Interestingly, the various degradation motifs appear to be recognised by different regions within the unfoldase. Some substrate classes (e.g. N-motif 1) strictly depend on interaction with the N-terminal domain, while other motif classes (e.g. C-motif 1, i.e. SsrA-tagged substrates) do not require this domain for direct recognition [50, 52, 69]. For example, λO (a replication protein of bacteriophage λ) carries an N-terminal degradation motif (N-motif 1, NH_2-TNTAKI), which is specifically recognised by the N-terminal domain of ClpX [52, 96, 99]. Indeed deletion of this domain (from ClpX) inhibits the ClpP-mediated degradation of λO [52], which is proposed to result from the low affinity of this class of substrate to the axial loops on ClpX. Tethering of this class of substrate, by the N-terminal domain, is likely to increase the effective concentration of the substrate, near the pore of ClpX. As a result, despite their low affinity to the pore loops, high affinity to the N-terminal domain promotes their engagement by the pore and, consequently, their efficient degradation. The N-terminal domain is also involved in the recognition of the adaptors proteins, SspB and UmuD, and substrate proteins such as MuA (C-motif 2, Fig. 1.7), which appear to share a conserved motif [50, 52, 54]. Importantly however, the adaptor proteins are not degraded by ClpXP, presumably because they are not recognised by the pore-1 motif of ClpX.

Other Degradation Tags

Currently, the substrate recognition motifs for ClpA are only poorly defined. The first ClpAP substrate to be identified was ClpA itself [100]. Interestingly, although the recognition motif within ClpA was originally proposed to be located at the

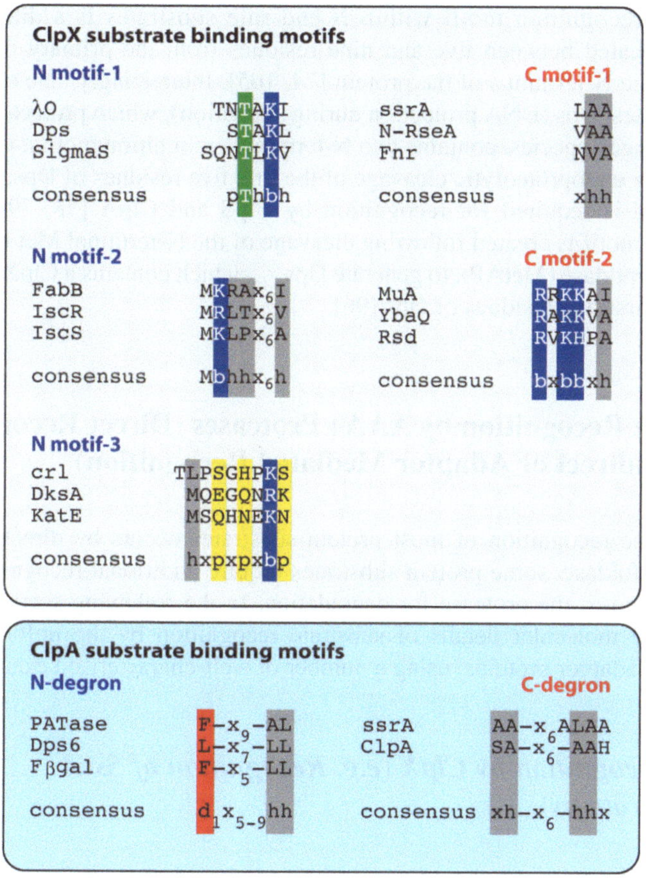

Fig. 1.7 Substrate-binding motifs for ClpX and ClpA. In general, AAA+proteases recognise either the N- or C-terminus of a substrate, as such several motifs have been defined for both ClpX and ClpA. ClpX Substrate recognition by can be divided into five broad groups, three N-terminal motifs (*N motif-1, -2 and -3*) and two C-terminal motifs (*C motif-1 and -2*). In contrast only two ClpA recognition motifs have been observed for ClpA (N-degron and C-degron)

N-terminus of ClpA, it was later shown to be C-terminal [101]. Interestingly, this motif within ClpA shares some similarity with the, well characterised, model degron – the SsrA tag (Fig. 1.7). ClpA has also been shown to recognise proteins via an N-terminal recognition motif but not an internal motif [102, 103]. The N-terminal recognition motifs can be classified into two groups, those that require the adaptor protein (ClpS) – N-end rule substrates, and those that do not. Currently, only a single substrate containing an N-terminal recognition sequence has been identified [104], and consequently a motif has not been defined. In contrast, several N-end rule substrates, both natural and model substrates have been identified and hence a motif for ClpA binding of these substrates has been proposed [74, 78].

The ClpA recognition motif within N-end rule substrates is a dihydrophobic element, located between five and nine residues from the primary destabilising residue at the N-terminus of the protein [74, 105]. Interestingly, one of the N-end rule substrates, Dps (DNA protection during starvation), which protects DNA from reactive oxygen species, contains two N-terminal recognition motifs. One motif is created after endoproteolytic cleavage of the first five residues of Dps, to generate Dps_{6-167} and is required for recognition by ClpS and ClpA [78, 79], the other N-terminal motif is created following cleavage of the N-terminal Met by methionine aminopeptidase (MetAP), to generate Dps_{2-167} which contains a ClpX (Nmotif-1) within the first five residues of Dps [96].

Substrate Recognition by AAA+ Proteases (Direct Recognition Versus Indirect or Adaptor Mediated Recognition)

Although the recognition of most protein substrates occurs by direct interaction with the unfoldase, some protein substrates require additional recognition factors to direct them to the protease for degradation. In the following sections, we will describe the molecular details of substrate recognition by the unfoldase and/or delivery by adaptor proteins, using a number of well-characterised examples.

Direct Recognition by ClpX (e.g. Recognition of SsrA Tagged Proteins)

In *E. coli*, the SsrA tag is composed of 11 amino acids (AANDENYALAA), however recognition of this tag by ClpX, only requires the last two alanines and the C-terminal α-carboxylate [106]. In contrast to some ClpX substrates (e.g. λO), recognition of the SsrA-tag by ClpX, does not involve the N-terminal domain. Consistent with the idea, removal of the N-domain of ClpX, did not alter the ClpP-mediated degradation of SsrA-tagged proteins [50, 52]. Rather, the SsrA-tag is specifically recognised by loops in, or near to, the axial pore of the AAA+ module. Indeed, three sets of pore loops in ClpX (RKH, pore-1 and pore-2, see Fig. 1.5) have been implicated in binding the SsrA tag [71, 107]. The RKH loops, as the name suggests, contains the tripeptide motif (RKH), which surrounds the entrance to the ClpX pore. The positively charged RKH loops are proposed to attract negatively charged sequences (i.e. the charged C-terminal α-carboxylate of the SsrA-tag) to the pore of ClpX [99]. Accordingly, mutations that reduce the positive charge of the RKH loop, reduced binding to SsrA-tagged proteins (or substrates containing a C motif-1), whilst simultaneously improved the binding of substrates containing a positively charged motif [99]. The pore-1 and pore-2 loops, in contrast to the RKH loop, interact with the two last alanine residues of the ssrA-tag [107, 108]. The pore-1 loop of ClpX contains the highly conserved GYVG motif, which plays a central role in substrate translocation across the pore and into the degradation chamber [59–61, 109]. Based on a number of mutations and series of crosslinking experiments, the pore-2 loops were shown to specifically interact with the terminal alanines of the SsrA-tag [108]. Interestingly, neither the RKH nor the pore-2 loops are conserved in human mitochondrial ClpX [108]. As such, human ClpX is unable to recognise proteins tagged with the *E. coli* SsrA tag. However, a crucial role for these loops in the recognition of SsrA-tagged was elegantly demonstrated by Sauer and colleagues by grafting the *E. coli* ClpX RKH and pore-2 loops onto human ClpX creating a chimeric ClpX protein [108]. Strikingly, when both the RKH loops and pore-2 loops from *E. coli* ClpX were grafted onto human ClpX, the resulting chimeric proteins was able, not only to recognize the SsrA-tagged substrates but also to deliver them to ClpP for degradation [108]. Interestingly, grafting of only the RKH or pore-2 loop, was insufficient to promote recognition of the SsrA-tag. Collectively, these results demonstrated the importance of both pore loops in the recognition of SsrA-tagged substrates.

Indirect Recognition (Adaptor Mediated Recognition)

As mentioned above, the recognition of some protein substrates by the unfoldase, either requires or is modulated by an additional component – known as an adaptor protein. In general, an adaptor protein acts as a bridge between the substrate and the unfoldase. As such, adaptor proteins invariably exhibit two separate activities; (i) substrate recognition and (ii) unfoldase docking, however in some cases the adaptor protein is also proposed to activate either the substrate or the unfoldase for delivery to the protease for degradation [42, 110, 111]. Typically, the adaptor protein is released and recycled in this process without being degraded, although in some cases the adaptor protein (e.g. MecA) is also degraded by the protease complex (i.e. ClpCP), which acts a negative feedback loop to control the turnover of the substrates delivered by this adaptor protein. To date, four adaptor proteins have been identified in *E. coli*, three of which (SspB, UmuD and RssB) deliver specific protein substrates to ClpXP [54, 112, 113] while a single adaptor protein (ClpS) is required for the delivery of a specific class of substrates to ClpAP [40, 114]. SspB increases the affinity of ClpX to SsrA tagged proteins [112]. RssB is essential in bacteria for ClpXP-mediated degradation of the stationary phase sigma factor, σ^S [113, 115, 116]. Interestingly, four adaptor proteins have also been identified in *B. subtilis*. However, in contrast to the adaptor proteins from *E. coli*, the vast majority of *B. subtilis* adaptor proteins (MecA, McsB and YpbH) function together with ClpC [45, 117–121], and only a single adaptor protein (YjbH) has been identified to function with ClpX [122, 123]. Surprisingly, with the exception of MecA and YpbH, the remaining adaptor proteins share little, to no, sequence homology and hence each adaptor protein is likely to function via a unique mechanism.

SspB (A Multi-functional Adaptor Protein)

SspB is certainly the best characterised ClpX adaptor protein and arguably the best characterised bacterial adaptor protein to be studied. It was first identified, in *E. coli* as a ribosome-interacting protein that specifically modulates the ClpXP-mediated turnover of SsrA-tagged proteins [112]. Subsequently, SspB was also shown to recognise and deliver another ClpX substrate (i.e. RseA) for ClpP-mediated degradation [76] for a recent review see [6]. Although the distribution of SspB homologs is largely limited to γ- and β-proteobacteria an ortholog of SspB, termed SspBα has also been identified *Caulobacter crescentus* and other α-proteobacteria [124, 125]. Interestingly, despite the poor sequence homology, the overall fold of SspBα is similar to *E. coli* SspB [124]. Nevertheless, in contrast to *E. coli* SspB, *C. crescentus* SspBα appears to be optimised for binding to the SsrA-tag. In the case of *E. coli* SspB, the protein is composed of two functional regions separated by a long unstructured segment (~40–50 residues long). The N-terminal region of SspB (~110–120 residues long) forms a dimeric module, which is involved in binding of the SsrA-tag [50, 126, 127]. This substrate-binding domain is tethered to ClpX, via a short motif (termed the ClpX-binding region (XBR)), located at the C-terminus of SspB [50, 127]. The XBR of SspB forms an anti-parallel β-sheet with the N-terminal ZBD of ClpX [53]. Indeed, it has been proposed that both XBRs (from the SspB dimer) interact simultaneously with two ZBDs on ClpX – a mode of attachment that places the SspB-bound cargo in an ideal position for interacting with the pore residues in the ClpX hexamer. Hence, SspB tethers the substrate to ClpX thereby increasing the local concentration of SsrA-tagged substrates near the ClpX pore [50, 53, 127–129] (Fig. 1.8). Importantly, both the unfoldase and the adaptor protein recognise exclusive regions within the SsrA tag (AANDENYALAA). The unfoldase recognises the AA motif (C motif-1) at the C-terminus of the SsrA tag (Fig. 1.7), while SspB binds towards the N-terminal end of the SsrA-tag (AANDxxY). In contrast, the ClpA binding motif within the SsrA-tag (AAxxxxxALA) overlaps with the SspB binding and as such SspB inhibits the ClpAP-mediated degradation of SsrA-tagged substrates [50, 130]. As a consequence, SspB-mediated tethering of the SsrA-tag to ClpX results in an increased affinity of the substrate for ClpX, and hence an improved rate of degradation [129]. As such, SspB is likely to play an important role in the delivery of substrates present at low concentrations as tethering to ClpX, effectively increases the local concentration of the substrate. Consistent with this substrate-tethering model, mutation of the ClpX recognition motif within the SsrA-tag (i.e. replacement of LAA with DAS, termed the DAS-tag), significantly reduce the ClpXP-mediated degradation of substrates bearing this tag, while, the addition of SspB improved the recognition and degradation of substrates bearing this modified DAS-tag by more than 100-fold [131].

RssB

In contrast to SspB, RssB is a dedicated adaptor protein that is uniquely responsible for the recognition of a single substrate – the general stress transcription factor SigmaS (σ^S, also referred to as RpoS). RssB (also known as SprE (stationary phase regulator)

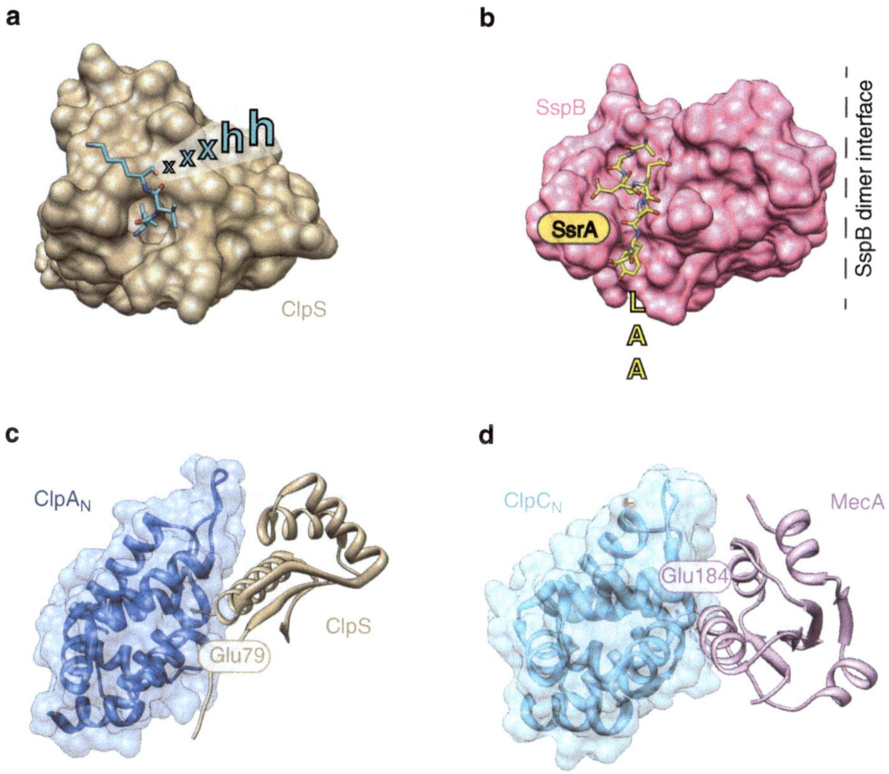

Fig. 1.8 Adaptor proteins. (**a**) ClpS (*tan*) contains a small substrate-binding pocket for the recognition of proteins bearing a primary destabilising residue at their N-terminus. This binding pocket exhibits exquisite specificity and not only forms a number of critical H-bonds with the α-amino group of the N-terminal residue of the substrate (*blue*), but also forms hydrophobic interactions with the side-chain of the N-terminal residue. The substrate extends away from ClpS and reaches towards the unfoldase, binding to ClpA is proposed to occur through the dihydrophobic (hh) element. (**b**) SspB (*pink*) forms a more permisscuous peptide-binding groove, which can accommodate peptides in different orientations. In the case of SsrA-tagged proteins, the substrate extends away from SspB and towards the pore loops of ClpX, which interact with the LAA motif. (**c–d**) Both ClpS (*tan*) and MecA (*pink*) interact with the N-terminal domain of ClpA (*blue*) and ClpC (*light blue*), respectively. An α-helix within the adaptor protein, contains a critical Glu residue which projects into the conserved pocket within the appropriate N-domain

in *E. coli* or MviA in *Salmonella typhimurium*) was first identified in *E. coli* using a genetic screen to discover genes involved in the RpoS expression and/or activity [115, 116]. RssB is a member of the two-component response regulator family and was the first family member to be shown to play a role in protein turnover. As a member of the response regulator family, RssB is phosphorylated on a highly conserved aspartate residue (D58). Phosphorylation of RssB at D58, stimulates binding of σ^S, resulting in the formation of a stable 1:1 complex [132]. Mutations that inhibit phosphorylation of RssB result in reduced binding to σ^S, and hence an increased stability of σ^S, both *in vitro* and *in vivo*. In contrast to SspB, which merely enhances the kinetics of substrate recognition, RssB is essential for the recognition of σ^S by ClpX [113]. Indeed it has been proposed that binding of RssB to σ^S triggers a conformational change in σ^S, which exposes a previously concealed ClpX recognition motif, however the mechanistic details of such a model are yet to be confirmed. For further details on the proteolytic control of the general stress response in bacteria refer to [7].

UmuD

The third and final, ClpX-specific adaptor protein in *E. coli*, is UmuD. In response to DNA damage, the first 24 residues of UmuD are auto-catalytically cleaved, in a RecA-dependent fashion. Following cleavage, the resulting protein (termed UmuD′), can form both homo- and hetero-oligomers [133]. As a heterodimer, UmuD/UmuD′ forms a component of the error-prone DNA polymerase V, which is able to bypass DNA lesions in the process of DNA replication and hence facilitates the cells recovery following DNA damage. Since this activity is necessary at times of DNA damage, but toxic under normal growth conditions, it is important that the cellular levels of UmuD′ (and hence UmuD/D′ oligomers) be carefully controlled during and after recovery. Indeed, this is elegantly achieved by the cell, as the N-terminal region of UmuD serves as a ClpX tethering sequence for delivery of UmuD′ to ClpXP when present in an UmuD/D′ complex [134]. Like the XBR of SspB, this region can bind to the ZBD of ClpX, but not as a degradation tag rather as a specific adaptor protein for the delivery of UmuD [54, 134].

Indirect Recognition by ClpA (Recognition of N-End Rule Substrates)

ClpS and the N-End Rule (A Specific ClpAP-Mediated Substrate)

In contrast to ClpX, which uses three different *E. coli* adaptor proteins, only a single adaptor protein (ClpS) has been identified for ClpA. Although ClpA appears to use only a single adaptor protein, this adaptor protein exhibits broad activity over its cognate unfoldase. Indeed, ClpS is able to regulate ClpA substrate selection, both negatively and positively [40, 114]. Originally identified as an inhibitor of ClpA auto-degradation, both *in vitro* and *in vivo*, and a negative regulator of ClpAP-mediated degradation of substrates bearing an SsrA-tag [40], ClpS was also shown to be an essential component of the N-end rule pathway [114].

The N-end rule pathway, originally identified in *Saccharomyces cerevisiae*, by Alexander Varshavsky's lab, is a highly conserved protein degradation pathway that is responsible for the recognition and degradation of proteins bearing a specific "destabilising" residue at the N-terminus [135, 136]. This pathway determines the half-life of a protein based on the N-terminal residue of that protein, which may be classified as "stabilising" or "destabilising". To date, this pathway has been identified in bacteria, plants and mammals and although the details of the various pathways differ, from one organism to the next, each pathway shares a number of common principals [135, 137–139]. In *E. coli*, like other organisms, the pathway is hierarchical, and destabilising residues can be separated into two classes (primary and secondary) [74]. Primary destabilising residues (L, F, W and Y) are recognised directly by the bacterial N-recognin, ClpS [114], while secondary destabilising residues (R, K and M) must first be converted to primary destabilising residues by the enzyme Leu/Phe-tRNA-protein transfersase (LFTR) before they are recognised by the adaptor protein [78, 140]. Interestingly, the first clue for a role of ClpS in the N-end rule pathway came from the structure of ClpS and comparison to the secondary structure of the human N-end rule recognition component (N-recognin), the E3-ligase, UBR1 [42, 141]. From this bioinformatic analysis, despite very low sequence homology, Lupas and colleagues proposed that ClpS was involved in the N-end rule pathway in bacteria [141]. Consistently, the crystal structure of ClpS (in complex with the N domain of ClpA) identified two conserved regions, one for interaction with the N-domain of ClpA and the other proposed to be involved in a substrate interaction [41, 42]. Subsequent biochemical and structural analysis confirmed that ClpS was indeed essential for the recognition of N-end rule substrates and that the second conserved region within ClpS was the N-degron binding site [142–144].

ClpS, like most characterised adaptor proteins is a small protein composed of two regions. The C-terminal domain of ClpS is the "workhorse" of the protein, it is responsible, not only for recognition of the substrate but also for docking to the N-terminal domain of ClpA [40–42, 114]. Despite both of these functions being located on the C-terminal domain of ClpS, this domain alone is neither sufficient for the inhibition of substrates bearing an SsrA-tag nor the delivery of N-end rule substrates [41, 110], suggesting that the N-terminal region of ClpS plays a crucial role in activation of ClpA. Hence in contrast to SspB, which merely modulates the affinity of ClpX for recognition of the SsrA-tag, the adaptor protein ClpS alters the substrate specificity of ClpA, by activating the unfoldase for recognition of N-degron bearing substrates. In summary, a substrate bearing an N-terminal primary destabilising residue, is bound by a small hydrophobic pocket on the surface of ClpS (Fig. 1.8). Importantly, this pocket exhibits exquisite specificity – it forms a number of important hydrogen bonds with both the α-amino group of the N-terminal residue and the carbonyl oxygen of the peptide bond, as well as several hydrophobic interactions with the side chain of the N-terminal amino acid [142–144]. Following recognition of the substrate by the adaptor protein, the substrate-ClpS complex docks to the N-terminal domain of ClpA [40–42]. Next, the N-terminal region is proposed to activate, an as yet undefined region of ClpA, for recognition of the N degron bearing substrate [110]. The unfoldase (ClpA), then recognises a hydrophobic region in the substrate approximately ~5–9 residues downstream of the primary destabilising residue (Fig. 1.7) [74, 78, 105].

MecA

Of the three known ClpC-adaptor proteins, MecA is currently the best characterised. It was first discovered in a genetic screen for repressors of competence development (which is a physiological state that permits *B. subtilis* cells to actively import DNA). In non-competent cells, the "competence" transcription factor (ComK) is recognised by MecA and targeted for degradation by ClpCP. Competence is triggered by the accumulation of a small peptide (ComS), which binds to MecA and thereby inhibits the MecA-dependent degradation of ComK by ClpCP [145]. Interestingly, MecA is not only involved in the development of competence through the regulated degradation of ComK but has also been proposed to be involved in general protein quality control, through the ClpCP-mediated degradation of misfolded and aggregated proteins [120]. Similar to most other adaptor proteins, MecA is composed of two regions an N-terminal domain, which is responsible for substrate recognition (i.e. ComK and ComS) and a C-terminal domain, which is required for docking to the unfoldase [146]. Interestingly, in contrast to other characterised adaptor proteins, docking of MecA (and hence substrate delivery to the protease) requires both the N-domain and the M-domain of ClpC [36, 48]. Despite the additional requirement for MecA binding to ClpC (i.e. to the M-domain), the mode of docking of MecA to the N-domain is strikingly similar to that of ClpS with the N-domain of ClpA [41, 42]. Indeed both adaptor proteins (ClpS and MecA) use a single α-helix to interact with the same region of the N-domain, and stabilise the complex by the formation of several H-bonds (Fig. 1.8). Interestingly, MecA is absent in cyanobacteria and ClpC was shown to cooperate with the adaptor protein ClpS [147]. Consistently, the distribution of MecA and ClpS appears to be mutually exclusive throughout evolution.

Substrate Processing by AAA+ Proteins

Substrate translocation is a basic mechanical process of all AAA+ proteins. This process is performed solely by the AAA+ module of the unfoldase, and like substrate binding has been extensively studied using both ClpA and ClpX as a model AAA+ protein. In recent years however, there have been many advances in this area of research by several researchers, including numerous contributions by the laboratory of Robert Sauer and Tania Baker to study the basic mechanism of action of ClpX.

Substrate Unfolding and Translocation

In order for a folded protein to enter the degradation chamber of ClpP, it must first be unfolded by the ATPase component. Although the pore of ClpX is large enough to simultaneously accommodate two or three peptide chains, most folded proteins are too large to enter [148]. As such, the narrow size of the ClpX pore prevents diffusion of folded proteins through ClpX and hence prevents the uncontrolled degradation of folded proteins by the ClpXP protease. Therefore protein substrates must first be unfolded, to enter the proteolytic chamber of ClpP. To achieve this, the unfoldase component converts the energy released from the binding and hydrolysis of ATP into a pulling force. This pulling force is responsible for the global unfolding of the substrate by threading it through the unfoldase pore, and into the degradation chamber of ClpP in a vectorial manner [95, 149].

The pulling force, generated by nucleotide-driven changes in the structure of ClpX is proposed to be transmitted to the substrate via movement of the pore-1 loops (Fig. 1.7). Indeed, the location of each pore-1 loop, within the central pore of the ClpX hexamer was shown to vary depending on the nucleotide bound state of the subunit [55]. As such, it has been proposed that ATP binding and hydrolysis drives conformational changes in the unfoldase, which result in movement of the pore-1 loops up and down the central pore [107, 150]. As these loops (in particular the highly conserved tyrosine residue) interact with the substrate's polypeptide chain, their movement along the pore provides the pulling force that is necessary for substrate unfolding and translocation. Do the subunits in a ClpX hexamer have to operate in a concerted fashion to promote successful unfolding and translocation of protein substrates? To examine this question Sauer and colleagues employed a method first used by the lab of Art Horwich, to study the role of individual subunits in GroEL [151]. In this case however, Sauer and colleagues created a single-chain hexamer of ClpX (lacking the N-terminal ZBD) by fusing six copies of *clpX* (lacking the sequence coding for the N-terminal ZBD) into a single gene [152]. This elegant experimental setup allowed the incorporation of specific mutations into various different ClpX subunits within the hexamer. Specifically, mutations in either the Walker B motif (E185Q) – which prevents ATP hydrolysis, or the sensor-2 motif (R370K) – which prevents both ATP hydrolysis and uncouples conformational changes linked to ClpP- and substrate-binding, were combined with wild type subunits and the degradation of SsrA-tagged substrates was examined [152]. Consequently, ClpX hexamers were created which contained either a single wild type subunit, or two wild type subunits and so on. Remarkably, the degradation rate increased linearly with the amount of the wild type subunits in a hexamer. For example, a ClpX hexamer that contained one wild type subunit led to degradation of an SsrA-tagged substrate 17% as fast as a wild type hexamer. In accordance, two wild type subunits in a hexamer resulted in a degradation rate that was 30% of that observed for a single-chain hexamer that was constructed of six wild type subunits. Moreover, in all cases, degradation was performed at a similar energetic cost (i.e. the amount of ATP hydrolyzed per substrate). These results indicated that even a single active subunit in a hexamer can promote efficient unfolding and translocation of substrates by ClpX and that concerted or sequential activity of multiple subunits is not essential for degradation.

Degradation of folded substrates proceeds much slower than degradation of unfolded substrates, suggesting that substrate unfolding is a rate-limiting step for proteolysis by AAA+proteases. This principle was elegantly demonstrated using the I27 domain of human titin as a substrate [92]. Titin-I27 was converted into a

ClpX substrate by creating a genetic fusion of titin-I27 with SsrA. The I27 domain of titin is extremely resistant to mechanical unfolding [153] and consistent with this, its degradation by ClpXP is relatively slow [92, 94]. Remarkably, an unfolded variant of titin-I27-SsrA can be obtained, simply by carboxymethylation of its two cysteine residues [92]. This simple chemical modification completely unfolds titin-I27 without altering its solubility. This permitted a direct comparison of the degradation kinetics of the substrate with respect to its folded state (i.e. either stably folded or unfolded). Interestingly, both the folded and unfolded substrates had a K_M of ~1 μM, similar to that observed for other SsrA tagged proteins. By contrast, the energetic cost for degradation of different substrates by ClpXP varied dramatically. For instance, the degradation of native titin-I27 required ~600 molecules of ATP, while the degradation of an unfolded mutant of titin-I27 only required ~100 molecules of ATP, suggesting that the cost of titin-I27 unfolding is ~500 ATP. Interestingly however, the rates of degradation do not correlate with the global thermodyamic stability of a substrate but rather seem to depend on the local stability of the region to which the recognition tag is attached [94, 154, 155]. In summary, the current model suggests that following binding, ClpX pulls on the degradation tag in an attempt to unfold the substrate. In some cases unfolding of the substrate may require multiple rounds of ATP hydrolysis until a power stroke of ClpX coincides with transient unfolding of a structural element near the degradation tag. When this occurs, ClpX can initiate substrate translocation and complete substrate unfolding very rapidly and with a high degree of cooperativity. As mentioned previously substrates may be recognised from either an internal site or from the N- or C-terminus of the protein [46]. Not surprisingly, substrate translocation may occur in either direction (i.e. from N-terminus to C-terminus or visa versa). Strikingly, single molecule experiments indicate that substrate unfolding eventually results from a single ClpX power stroke [154]. Following the initial unfolding event, substrate translocation proceeds rapidly and without considerable specificity [156]. Indeed, ClpX was shown to efficiently translocate a variety of different polymers, including homopolymeric tracts of glycine, proline and lysine non-amino-acid aliphatic chains. In addition, ClpX can carry out translocation of a polypeptide from the N-terminus to the C-terminus as efficiently as in the opposite direction.

References

1. Sauer RT, Baker TA (2011) AAA+proteases: ATP-fueled machines of protein destruction. Annu Rev Biochem 80:587–612
2. Samanovic M, Li H, Darwin KH (2013) The Pup-proteasome system of Mycobacterium tuberculosis. In: Dougan DA (ed) Regulated proteolysis in microorganisms. Springer, Subcell Biochem 66:267–295
3. Molière N, Turgay K (2013) General and regulatory proteolysis in Bacillus subtilis. In: Dougan DA (ed) Regulated proteolysis in microorganisms. Springer, Subcell Biochem 66:73–103
4. Gur E (2013) The Lon AAA+protease. In: Dougan DA (ed) Regulated proteolysis in microorganisms. Springer, Subcell Biochem 66:35–51

5. Okuno T, Ogura T (2013) FtsH protease-mediated regulation of various cellular functions. In: Dougan DA (ed) Regulated proteolysis in microorganisms. Springer, Subcell Biochem 66:53–69
6. Barchinger SE, Ades SE (2013) Regulated proteolysis: control of the Escherichia coli σE-dependent cell envelope stress response. In: Dougan DA (ed) Regulated proteolysis in microorganisms. Springer, Subcell Biochem 66:129–160
7. Micevski D, Dougan DA (2013) Proteolytic regulation of stress response pathways in Escherichia coli. In: Dougan DA (ed) Regulated proteolysis in microorganisms. Springer, Subcell Biochem 66:105–128
8. Frees D, Brøndsted L, Ingmer H (2013) Bacterial proteases and virulence. In: Dougan DA (ed) Regulated proteolysis in microorganisms. Springer, Subcell Biochem 66:161–192
9. Kirstein-Miles J, Morimoto RI (2010) Caenorhabditis elegans as a model system to study intercompartmental proteostasis: interrelation of mitochondrial function, longevity, and neurodegenerative diseases. Dev Dyn 239(5):1529–1538
10. Truscott KN, Bezawork-Geleta A, Dougan DA (2011) Unfolded protein responses in bacteria and mitochondria: a central role for the ClpXP machine. IUBMB Life 63(11):955–963
11. Kwasniak M, Pogorzelec L, Migdal I, Smakowska E et al (2012) Proteolytic system of plant mitochondria. Physiol Plant 145(1):187–195
12. Truscott KN, Lowth BR, Strack PR, Dougan DA (2010) Diverse functions of mitochondrial AAA+proteins: protein activation, disaggregation, and degradation. Biochem Cell Biol 88(1):97–108
13. van Dyck L, Dembowski M, Neupert W, Langer T (1998) Mcx1p, a ClpX homologue in mitochondria of Saccharomyces cerevisiae. FEBS Lett 438(3):250–254
14. Voos W, Ward L, Truscott KN (2013) The role of AAA+proteases in mitochondrial protein biogenesis, homeostasis and activity control. In: Dougan DA (ed) Regulated proteolysis in microorganisms. Springer, Subcell Biochem 66:223–263
15. Clarke AK (2012) The chloroplast ATP-dependent Clp protease in vascular plants – new dimensions and future challenges. Physiol Plant 145(1):235–244
16. Olinares PD, Kim J, van Wijk KJ (2011) The Clp protease system; a central component of the chloroplast protease network. Biochim Biophys Acta 1807(8):999–1011
17. Bochtler M, Hartmann C, Song HK, Bourenkov GP et al (2000) The structures of HslU and the ATP-dependent protease HslU-HslV. Nature 403(6771):800–805
18. Sousa MC, Trame CB, Tsuruta H, Wilbanks SM et al (2000) Crystal and solution structures of an HslUV protease-chaperone complex. Cell 103(4):633–643
19. Maglica Z, Kolygo K, Weber-Ban E (2009) Optimal efficiency of ClpAP and ClpXP chaperone-proteases is achieved by architectural symmetry. Structure 17(4):508–516
20. Wang J, Hartling JA, Flanagan JM (1997) The structure of ClpP at 2.3 A resolution suggests a model for ATP-dependent proteolysis. Cell 91(4):447–456
21. Gribun A, Kimber MS, Ching R, Sprangers R et al (2005) The ClpP double ring tetradecameric protease exhibits plastic ring-ring interactions, and the N termini of its subunits form flexible loops that are essential for ClpXP and ClpAP complex formation. J Biol Chem 280(16):16185–16196
22. Sprangers R, Gribun A, Hwang PM, Houry WA et al (2005) Quantitative NMR spectroscopy of supramolecular complexes: dynamic side pores in ClpP are important for product release. Proc Natl Acad Sci U S A 102(46):16678–16683
23. Maurizi MR, Clark WP, Katayama Y, Rudikoff S et al (1990) Sequence and structure of Clp P, the proteolytic component of the ATP-dependent Clp protease of Escherichia coli. J Biol Chem 265(21):12536–12545
24. Arribas J, Castano JG (1993) A comparative study of the chymotrypsin-like activity of the rat liver multicatalytic proteinase and the ClpP from Escherichia coli. J Biol Chem 268(28): 21165–21171
25. Thompson MW, Maurizi MR (1994) Activity and specificity of Escherichia coli ClpAP protease in cleaving model peptide substrates. J Biol Chem 269(27):18201–18208

26. Jennings LD, Lun DS, Medard M, Licht S (2008) ClpP hydrolyzes a protein substrate processively in the absence of the ClpA ATPase: mechanistic studies of ATP-independent proteolysis. Biochemistry 47(44):11536–11546
27. Effantin G, Ishikawa T, De Donatis GM, Maurizi MR et al (2010) Local and global mobility in the ClpA AAA+chaperone detected by cryo-electron microscopy: functional connotations. Structure 18(5):553–562
28. Jennings LD, Bohon J, Chance MR, Licht S (2008) The ClpP N-terminus coordinates substrate access with protease active site reactivity. Biochemistry 47(42):11031–11040
29. Li DH, Chung YS, Gloyd M, Joseph E et al (2010) Acyldepsipeptide antibiotics induce the formation of a structured axial channel in ClpP: a model for the ClpX/ClpA-bound state of ClpP. Chem Biol 17(9):959–969
30. Leung E, Datti A, Cossette M, Goodreid J et al (2011) Activators of cylindrical proteases as antimicrobials: identification and development of small molecule activators of ClpP protease. Chem Biol 18(9):1167–1178
31. Lee BG, Park EY, Lee KE, Jeon H et al (2010) Structures of ClpP in complex with acyldepsipeptide antibiotics reveal its activation mechanism. Nat Struct Mol Biol 17(4):471–478
32. Kirstein J, Hoffmann A, Lilie H, Schmidt R et al (2009) The antibiotic ADEP reprogrammes ClpP, switching it from a regulated to an uncontrolled protease. EMBO Mol Med 1(1):37–49
33. Dougan DA (2011) Chemical activators of ClpP: turning Jekyll into Hyde. Chem Biol 18(9):1072–1074
34. Brotz-Oesterhelt H, Beyer D, Kroll HP, Endermann R et al (2005) Dysregulation of bacterial proteolytic machinery by a new class of antibiotics. Nat Med 11(10):1082–1087
35. Sass P, Josten M, Famulla K, Schiffer G et al (2011) Antibiotic acyldepsipeptides activate ClpP peptidase to degrade the cell division protein FtsZ. Proc Natl Acad Sci U S A 108(42):17474–17479
36. Kirstein J, Schlothauer T, Dougan DA, Lilie H et al (2006) Adaptor protein controlled oligomerization activates the AAA+protein ClpC. EMBO J 25(7):1481–1491
37. Lowth BR, Kirstein-Miles J, Saiyed T, Brotz-Oesterhelt H et al (2012) Substrate recognition and processing by a Walker B mutant of the human mitochondrial AAA+protein CLPX. J Struct Biol 179(2):193–201
38. Kang SG, Dimitrova MN, Ortega J, Ginsburg A et al (2005) Human mitochondrial ClpP is a stable heptamer that assembles into a tetradecamer in the presence of ClpX. J Biol Chem 280(42):35424–35432
39. Alexopoulos JA, Guarne A, Ortega J (2012) ClpP: a structurally dynamic protease regulated by AAA+proteins. J Struct Biol 179(2):202–210
40. Dougan DA, Reid BG, Horwich AL, Bukau B (2002) ClpS, a substrate modulator of the ClpAP machine. Mol Cell 9(3):673–683
41. Guo F, Esser L, Singh SK, Maurizi MR et al (2002) Crystal structure of the heterodimeric complex of the adaptor, ClpS, with the N-domain of the AAA+chaperone, ClpA. J Biol Chem 277(48):46753–46762
42. Zeth K, Ravelli RB, Paal K, Cusack S et al (2002) Structural analysis of the adaptor protein ClpS in complex with the N-terminal domain of ClpA. Nat Struct Biol 9(12):906–911
43. Erbse AH, Wagner JN, Truscott KN, Spall SK et al (2008) Conserved residues in the N-domain of the AAA+chaperone ClpA regulate substrate recognition and unfolding. FEBS J 275(7):1400–1410
44. Lo JH, Baker TA, Sauer RT (2001) Characterization of the N-terminal repeat domain of Escherichia coli ClpA-A class I Clp/HSP100 ATPase. Protein Sci 10(3):551–559
45. Kirstein J, Dougan DA, Gerth U, Hecker M et al (2007) The tyrosine kinase McsB is a regulated adaptor protein for ClpCP. EMBO J 26(8):2061–2070
46. Kirstein J, Moliere N, Dougan DA, Turgay K (2009) Adapting the machine: adaptor proteins for Hsp100/Clp and AAA+proteases. Nat Rev Microbiol 7(8):589–599
47. Kojetin DJ, McLaughlin PD, Thompson RJ, Dubnau D et al (2009) Structural and motional contributions of the Bacillus subtilis ClpC N-domain to adaptor protein interactions. J Mol Biol 387(3):639–652

48. Wang F, Mei Z, Qi Y, Yan C et al (2011) Structure and mechanism of the hexameric MecA-ClpC molecular machine. Nature 471(7338):331–335
49. Donaldson LW, Wojtyra U, Houry WA (2003) Solution structure of the dimeric zinc binding domain of the chaperone ClpX. J Biol Chem 278(49):48991–48996
50. Dougan DA, Weber-Ban E, Bukau B (2003) Targeted delivery of an ssrA-tagged substrate by the adaptor protein SspB to its cognate AAA+ protein ClpX. Mol Cell 12(2):373–380
51. Thibault G, Houry WA (2012) Role of the N-terminal domain of the chaperone ClpX in the recognition and degradation of lambda phage protein O. J Phys Chem B 116(23): 6717–6724
52. Wojtyra UA, Thibault G, Tuite A, Houry WA (2003) The N-terminal zinc binding domain of ClpX is a dimerization domain that modulates the chaperone function. J Biol Chem 278(49):48981–48990
53. Park EY, Lee BG, Hong SB, Kim HW et al (2007) Structural basis of SspB-tail recognition by the zinc binding domain of ClpX. J Mol Biol 367(2):514–526
54. Neher SB, Sauer RT, Baker TA (2003) Distinct peptide signals in the UmuD and UmuD′ subunits of UmuD/D′ mediate tethering and substrate processing by the ClpXP protease. Proc Natl Acad Sci U S A 100(23):13219–13224
55. Glynn SE, Martin A, Nager AR, Baker TA et al (2009) Structures of asymmetric ClpX hexamers reveal nucleotide-dependent motions in a AAA+ protein-unfolding machine. Cell 139(4):744–756
56. Kim DY, Kim KK (2003) Crystal structure of ClpX molecular chaperone from Helicobacter pylori. J Biol Chem 278(50):50664–50670
57. Neuwald AF, Aravind L, Spouge JL, Koonin EV (1999) AAA+: a class of chaperone-like ATPases associated with the assembly, operation, and disassembly of protein complexes. Genome Res 9(1):27–43
58. Hinnerwisch J, Fenton WA, Furtak KJ, Farr GW et al (2005) Loops in the central channel of ClpA chaperone mediate protein binding, unfolding, and translocation. Cell 121(7):1029–1041
59. Okuno T, Yamanaka K, Ogura T (2006) Characterization of mutants of the Escherichia coli AAA protease, FtsH, carrying a mutation in the central pore region. J Struct Biol 156(1):109–114
60. Schlieker C, Weibezahn J, Patzelt H, Tessarz P et al (2004) Substrate recognition by the AAA+ chaperone ClpB. Nat Struct Mol Biol 11(7):607–615
61. Yamada-Inagawa T, Okuno T, Karata K, Yamanaka K et al (2003) Conserved pore residues in the AAA protease FtsH are important for proteolysis and its coupling to ATP hydrolysis. J Biol Chem 278(50):50182–50187
62. Turgay K, Persuh M, Hahn J, Dubnau D (2001) Roles of the two ClpC ATP binding sites in the regulation of competence and the stress response. Mol Microbiol 42(3):717–727
63. Singh SK, Maurizi MR (1994) Mutational analysis demonstrates different functional roles for the two ATP-binding sites in ClpAP protease from Escherichia coli. J Biol Chem 269(47):29537–29545
64. Kress W, Mutschler H, Weber-Ban E (2009) Both ATPase domains of ClpA are critical for processing of stable protein structures. J Biol Chem 284(45):31441–31452
65. Hersch GL, Burton RE, Bolon DN, Baker TA et al (2005) Asymmetric interactions of ATP with the AAA+ ClpX6 unfoldase: allosteric control of a protein machine. Cell 121(7):1017–1027
66. Grimaud R, Kessel M, Beuron F, Steven AC et al (1998) Enzymatic and structural similarities between the Escherichia coli ATP-dependent proteases, ClpXP and ClpAP. J Biol Chem 273(20):12476–12481
67. Joshi SA, Hersch GL, Baker TA, Sauer RT (2004) Communication between ClpX and ClpP during substrate processing and degradation. Nat Struct Mol Biol 11(5):404–411
68. Kim YI, Levchenko I, Fraczkowska K, Woodruff RV et al (2001) Molecular determinants of complex formation between Clp/Hsp100 ATPases and the ClpP peptidase. Nat Struct Biol 8(3):230–233
69. Singh SK, Rozycki J, Ortega J, Ishikawa T et al (2001) Functional domains of the ClpA and ClpX molecular chaperones identified by limited proteolysis and deletion analysis. J Biol Chem 276(31):29420–29429

70. Bewley MC, Graziano V, Griffin K, Flanagan JM (2006) The asymmetry in the mature amino-terminus of ClpP facilitates a local symmetry match in ClpAP and ClpXP complexes. J Struct Biol 153(2):113–128
71. Martin A, Baker TA, Sauer RT (2007) Distinct static and dynamic interactions control ATPase-peptidase communication in a AAA+ protease. Mol Cell 27(1):41–52
72. Ishihama Y, Schmidt T, Rappsilber J, Mann M et al (2008) Protein abundance profiling of the Escherichia coli cytosol. BMC Genomics 9:102
73. Schrader EK, Harstad KG, Matouschek A (2009) Targeting proteins for degradation. Nat Chem Biol 5(11):815–822
74. Dougan DA, Truscott KN, Zeth K (2010) The bacterial N-end rule pathway: expect the unexpected. Mol Microbiol 76(3):545–558
75. Karzai AW, Roche ED, Sauer RT (2000) The SsrA-SmpB system for protein tagging, directed degradation and ribosome rescue. Nat Struct Biol 7(6):449–455
76. Flynn JM, Levchenko I, Sauer RT, Baker TA (2004) Modulating substrate choice: the SspB adaptor delivers a regulator of the extracytoplasmic-stress response to the AAA+ protease ClpXP for degradation. Genes Dev 18(18):2292–2301
77. Neher SB, Flynn JM, Sauer RT, Baker TA (2003) Latent ClpX-recognition signals ensure LexA destruction after DNA damage. Genes Dev 17(9):1084–1089
78. Ninnis RL, Spall SK, Talbo GH, Truscott KN et al (2009) Modification of PATase by L/F-transferase generates a ClpS-dependent N-end rule substrate in Escherichia coli. EMBO J 28(12):1732–1744
79. Schmidt R, Zahn R, Bukau B, Mogk A (2009) ClpS is the recognition component for Escherichia coli substrates of the N-end rule degradation pathway. Mol Microbiol 72(2):506–517
80. Lange R, Hengge-Aronis R (1994) The cellular concentration of the sigma S subunit of RNA polymerase in Escherichia coli is controlled at the levels of transcription, translation, and protein stability. Genes Dev 8(13):1600–1612
81. Lies M, Maurizi MR (2008) Turnover of endogenous SsrA-tagged proteins mediated by ATP-dependent proteases in Escherichia coli. J Biol Chem 283(34):22918–22929
82. Moore SD, Sauer RT (2005) Ribosome rescue: tmRNA tagging activity and capacity in Escherichia coli. Mol Microbiol 58(2):456–466
83. Moore SD, Sauer RT (2007) The tmRNA system for translational surveillance and ribosome rescue. Annu Rev Biochem 76:101–124
84. Keiler KC, Waller PR, Sauer RT (1996) Role of a peptide tagging system in degradation of proteins synthesized from damaged messenger RNA. Science 271(5251):990–993
85. Oh BK, Chauhan AK, Isono K, Apirion D (1990) Location of a gene (ssrA) for a small, stable RNA (10Sa RNA) in the Escherichia coli chromosome. J Bacteriol 172(8):4708–4709
86. Komine Y, Kitabatake M, Yokogawa T, Nishikawa K et al (1994) A tRNA-like structure is present in 10Sa RNA, a small stable RNA from Escherichia coli. Proc Natl Acad Sci U S A 91(20):9223–9227
87. Tu GF, Reid GE, Zhang JG, Moritz RL et al (1995) C-terminal extension of truncated recombinant proteins in Escherichia coli with a 10Sa RNA decapeptide. J Biol Chem 270(16):9322–9326
88. Gottesman S, Roche E, Zhou Y, Sauer RT (1998) The ClpXP and ClpAP proteases degrade proteins with carboxy-terminal peptide tails added by the SsrA-tagging system. Genes Dev 12(9):1338–1347
89. Gur E, Sauer RT (2008) Evolution of the ssrA degradation tag in Mycoplasma: specificity switch to a different protease. Proc Natl Acad Sci U S A 105(42):16113–16118
90. Herman C, Prakash S, Lu CZ, Matouschek A et al (2003) Lack of a robust unfoldase activity confers a unique level of substrate specificity to the universal AAA protease FtsH. Mol Cell 11(3):659–669
91. Farrell CM, Grossman AD, Sauer RT (2005) Cytoplasmic degradation of ssrA-tagged proteins. Mol Microbiol 57(6):1750–1761

92. Kenniston JA, Baker TA, Fernandez JM, Sauer RT (2003) Linkage between ATP consumption and mechanical unfolding during the protein processing reactions of an AAA+ degradation machine. Cell 114(4):511–520
93. Kenniston JA, Baker TA, Sauer RT (2005) Partitioning between unfolding and release of native domains during ClpXP degradation determines substrate selectivity and partial processing. Proc Natl Acad Sci U S A 102(5):1390–1395
94. Kenniston JA, Burton RE, Siddiqui SM, Baker TA et al (2004) Effects of local protein stability and the geometric position of the substrate degradation tag on the efficiency of ClpXP denaturation and degradation. J Struct Biol 146(1–2):130–140
95. Weber-Ban EU, Reid BG, Miranker AD, Horwich AL (1999) Global unfolding of a substrate protein by the Hsp100 chaperone ClpA. Nature 401(6748):90–93
96. Flynn JM, Neher SB, Kim YI, Sauer RT et al (2003) Proteomic discovery of cellular substrates of the ClpXP protease reveals five classes of ClpX-recognition signals. Mol Cell 11(3):671–683
97. Neher SB, Villen J, Oakes EC, Bakalarski CE et al (2006) Proteomic profiling of ClpXP substrates after DNA damage reveals extensive instability within SOS regulon. Mol Cell 22(2):193–204
98. Gonciarz-Swiatek M, Wawrzynow A, Um SJ, Learn BA et al (1999) Recognition, targeting, and hydrolysis of the lambda O replication protein by the ClpP/ClpX protease. J Biol Chem 274(20):13999–14005
99. Farrell CM, Baker TA, Sauer RT (2007) Altered specificity of a AAA+ protease. Mol Cell 25(1):161–166
100. Gottesman S, Clark WP, Maurizi MR (1990) The ATP-dependent Clp protease of Escherichia coli. Sequence of clpA and identification of a Clp-specific substrate. J Biol Chem 265(14):7886–7893
101. Maglica Z, Striebel F, Weber-Ban E (2008) An intrinsic degradation tag on the ClpA C-terminus regulates the balance of ClpAP complexes with different substrate specificity. J Mol Biol 384(2):503–511
102. Hoskins JR, Singh SK, Maurizi MR, Wickner S (2000) Protein binding and unfolding by the chaperone ClpA and degradation by the protease ClpAP. Proc Natl Acad Sci U S A 97(16):8892–8897
103. Hoskins JR, Wickner S (2006) Two peptide sequences can function cooperatively to facilitate binding and unfolding by ClpA and degradation by ClpAP. Proc Natl Acad Sci U S A 103(4):909–914
104. Hoskins JR, Kim SY, Wickner S (2000) Substrate recognition by the ClpA chaperone component of ClpAP protease. J Biol Chem 275(45):35361–35367
105. Wang KH, Oakes ES, Sauer RT, Baker TA (2008) Tuning the strength of a bacterial N-end rule degradation signal. J Biol Chem 283(36):24600–24607
106. Kim YI, Burton RE, Burton BM, Sauer RT et al (2000) Dynamics of substrate denaturation and translocation by the ClpXP degradation machine. Mol Cell 5(4):639–648
107. Martin A, Baker TA, Sauer RT (2008) Pore loops of the AAA+ ClpX machine grip substrates to drive translocation and unfolding. Nat Struct Mol Biol 15(11):1147–1151
108. Martin A, Baker TA, Sauer RT (2008) Diverse pore loops of the AAA+ ClpX machine mediate unassisted and adaptor-dependent recognition of ssrA-tagged substrates. Mol Cell 29(4):441–450
109. Park E, Rho YM, Koh OJ, Ahn SW et al (2005) Role of the GYVG pore motif of HslU ATPase in protein unfolding and translocation for degradation by HslV peptidase. J Biol Chem 280(24):22892–22898
110. Hou JY, Sauer RT, Baker TA (2008) Distinct structural elements of the adaptor ClpS are required for regulating degradation by ClpAP. Nat Struct Mol Biol 15(3):288–294
111. Roman-Hernandez G, Hou JY, Grant RA, Sauer RT et al (2011) The ClpS adaptor mediates staged delivery of N-end rule substrates to the AAA+ ClpAP protease. Mol Cell 43(2):217–228
112. Levchenko I, Seidel M, Sauer RT, Baker TA (2000) A specificity-enhancing factor for the ClpXP degradation machine. Science 289(5488):2354–2356

113. Zhou Y, Gottesman S, Hoskins JR, Maurizi MR et al (2001) The RssB response regulator directly targets sigma(S) for degradation by ClpXP. Genes Dev 15(5):627–637
114. Erbse A, Schmidt R, Bornemann T, Schneider-Mergener J et al (2006) ClpS is an essential component of the N-end rule pathway in Escherichia coli. Nature 439(7077):753–756
115. Muffler A, Fischer D, Altuvia S, Storz G et al (1996) The response regulator RssB controls stability of the sigma(S) subunit of RNA polymerase in Escherichia coli. EMBO J 15(6):1333–1339
116. Pratt LA, Silhavy TJ (1996) The response regulator SprE controls the stability of RpoS. Proc Natl Acad Sci U S A 93(6):2488–2492
117. Elsholz AK, Turgay K, Michalik S, Hessling B et al (2012) Global impact of protein arginine phosphorylation on the physiology of Bacillus subtilis. Proc Natl Acad Sci U S A 109(19):7451–7456
118. Fuhrmann J, Schmidt A, Spiess S, Lehner A et al (2009) McsB is a protein arginine kinase that phosphorylates and inhibits the heat-shock regulator CtsR. Science 324(5932):1323–1327
119. Persuh M, Mandic-Mulec I, Dubnau D (2002) A MecA paralog, YpbH, binds ClpC, affecting both competence and sporulation. J Bacteriol 184(8):2310–2313
120. Schlothauer T, Mogk A, Dougan DA, Bukau B et al (2003) MecA, an adaptor protein necessary for ClpC chaperone activity. Proc Natl Acad Sci U S A 100(5):2306–2311
121. Turgay K, Hahn J, Burghoorn J, Dubnau D (1998) Competence in Bacillus subtilis is controlled by regulated proteolysis of a transcription factor. EMBO J 17(22):6730–6738
122. Chan CM, Garg S, Lin AA, Zuber P (2012) Geobacillus thermodenitrificans YjbH recognizes the C-terminal end of Bacillus subtilis Spx to accelerate Spx proteolysis by ClpXP. Microbiology 158(Pt 5):1268–1278
123. Nakano MM, Nakano S, Zuber P (2002) Spx (YjbD), a negative effector of competence in Bacillus subtilis, enhances ClpC-MecA-ComK interaction. Mol Microbiol 44(5):1341–1349
124. Chien P, Grant RA, Sauer RT, Baker TA (2007) Structure and substrate specificity of an SspB ortholog: design implications for AAA+ adaptors. Structure 15(10):1296–1305
125. Lessner FH, Venters BJ, Keiler KC (2007) Proteolytic adaptor for transfer-messenger RNA-tagged proteins from alpha-proteobacteria. J Bacteriol 189(1):272–275
126. Levchenko I, Grant RA, Wah DA, Sauer RT et al (2003) Structure of a delivery protein for an AAA+ protease in complex with a peptide degradation tag. Mol Cell 12(2):365–372
127. Wah DA, Levchenko I, Rieckhof GE, Bolon DN et al (2003) Flexible linkers leash the substrate binding domain of SspB to a peptide module that stabilizes delivery complexes with the AAA+ ClpXP protease. Mol Cell 12(2):355–363
128. Bolon DN, Grant RA, Baker TA, Sauer RT (2004) Nucleotide-dependent substrate handoff from the SspB adaptor to the AAA+ ClpXP protease. Mol Cell 16(3):343–350
129. Bolon DN, Wah DA, Hersch GL, Baker TA et al (2004) Bivalent tethering of SspB to ClpXP is required for efficient substrate delivery: a protein-design study. Mol Cell 13(3):443–449
130. Flynn JM, Levchenko I, Seidel M, Wickner SH et al (2001) Overlapping recognition determinants within the ssrA degradation tag allow modulation of proteolysis. Proc Natl Acad Sci U S A 98(19):10584–10589
131. McGinness KE, Baker TA, Sauer RT (2006) Engineering controllable protein degradation. Mol Cell 22(5):701–707
132. Becker G, Klauck E, Hengge-Aronis R (1999) Regulation of RpoS proteolysis in Escherichia coli: the response regulator RssB is a recognition factor that interacts with the turnover element in RpoS. Proc Natl Acad Sci U S A 96(11):6439–6444
133. Jonczyk P, Nowicka A (1996) Specific in vivo protein-protein interactions between Escherichia coli SOS mutagenesis proteins. J Bacteriol 178(9):2580–2585
134. Gonzalez M, Rasulova F, Maurizi MR, Woodgate R (2000) Subunit-specific degradation of the UmuD/D′ heterodimer by the ClpXP protease: the role of trans recognition in UmuD′ stability. EMBO J 19(19):5251–5258
135. Varshavsky A (1996) The N-end rule: functions, mysteries, uses. Proc Natl Acad Sci U S A 93(22):12142–12149

136. Varshavsky A (2011) The N-end rule pathway and regulation by proteolysis. Protein Sci 20(8):1298–1345
137. Dougan DA, Micevski D, Truscott KN (2012) The N-end rule pathway: from recognition by N-recognins, to destruction by AAA+ proteases. Biochim Biophys Acta 1823(1):83–91
138. Graciet E, Wellmer F (2010) The plant N-end rule pathway: structure and functions. Trends Plant Sci 15(8):447–453
139. Mogk A, Schmidt R, Bukau B (2007) The N-end rule pathway for regulated proteolysis: prokaryotic and eukaryotic strategies. Trends Cell Biol 17(4):165–172
140. Shrader TE, Tobias JW, Varshavsky A (1993) The N-end rule in Escherichia coli: cloning and analysis of the leucyl, phenylalanyl-tRNA-protein transferase gene aat. J Bacteriol 175(14):4364–4374
141. Lupas AN, Koretke KK (2003) Bioinformatic analysis of ClpS, a protein module involved in prokaryotic and eukaryotic protein degradation. J Struct Biol 141(1):77–83
142. Roman-Hernandez G, Grant RA, Sauer RT, Baker TA (2009) Molecular basis of substrate selection by the N-end rule adaptor protein ClpS. Proc Natl Acad Sci U S A 106(22):8888–8893
143. Schuenemann VJ, Kralik SM, Albrecht R, Spall SK et al (2009) Structural basis of N-end rule substrate recognition in Escherichia coli by the ClpAP adaptor protein ClpS. EMBO Rep 10(5):508–514
144. Wang KH, Roman-Hernandez G, Grant RA, Sauer RT et al (2008) The molecular basis of N-end rule recognition. Mol Cell 32(3):406–414
145. Turgay K, Hamoen LW, Venema G, Dubnau D (1997) Biochemical characterization of a molecular switch involving the heat shock protein ClpC, which controls the activity of ComK, the competence transcription factor of Bacillus subtilis. Genes Dev 11(1):119–128
146. Persuh M, Turgay K, Mandic-Mulec I, Dubnau D (1999) The N- and C-terminal domains of MecA recognize different partners in the competence molecular switch. Mol Microbiol 33(4):886–894
147. Andersson FI, Blakytny R, Kirstein J, Turgay K et al (2006) Cyanobacterial ClpC/HSP100 protein displays intrinsic chaperone activity. J Biol Chem 281(9):5468–5475
148. Burton RE, Siddiqui SM, Kim YI, Baker TA et al (2001) Effects of protein stability and structure on substrate processing by the ClpXP unfolding and degradation machine. EMBO J 20(12):3092–3100
149. Reid BG, Fenton WA, Horwich AL, Weber-Ban EU (2001) ClpA mediates directional translocation of substrate proteins into the ClpP protease. Proc Natl Acad Sci U S A 98(7):3768–3772
150. Wang J, Song JJ, Seong IS, Franklin MC et al (2001) Nucleotide-dependent conformational changes in a protease-associated ATPase HslU. Structure 9(11):1107–1116
151. Farr GW, Furtak K, Rowland MB, Ranson NA et al (2000) Multivalent binding of nonnative substrate proteins by the chaperonin GroEL. Cell 100(5):561–573
152. Martin A, Baker TA, Sauer RT (2005) Rebuilt AAA+ motors reveal operating principles for ATP-fuelled machines. Nature 437(7062):1115–1120
153. Li H, Carrion-Vazquez M, Oberhauser AF, Marszalek PE et al (2000) Point mutations alter the mechanical stability of immunoglobulin modules. Nat Struct Biol 7(12):1117–1120
154. Aubin-Tam ME, Olivares AO, Sauer RT, Baker TA et al (2011) Single-molecule protein unfolding and translocation by an ATP-fueled proteolytic machine. Cell 145(2):257–267
155. Lee C, Schwartz MP, Prakash S, Iwakura M et al (2001) ATP-dependent proteases degrade their substrates by processively unraveling them from the degradation signal. Mol Cell 7(3):627–637
156. Barkow SR, Levchenko I, Baker TA, Sauer RT (2009) Polypeptide translocation by the AAA+ ClpXP protease machine. Chem Biol 16(6):605–612

Chapter 2
The Lon AAA+ Protease

Eyal Gur

Abstract As the first ATP-dependent protease to be identified, Lon holds a special place in the history of cellular biology. In fact, the concept of ATP-dependent protein degradation was established through the findings that led to the discovery of Lon. Therefore, this chapter begins with a historical perspective, describing the milestones that led to the discovery of Lon and ATP-dependent proteolysis, starting from the early findings in the 1960s until the demonstration of Lon's ATP-dependent proteolytic activity *in vitro*, in 1981. Most of our knowledge on Lon derives from studies of the *Escherichia coli* Lon ortholog, and, therefore, most of this chapter relates to this particular enzyme. Nonetheless, Lon is not only found in most bacterial species, it is also found in Archaea and in the mitochondrion and chloroplast of eukaryotic cells. Therefore many of the conclusions gained from studies on the *E. coli* enzyme are relevant to Lon proteases in other organisms. Lon, more than any other bacterial or organellar protease, is associated with the degradation of misfolded proteins and protein quality control. In addition, Lon also degrades many regulatory proteins that are natively folded, thus it also plays a prominent role in regulation of physiological processes. Throughout the years, many Lon substrates have been identified, confirming its role in the regulation of diverse cellular processes, including cell division, DNA replication, differentiation, and adaptation to stress conditions. Some examples of these functions are described and discussed here, as is the role of Lon in the degradation of misfolded proteins and in protein quality control. Finally, this chapter deals with the exquisite sensitivity of protein degradation inside a cell. How can a protease distinguish so many substrates from cellular proteins that

E. Gur (✉)
Life Sciences Department, Ben-Gurion University of the Negev,
Beer-Sheva 84105, Israel

Life Sciences Department, The National Institute for Biotechnology in the Negev,
Beer-Sheva 84105, Israel
e-mail: gure@bgu.ac.il

D.A. Dougan (ed.), *Regulated Proteolysis in Microorganisms*, Subcellular Biochemistry 66, DOI 10.1007/978-94-007-5940-4_2, © Springer Science+Business Media Dordrecht 2013

should not be degraded? Can the specificity of a protease be regulated according to the physiological needs of a cell? This chapter thus broadly discusses the substrate specificity of Lon and its allosteric regulation.

From the *lac operon* to the Discovery of ATP-Dependent Proteolysis

In 1961 Francois Jacob and Jacques Monod made a major breakthrough in the biological sciences, as they presented their model of the *lac operon* and established the first paradigm for regulation of gene expression [1]. It was a brilliant, simple and coherent model that was based on negative regulation, i.e., on a repressor that binds to an operator region on the DNA and controls the expression of downstream genes. The model was readily accepted as a general mechanism for gene regulation in living organisms. Indeed, the basic principles that are exemplified by the *lac operon* model are still relevant today, 50 years after their discovery. However, Jacob and Monod extrapolated their interpretation too far, proposing that negative regulation is the *only* mechanism for gene regulation. The beauty and simplicity of their model and the dominant character of its discoverers (mostly of Jacques Monod) persuaded many scientists to accept the model as it stood, thereby discouraging investigation of alternative mechanisms for regulation of gene expression. Too many scientists forced negative regulation on their results at times when other hypotheses should have been raised. Such was the case with the initial results that led eventually to the discovery of ATP-dependent proteolysis, as described below.

About 3 years after the *lac operon* model was published, a peculiar bacterial phenotype – the mucoid phenotype – was reported by Alvin Markovitz [2]. *Escherichia coli* K12 mucoid mutants appeared watery and slimy on agar plates, and further investigation indicated that the phenotype resulted from increased synthesis of the cell's polysaccharide capsule [2]. These mutants exhibited constitutive expression of *capsule synthesis* (*cps*) genes as a result of a mutation in a gene that was termed initially *R* and later *capR*. These mutants were actually *lon* mutants and the *capR* gene product was the Lon AAA+ protease, but it took 20 more years to realize that. Initially, it was hypothesized that, in accordance with the negative control paradigm for regulation of gene expression, that the *capR* gene encodes a repressor that negatively regulates the expression of *cps* genes. Mutations in the repressor gene, it was thought, led to constitutive expression of the *cps* genes and initially, the available genetic data indeed supported negative control as a mechanism for *cps* gene expression by the R regulator (Table 2.1). For example, in both the *lac* system and the *cps* system, regulator deficient cells exhibited constitutive expression of the regulated genes, and in both systems the wild type allele was dominant over the mutant allele. The similarities were compelling! However, gradually data accumulated raising concerns regarding the plausibility of negative control in the *cps* system (discussed below).

Table 2.1 Similarities between the lactose and capsule-synthesis systems

Lactose system		Capsule synthesis system	
Genotype	Phenotype	Genotype	Phenotype
lacI$^+$	Wild Type Inducible *lacZ* expression	R^+	Wild Type Normal colonies due to uninduced expression of capsule synthesis (*cps*) genes
	Repressed expression in the absence of an inducer		Inducer not identified
lacI$^-$	Constitutive *lacZ* expression due to lack of repression	R^-	Mucoid colonies due to constitutive expression of *cps* genes
lacI$^-$/*lacI*$^+$	W.T. *lacI*$^+$ is dominant over *lacI*$^-$	R^-/R^+	W.T. R^+ is dominant over R^-

The constitutive expression of capsule synthesis genes in R^- mutants and the dominance of the wild type allele, R^+, over R^- mistakenly suggested, by similarity to the lactose system, that R is a repressor of capsule synthesis gene expression

At roughly the same time as the discovery of the mucoid phenotype, another, completely different phenotype was discovered. These *E. coli* mutants were found to be much more sensitive to ionizing irradiation than wild type cells [3]. Consequently, these mutants failed to recover from X-ray or UV irradiation. Close inspection of the irradiated mutants under a microscope revealed that they were much longer than normal cells and contained multiple nucleoids [3–5]. It was found that these mutants were unable to form a septum during cell division, and therefore, increased in cell mass but could not divide. The responsible gene was termed *lon* because of the Long phenotype associated with the mutation [5]. It was apparent that these mutants were also mucoid [4, 5], and genetic analysis revealed that *lon* and *capR* were actually the same gene [6]. Thus, the *lon* mutation had a pleiotropic effect, as it caused two, apparently unrelated phenotypes (i.e., sensitivity to ionizing irradiation and the formation of mucoid colonies). This complicated the hypothesis of negative regulation by Lon (i.e., repression of *cps* gene synthesis), since repressor mutations did not cause pleiotropic effects in either the *lac* system or in other genetic systems that were known at that time to be regulated by repression (i.e. tryptophan biosynthesis and bacteriophage lambda development). Rather, they were associated with phenotypes that were specifically related to a single metabolic or developmental pathway (e.g. lactose catabolism). Nonetheless, the pleiotropic effect of the *lon* mutations could be reconciled with negative regulation by repression, assuming that *lon* encodes a global transcription regulator that affects several, apparently unrelated, pathways. Markovitz rationalized that if CapR is indeed a transcriptional regulator, it should have a DNA binding activity like the *lac* and lambda repressors [7]. Therefore, he purified CapR and examined its ability to bind DNA. Astonishingly, it was found that CapR was indeed a DNA-binding protein [7]. However, it bound

DNA non-specifically – a property that is not expected from a transcriptional regulator. Among all the *E. coli* proteases, Lon is the only DNA-binding protein, and today, more than 30 years following the discovery of this activity, the physiological relevance of DNA-binding, by Lon, remains unclear.

Almost 10 years later, following the discoveries of the mucoid and irradiation sensitivity phenotypes of *lon* mutant strains, an additional phenotype of *lon*⁻ cells was identified. This discovery, came from an unexpected direction, albeit one still very much related to the *lac* system of *E. coli*. Brenner and colleagues noticed that nonsense mutations in *lacZ* reduced expression of the two downstream genes in the *lac operon*, *lacY* and *lacA* [8]. This phenomenon, termed polarity, was due to the indirect effect of these nonsense mutations. As a result of nonsense mutations in the *lacZ* open reading frame, translation of the *lacZ* mRNA is terminated prematurely, allowing the naked mRNA to form secondary structures that terminate the transcription of the *lac operon* and hence expression of the *lacYA* genes. The polar effect of nonsense mutations in operons was thus established. Zipser and Goldschmidt then took Brenner's system one step further, and instead of testing the expression level of *lacY* and *lacA* in the *lacZ* nonsense mutants they monitored the expression of β-galactosidase. They discovered that following induction of the *lac operon* with IPTG, the cytoplasmic concentration of full-length wild type β-galactosidase was significantly higher than the levels of the β-galactosidase nonsense fragments, under similar conditions [9, 10]. Strikingly, the low cytoplasmic level of the fragments, was a result of their degradation *in vivo*, as revealed by pulse-chase experiments [11]. Around this time, intracellular protein degradation was also reported in other systems [12, 13]. For instance, it was found that a defective mutant of the *lac* repressor was rapidly degraded [13]. Similarly, proteins in puromycin-treated cells were unstable and rapidly degraded [12]. From these observations, it became apparent that aberrant proteins were removed from cells by proteolysis. But what was the molecular mechanism of intracellular protein degradation? Which enzymes were responsible for this activity? To address these questions, David Zipser sought to isolate mutants that were defective in the proteolysis of aberrant proteins. To this end, an elegant mutant selection system was developed, based on complementation of β-galactosidase activity [14]. A peculiar phenomenon was discovered a few years earlier by Monod and colleagues in which, the enzyme activity of a *lacZ* mutant that carried a deletion near the 5′ end of the gene, could be recovered by *in trans* expression of an N-terminal fragment of β-galactosidase. The two β-galactosidase fragments associated, forming an active enzyme [15]. Thus, the N-terminal fragment was termed the α-donor, the deletion-containing enzyme was designated the α-acceptor, and the process was termed α-complementation: A concept that is commonly used today for gene cloning, as the basic principle of blue-white screening systems in *E. coli*. Since α-donors are aberrant β-galactosidase fragments, they are rapidly degraded *in vivo*, and hence their cytoplasmic concentration is very low. Zipser and colleagues identified α-donor and α-acceptor pairs that do not associate sufficiently well *in vivo* to support growth on a medium containing lactose as the sole carbon source. They hypothesized that in proteolysis-deficient cells elevated concentrations of the α-donor should promote association with the α-acceptor.

They, therefore, looked for *E. coli* mutants that were able to grow on lactose as the sole carbon source. Such mutants were indeed isolated and, as expected, were defective in the degradation of nonsense β-galactosidase fragments [14]. Accordingly, these mutants were designated *deg*, and they all mapped to the same chromosomal locus [14]. Surprisingly, these mutants were mucoid, and it soon became evident that *deg* mutants were actually *lon* mutants and that the Deg phenotype is yet another phenotype of the *lon* mutation in addition to the mucoid phenotype and irradiation sensitivity [16]. But what was the molecular mechanism behind these phenotypes? A major step forward in answering this question came from a completely different experimental approach.

While studying the degradation of abnormal proteins in mammalian and bacterial cells Goldberg and colleagues discovered that intracellular degradation of abnormal proteins was ATP dependent [17–19]. This was an unexpected finding, as the cleavage of a peptide bond is a "down-hill" reaction that does not require coupling with ATP hydrolysis. Indeed, intestinal proteases, such as trypsin and chymotrypsin, do not require ATP for their activity. Goldberg and colleagues decided to purify the ATP-dependent protease from *E. coli*. They fractionated lysates of *E. coli* cells and assayed each fraction for degradation of ^{14}C-labelled α-casein both in the presence and absence of ATP. Several proteases were identified and designated Do, Re, Mi, Fa and so on [20]. One of them, protease La, was the first ATP-dependent protease to be purified. Soon afterwards, Chung and Goldberg [21] showed that La was the polypeptide product of the *lon* gene. Simultaneously, the group of Markovitz reported that the CapR DNA-binding protein – previously thought to be a transcriptional regulator – had ATP-dependent protease activity [22]. Over the years, these names – La and CapR – were abandoned and the protease was named Lon, after its gene.

The Mucoid and Irradiation-Sensitivity Phenotypes – Regulation Defects in *E. coli lon* Mutants

Following the realization that *lon* encodes an ATP-dependent protease, the molecular mechanism underlying the two known *lon* phenotypes – mucoid and irradiation sensitivity – were investigated. At this time, the mucoid phenotype was known to result from over-expression of capsule synthesis genes. It was found that capsule synthesis was regulated by a complex set of transcriptional regulators, one of which was a transcription activator, termed RcsA [23], which was later shown to be a substrate of Lon [24]. Under normal growth conditions, capsule synthesis is repressed as a result of the rapid degradation of RcsA by Lon. Therefore, in *lon* mutants RcsA accumulates, leading to transcriptional activation of *cps* genes and, as a result, the over production of capsule polysaccharides.

Irradiation sensitivity of *lon* mutants is caused by increased cytoplasmic concentration of a different type of regulator (see below). At the time that the irradiation sensitivity of *lon* mutants was identified, suppressor strains of this phenotype were

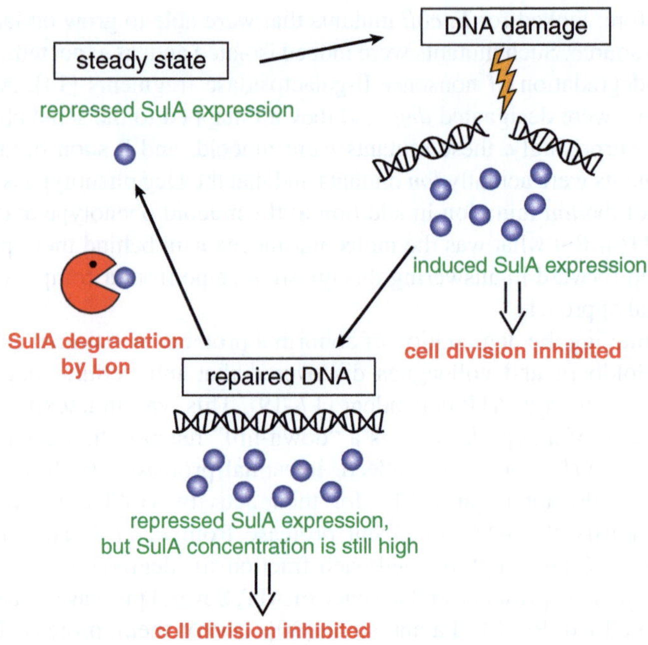

Fig. 2.1 SulA proteolysis and its role in the SOS response. Under normal growth conditions, the cytoplasmic concentration of SulA is very low due to both repressed *sulA* expression and Lon degradation. Following DNA damage, however, SulA expression is induced, and its cytoplasmic concentration increases markedly. At high concentrations, SulA blocks cell division by binding to the septum-forming protein, FtsZ. After DNA repair, resumption of cell division depends on repressed SulA expression and on degradation by Lon. The elongated phenotype of *lon* mutants following ionizing irradiation results from their inability to degrade SulA

isolated in an attempt to understand the cause of this sensitivity. These suppressor mutations were termed *sul* (supressor of *lon*) [25, 26]. One of these supressors, *sulA*, was later found to encode an inhibitor of cell division and a substrate of Lon [27, 28]. Under normal growth conditions, transcription of *sulA* is tightly controlled by the repressor, LexA – a transcriptional repressor of a group of genes that are responsible for recovery from, and repair of, DNA damage [29]. Together, these genes comprise the SOS regulon, and their transcription is induced as a result of LexA destruction in response to DNA damage [30, 31]. The role of SulA is to block cell division after DNA damage in order to prevent inheritance of chromosomal aberrations from mother cells [27]. This is achieved by the interaction of SulA with FtsZ, the protein that comprises the Z-ring during septation [27, 32, 33]. After repair of the damaged DNA, transcription of SOS genes is again repressed, and degradation of SulA by Lon allows the resumption of cell division (Fig. 2.1). When *lon* mutants are irradiated, DNA damage results in induction of *sulA* as part of the SOS response. However, unlike wild type cells, *lon* mutants fail to deplete SulA from the cytoplasm following DNA repair and, as a consequence, are unable to resume cell division.

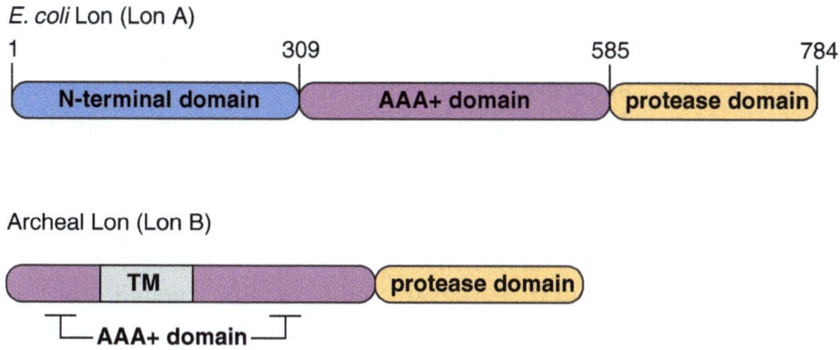

Fig. 2.2 Domain organization of Lon proteases. Shown are the two Lon types found in bacteria (predominantly Lon A) and in archaea (Lon B). TM refers to the trans-membrane anchor that is inserted in between the large and the small AAA+subdomains of Lon B types

Since the identification of SulA and RcsA as Lon substrates and since the elucidation of the regulatory defects that are responsible for the mucoid and irradiation sensitivity phenotypes, many more regulatory proteins have been identified as substrates of Lon (and of related proteases) [34]. These proteins include transcription regulators and enzymes that are involved in regulation of a variety of cellular processes, including virulence, sporulation and adaptation to stress conditions (see also [35–37]). Indeed, ATP-dependent proteolysis plays a prominent regulatory role in all living cells.

Structure and Function of Lon Proteases

Lon proteases are homohexameric ring-shaped complexes that are encoded by a single gene. Each Lon polypeptide chain contains both a protease and an AAA+ domain. The proteolytic active site of Lon proteases is composed of a Ser-Lys catalytic dyad [38, 39], an arrangement that differs from the Ser-His-Asp catalytic triad that is found in serine proteases, such as trypsin, chymotrypsin, elastase, and AAA+ proteases, such as ClpXP and the proteasome [40].

Based on sequence homology and structural features, Lon proteases may be classified into two distinct types (Fig. 2.2). Lon A (found in most bacteria) contains a large N-terminal domain (~300 amino acids), an AAA+ domain, and a C-terminal protease domain. Lon B (found predominantly in Archaea) lacks the N-terminal domain, but contains a trans-membrane region that is inserted between the large and small AAA+ subdomains [41]. In the folded structure, the trans-membrane region protrudes from the apical side of the protease [42], such that the pore region of the protease faces the inner side of the cell membrane. In most bacteria, FtsH (a membrane-bound protease) is responsible for intracellular degradation of membrane-bound proteins. Archaea however, lack FtsH and instead Lon B acts as their membrane-bound protease. Lon B is found also in some bacterial species; for example, *Bacillus subtilis*, *Pseudomonas aeruginosa* and *Thermotoga maritime* possess both types of Lon proteases in the cytoplasm.

The information on the structure and subunit stoichiometry of Lon proteases has accumulated gradually over the years. Initially, based on size-exclusion chromatography, it was hypothesized that Lon forms either tetramers or pentamers [21]. Later, an open-ring heptameric oligomerization of the protease was proposed, based on cryo electron microscopy (cryo-EM) studies of mitochondrial Lon from *Saccharomyces cerevisiae* [43]. However, non-hexameric subunit stoichiometry became improbable with the realization that AAA+ proteins form hexamers [44], suggesting that earlier interpretations of Lon oligomerization were mistaken.

The first evidence that Lon exists as a hexamer came from sedimentation analysis of the *Mycobacterium smegmatis* Lon A protease [45]. These studies further indicated that magnesium ions are necessary for hexamerization of Lon. These conclusions were later supported by a more recent cryo-EM study of *E. coli* Lon, clearly showing a magnesium-dependent hexamerization of the protease [46]. Lon hexamerization was also evident from the first Lon crystal structure. However, Lon is infamous for being recalcitrant to crystallization, specifically, it becomes insoluble in the presence of magnesium ions at the high protein concentrations. In contrast, isolated domains of *E. coli* Lon could be crystallized, and the first published X-ray structure of Lon was of the protease domain of the *E. coli* enzyme [39]. This structure showed a ring-shaped hexamer, in which the active site (Ser679) was completely exposed. This architecture challenged the basic principle of compartmentalized proteolysis by AAA+ proteases. Indeed, the three-dimensional structures of related AAA+ proteases demonstrated that the proteolytic active sites were sequestered in a closed chamber [47, 48]. It was not clear, therefore, whether the observed architecture of the *E. coli* Lon protease domain only represented the spatial rearrangement of the domain in the absence of the AAA+ and N-terminal domains or whether it truly represented the architecture of the domain, in the context of the full-length protein. Nonetheless, this structure clearly demonstrated that the activity of Lon is based on a Ser-Lys catalytic dyad, as indicated by the proximity of Lys722 to the active site Ser679 (mutated to Ala in the protein used for crystallisation) and the absence of other potential catalytic side chains. Two other protease domain structures, those of *Methanococcus jannaschii* and *Archaeoglobus fulgidus* (both Archaea), revealed similar oligomerization and domain architectures, with certain variations in the organisation of the active site residues [49, 50].

In addition to the structure of the protease domain, structures of two subdomains of the *E. coli* Lon are now available. One is the small sub-domain of the AAA+ domain and the second is part of the N-terminal domain (residues 1–119 out of a total length of about 309 residues) [51, 52]. Notably, however, a solved structure of a full-length Lon A is still missing, thereby preventing a better understanding of many function-related issues. Major progress was recently achieved with the publication of two Lon structures: that of the archeal *Thermococcus onnurineus* Lon B (*Ton* Lon; Fig. 2.3) and that of the bacterial *B. subtilis* Lon A [42, 53]. A structure was solved for most of the N-terminal domain of the *B. subtilis* Lon (discussed

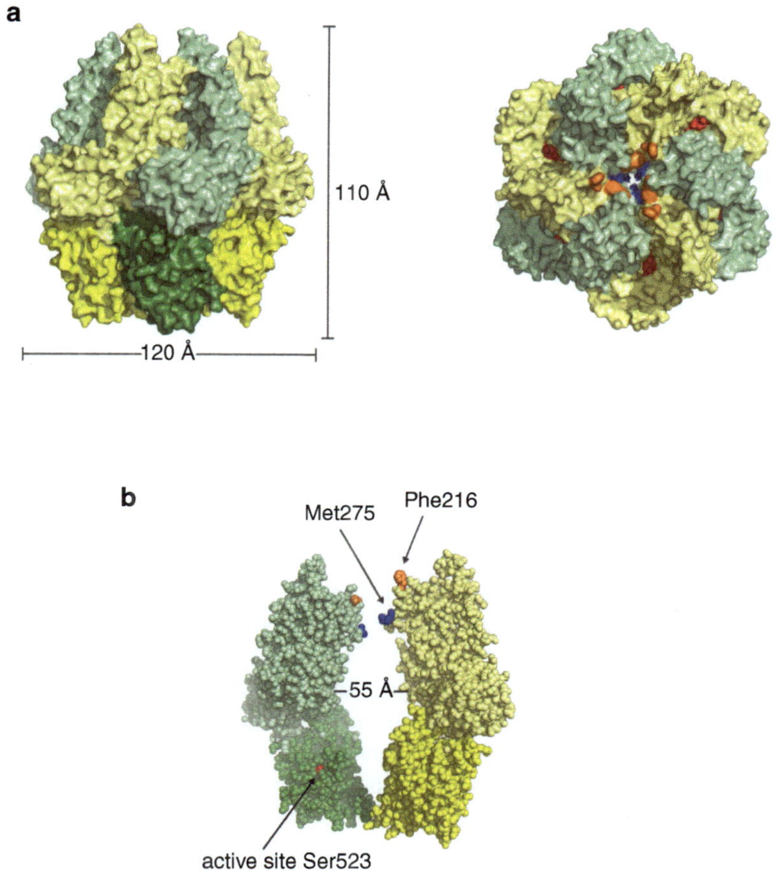

Fig. 2.3 Crystal structure of an archaeal Lon protease. (**a**) Surface representations of a *side-view* (*left*) and *top-view* (*right*) of the *Ton* Lon protease (3K1J). In the side-view representation, the protease domain of each monomer is shown in *dark green* or *dark yellow*, and the AAA+domain of each monomer is shown in *light green* or *light yellow*. In the *top-view* representation, ADP molecules are shown in *red*. Pore loop residues, Phe216 (*in orange*) and Met275 (*in blue*) are also shown. (**b**) The degradation chamber is shown by a cut-away side view of two opposing subunits

below) and separately for a large fragment comprising the AAA+ and protease domains. In this structure, these domains are arranged in an open hexameric spiral, probably reflecting an artifactual arrangement in the absence of the N-domain. Nonetheless, this structure was a major step forward in visualizing domain orientation in Lon proteases and the formation of a closed degradation chamber. The *Ton* Lon structure shown in Fig. 2.3 represents a full-length protease, with the exception of the trans-membrane hydrophobic anchor, which was removed from the protein used for crystallisation. The trans-membrane anchor is proposed to be joined, via two flexible linkers, to the top of the folded oligomeric protease. This arrangement allows the pore entrance of the protease to face the membrane, from where it captures its substrates. *Ton* Lon assumes a spherical shape (Fig. 2.3), with the proteolytic active sites hidden in a degradation chamber. The apical half-sphere of this degradation chamber is composed of the AAA+ domain, and its basal half-sphere is composed of the protease domain. Together, the two oligomeric domains form a closed sphere of ~55 Å at its widest part. A narrow pore for substrate entrance from the apical part is obvious, and so is an opening at the bottom of the chamber for the exit of degradation products. This structure shows, for the first time, the orientation of Lon pore loop residues, Phe216 and Met275 (Fig. 2.3). These protrude from the apical opening, thus forming an axial portal that is most likely involved in substrate gating and processing (see [54] for further information on pore loops of AAA+ proteases). Indeed, a Phe216Ala mutant of *Ton* Lon failed to degrade a protein substrate, suggesting a role for Phe216 in substrate binding and translocation into the protease domain [42, 53]. By contrast, a Met275Ala mutant retained its proteolytic activity, indicating that it does not play a direct role in substrate processing. The location however, of Met275 at the entrance of the protease chamber may suggest a gating role for this residue, to prevent random protein degradation.

The oligomeric protease domain of *Ton* Lon is arranged in almost perfect six fold symmetry. In contrast, the AAA+ domain forms asymmetric hexamers, despite similar AAA+ conformations in each monomer. This asymmetric arrangement results from two alternative rotation angles of the monomeric AAA+ domain relative to the monomeric protease domain of the same polypeptide chain. The two conformers have different affinities for nucleotides, an observation that is consistent with a previous biochemical analysis of nucleotide binding by *E. coli* Lon [55]. The rotation-based asymmetry and the resulting changes in nucleotide affinity closely resemble the domain-rotation and nucleotide binding observed for ClpX and may reflect a common mechanism for substrate processing by the two AAA+ proteases [56, 57]. Asymmetric nucleotide transactions have also been reported for HslUV, based on biochemical studies [58]. In contrast, asymmetric domain arrangements were not observed in the crystal structure of HslUV protease [47].

Ton Lon and all other Lon B proteases lack the N-terminal domain that is found in Lon A types. It was found that Lon variants that carry a range of deletions in their N-terminal domain have lost their ability to bind substrates but retain their ATPase activity and can hydrolyze small peptides [59]. These observations indicated that the large N-terminal domains of Lon A proteases play an important role in substrate binding. Currently, there is no solved structure of a full-length Lon A or of an oligomeric form of the N-terminal domain. There are, however, known structures for parts of a monomeric N-terminal domain [52, 53]. The largest segment of the N-terminal domain with a known structure (in its monomeric form) is that of the *B. subtilis* Lon A [53]. Indeed, a structure has been solved for residues 1–209 and 246–300 out of the ~300-residue *B. subtilis* Lon N-terminal domain. In this structure, there are three hydrophobic patches, each a candidate for substrate binding, and these patches can explain the affinity of Lon A proteases for misfolded proteins. Clearly, important information regarding substrate binding to the N-terminal domain is still missing, and the determination of an oligomeric N-terminal domain structure (if not of a full-length Lon A hexamer) and mutational analysis are much in need.

PinA: A Specific Protein Inhibitor of the *E. coli* Lon

As described above, prior to the concept of ATP-dependent proteolysis and before the discovery of Lon, researchers noticed that nonsense fragments and, more generally, abnormal proteins were rapidly degraded in *E. coli* [11]. In sharp contrast to this observation, it was also noticed that nonsense fragments of bacteriophage T4 proteins were quite stable in *E. coli* following infection [60]. It was subsequently shown that T4 infection shuts down not only the degradation of the phage's nonsense fragments but also degradation of other abnormal proteins in the host cell [61]. The T4 gene that is responsible for the proteolysis-arrest was mapped and termed *pin* (proteolysis inhibition). It was later found that *pin* specifically inhibits Lon and no other ATP-dependent protease [62], but the importance of Lon inhibition for the development of T4 bacteriophage remains unclear.

PinA (18,816 Da), the product of the *pin* gene, was purified and studied biochemically [63, 64]. It binds Lon tightly, as indicated by a dissociation constant (K_D) of ~10 nM. Following binding, PinA inhibits ATP hydrolysis by Lon. By contrast, it does not inhibit the peptidase activity of Lon, nor does it inhibit the ability of substrates or ATP binding to stimulate this activity. Therefore, PinA inhibits the AAA+ domain but does not prevent substrate binding. Indeed, kinetic measurements of Lon degradation in the presence of increasing PinA concentrations exhibited a V_{max} (rather than K_M) effect. These measurements indicated that PinA does not compete with the substrate for Lon binding, but rather inhibits the catalytic activity of the protease. Currently, however, neither the PinA binding site in Lon, nor the PinA residues that interact with the protease are known, and at present there are no structural data that can shed light on this unique interaction. It will be interesting to learn the fine details of nature's design of an AAA+ protease inhibitor.

Degradation of Misfolded Proteins by Lon and Principles of Substrate Recognition

Proper folding of proteins is required for the correct functioning of all cells. However, a protein cannot maintain proper folding forever, and even the most stable proteins eventually lose their tertiary structure. The occurrence of protein misfolding events is dramatically accelerated under certain conditions, such as elevated temperatures, unsuitable pH and exposure to oxidative reagents. Such conditions endanger the cell, and above a protein "misfolding" threshold, lead to its death. Therefore, all cells possess molecular mechanisms that promote survival under protein misfolding stresses. These mechanisms are collectively referred to as "protein quality control" and are predominantly characterized by the activity of chaperones and ATP-dependent proteases [65]. Various chaperones act to prevent unfolding, refold misfolded species, and disassemble protein aggregates. Proteases, on the other hand, prevent accumulation of misfolded proteins by degrading them.

In bacteria and in the mitochondria of eukaryotes, Lon is the major protein quality control protease and, as such, is responsible for most of the ATP-dependent degradation of misfolded proteins [34]. It degrades a broad range of different proteins and yet it maintains selectivity. Indeed, broad specificity and strict selectivity, two apparently conflicting qualities, are remarkably combined in bacterial proteases. How does Lon probe the folding state of a protein, efficiently distinguishing between misfolded and folded species and eliminating only the misfolded ones? It appears that the selectivity of Lon relies on its interaction with sequences that are often found in the hydrophobic core of proteins but very rarely on the surface [66]. Consequently, most folded proteins are resistant to degradation by Lon simply because they do not expose Lon recognition sequences. In misfolded proteins, however, these sequences are readily available for interaction with Lon and promote degradation of the substrate. Characterization of Lon recognition sequences has shown them to be composed of short hydrophobic polypeptide stretches of about 15 residues that contain aromatic residues [66]. In contrast, negatively charged residues severely hamper Lon binding. Within these constraints, much freedom is allowed, thereby ensuring the unusual broad specificity of the protease. Indeed, Lon recognition sequences vary dramatically in their amino acid composition. Owing to this variability, Lon is able to recognize a vast number of different sequences and substrates. At the same time, the rare occurrence of these Lon recognition sequences on the surface of a protein, make them ideal for recognition by a protein quality control protease.

Recognition of Natively Folded Substrates and Allosteric Activation of Lon by Its Substrates

In addition to misfolded proteins, Lon also degrades many natively folded regulatory proteins [38]. These natively folded proteins usually carry a degradation tag at their amino or carboxy terminus. As the ends of proteins are often accessible, proteins that are tagged at their termini may be recognized by Lon in the absence of unfolding. The best-studied example of recognition of a regulatory protein by Lon is that of *E. coli* SulA. SulA carries a Lon degradation tag at its C-terminus [67]. The tag sequence is moderately hydrophobic and contains a histidine at the C-terminus and a penultimate tyrosine. These two residues play a critical role in Lon binding, as has been shown in site-directed mutagenesis experiments [68, 69]. It is evident that the SulA tag is autonomous in its ability to direct efficient recognition by Lon, as a peptide that is comprised of the last 20 amino acids of SulA (sul20C) is efficiently degraded by Lon. In addition, the attachment of these 20 amino acids of SulA to the C-terminus of a model protein is sufficient to generate a substrate that is rapidly degraded by Lon [68].

In a recent study, that compared the degradation kinetics of a model substrate that was tagged by either the SulA tag (sul20C) or by a degradation tag of the same length that was derived from β-galactosidase (β20) revealed surprising rate differences [68]. Oddly, the sul20C tagged protein was degraded about ten-fold

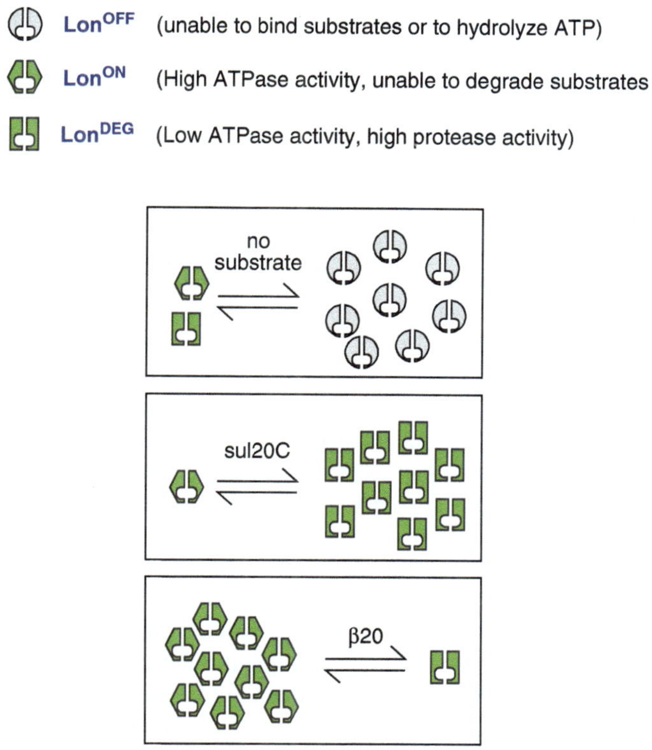

Fig. 2.4 An allostery-based model for Lon stimulation by its substrates. The model is based on three alternative Lon conformations that exist in equilibrium. The sul20C tag shifts the equilibrium to the Lon^{DEG} conformation, resulting in rapid and efficient degradation of the bound substrates. The β20 tag shifts the equilibrium to the Lon^{ON} conformation, simulating a high rate of ATP hydrolysis but not degradation. Lon^{OFF} dominates in the absence of substrate

faster at substrate saturation. In other words, the V_{max} for the degradation of the sul20C-tagged substrate was dramatically higher than that of the β20-tagged substrate. This observation, suggested that the magnitude of the catalytic activity of Lon – whether, high or low – is determined by the degradation tag. Moreover, the rapid degradation of the sul20C-tagged substrate required considerably less ATP than degradation of the β20-tagged substrate. Therefore, Lon degraded the sul20C-tagged substrate not only more rapidly but also more efficiently in terms of energy consumption. How can the degradation tag of the substrate determine the operation mode of Lon? A tag-induced allosteric transition model provided the answer to this question. According to this model, which is based on the Monod-Wyman-Changeux (MWC) model for enzyme allostery [70], Lon can adopt three inter-convertible conformations that exist in equilibrium (Fig. 2.4). One conformation, Lon^{OFF}, can neither bind substrates nor hydrolyze ATP. The second conformation, Lon^{ON}, is proteolytically inactive but hydrolyzes ATP at a high rate. The third conformation,

Lon^{DEG}, is an active protease that has a low rate of ATP hydrolysis. In the absence of a substrate, the dominating conformation is Lon^{OFF}, and the low rate of ATP hydrolysis by Lon [66, 68] results from the low abundance of Lon^{ON} and Lon^{DEG}. In the presence of a substrate, either Lon^{ON} or Lon^{DEG} predominate. Importantly, the model postulates that *sul20C*-type tags bind preferentially to the Lon^{DEG} conformation, whereas *β20*-type tags bind preferentially to the Lon^{ON} conformation. Consequently, substrates that carry *sul20C*-type tags shift the equilibrium in favour of the Lon^{DEG} conformation, leading to their rapid degradation. In contrast, substrates that carry *β20*-type tags shift the equilibrium in favour of the Lon^{ON} conformation, and the slow degradation rate observed for these substrates is due to the low abundance of substrate-bound enzymes in the Lon^{DEG} conformation. In accordance with biochemical data, the allostery-based model postulates at least two substrate-binding sites in Lon. As a result, cooperative binding becomes an intrinsic property of the model, as proposed by the MWC model. Indeed, cooperativity in substrate binding is a hallmark of Lon degradation kinetics [55, 66, 68]. Currently, there are no adequate structural data to support this allostery-based model. Support for this model would require a comparison between a full-length Lon A structure and substrate-bound protease structures, but to date neither has been determined.

Is there a physiological rationale to the alternative activity modes of Lon degradation? In some instances, it is easy to comprehend how adjustment of the degradation rate can contribute to physiological needs, as in the case of SulA degradation. As described earlier in this chapter, SulA is an inhibitor of cell division whose expression is induced in response to DNA damage as part of the SOS response. After the repair of DNA damage, resumption of cell cycle depends on SulA degradation by Lon (Fig. 2.1). The faster SulA is degraded after DNA repair, the faster the bacteria can resume normal growth. It therefore makes sense that the SulA degradation tag induces Lon to act in a rapid degradation mode. Are there, however, cases in which slow degradation is preferred? This is not clear, but one can speculate that in the case of degradation of misfolded proteins, the cell can benefit if Lon does not automatically degrade the substrate, as "binding and release" without degradation may give the misfolded protein a second chance to fold correctly. Since protein refolding is much more cost-effective to the cell, in terms of energy consumption, than degradation and re-synthesis, it is tempting to speculate that Lon can act *in vivo* not only as a protease, but also as a chaperone. Evidence for such as activity is, however, still eagerly awaited.

References

1. Jacob F, Monod J (1961) Genetic regulatory mechanisms in the synthesis of proteins. J Mol Biol 3:318–356
2. Markovitz A (1964) Regulatory mechanisms for synthesis of capsular polysaccharide in mucoid mutants of Escherichia coli K12. Proc Natl Acad Sci U S A 51:239–246
3. Adler HI, Hardigree AA (1964) Cell elongation in strains of Escherichia coli. J Bacteriol 87(5):1240–1242

4. Adler HI, Hardigree AA (1964) Analysis of a gene controlling cell division and sensitivity to radiation in Escherichia coli. J Bacteriol 87:720–726
5. Howard-Flanders P, Simson E, Theriot L (1964) A locus that controls filament formation and sensitivity to radiation in Escherichia coli K-12. Genetics 49:237–246
6. Bush JW, Markovitz A (1973) The genetic basis for mucoidy and radiation sensitivity in capR (lon) mutants of E. coli K-12. Genetics 74(2):215–225
7. Zehnbauer BA, Foley EC, Henderson GW, Markovitz A (1981) Identification and purification of the Lon+(capR+) gene product, a DNA-binding protein. Proc Natl Acad Sci U S A 78(4):2043–2047
8. Newton WA, Beckwith JR, Zipser D, Brenner S (1965) Nonsense mutants and polarity in the lac operon of Escherichia coli. J Mol Biol 14(1):290–296
9. Morrison SL, Zipser D (1970) Polypeptide products of nonsense mutations. I. Termination fragments from nonsense mutations in the Z gene of the lac operon of Escherichia coli. J Mol Biol 50(2):359–371
10. Morrison SL, Zipser D, Goldschmidt R (1971) Polypeptide products of nonsense mutations. II. Minor fragments produced by nonsense mutations in the z gene of the lactose operon of Escherichia coli. J Mol Biol 60(3):485–497
11. Goldschmidt R (1970) In vivo degradation of nonsense fragments in E. coli. Nature 228(5277):1151–1154
12. Goldberg AL (1972) Degradation of abnormal proteins in Escherichia coli (protein breakdown-protein structure-mistranslation-amino acid analogs-puromycin). Proc Natl Acad Sci U S A 69(2):422–426
13. Platt T, Miller JH, Weber K (1970) In vivo degradation of mutant lac repressor. Nature 228(5277):1154–1156
14. Bukhari AI, Zipser D (1973) Mutants of Escherichia coli with a defect in the degradation of nonsense fragments. Nat New Biol 243(129):238–241
15. Ullmann A, Jacob F, Monod J (1967) Characterization by in vitro complementation of a peptide corresponding to an operator-proximal segment of the beta-galactosidase structural gene of Escherichia coli. J Mol Biol 24(2):339–343
16. Gottesman S, Zipser D (1978) Deg phenotype of Escherichia coli lon mutants. J Bacteriol 133(2):844–851
17. Etlinger JD, Goldberg AL (1977) A soluble ATP-dependent proteolytic system responsible for the degradation of abnormal proteins in reticulocytes. Proc Natl Acad Sci U S A 74(1):54–58
18. Goldberg AL, Strnad NP, Swamy KH (1979) Studies of the ATP dependence of protein degradation in cells and cell extracts. CIBA Found Symp 75:227–251
19. Murakami K, Voellmy R, Goldberg AL (1979) Protein degradation is stimulated by ATP in extracts of Escherichia coli. J Biol Chem 254(17):8194–8200
20. Swamy KH, Goldberg AL (1981) E. coli contains eight soluble proteolytic activities, one being ATP dependent. Nature 292(5824):652–654
21. Chung CH, Goldberg AL (1981) The product of the lon (capR) gene in Escherichia coli is the ATP-dependent protease, protease La. Proc Natl Acad Sci U S A 78(8):4931–4935
22. Charette MF, Henderson GW, Markovitz A (1981) ATP hydrolysis-dependent protease activity of the lon (capR) protein of Escherichia coli K-12. Proc Natl Acad Sci U S A 78(8):4728–4732
23. Gottesman S, Trisler P, Torres-Cabassa A (1985) Regulation of capsular polysaccharide synthesis in Escherichia coli K-12: characterization of three regulatory genes. J Bacteriol 162(3):1111–1119
24. Torres-Cabassa AS, Gottesman S (1987) Capsule synthesis in Escherichia coli K-12 is regulated by proteolysis. J Bacteriol 169(3):981–989
25. Donch J, Chung YS, Greenberg J (1969) Locus for radiation resistance in Escherichia coli strain B-r. Genetics 61(2):363–370
26. Johnson BF, Greenberg J (1975) Mapping of sul, the suppressor of lon in Escherichia coli. J Bacteriol 122(2):570–574
27. Gottesman S, Halpern E, Trisler P (1981) Role of sulA and sulB in filamentation by lon mutants of Escherichia coli K-12. J Bacteriol 148(1):265–273

28. Mizusawa S, Gottesman S (1983) Protein degradation in Escherichia coli: the lon gene controls the stability of sulA protein. Proc Natl Acad Sci U S A 80(2):358–362
29. Butala M, Zgur-Bertok D, Busby SJ (2009) The bacterial LexA transcriptional repressor. Cell Mol Life Sci 66(1):82–93
30. d'Ari R (1985) The SOS system. Biochimie 67(3–4):343–347
31. Sutton MD, Smith BT, Godoy VG, Walker GC (2000) The SOS response: recent insights into umuDC-dependent mutagenesis and DNA damage tolerance. Annu Rev Genet 34:479–497
32. Bi E, Lutkenhaus J (1993) Cell division inhibitors SulA and MinCD prevent formation of the FtsZ ring. J Bacteriol 175(4):1118–1125
33. Cordell SC, Robinson EJ, Lowe J (2003) Crystal structure of the SOS cell division inhibitor SulA and in complex with FtsZ. Proc Natl Acad Sci U S A 100(13):7889–7894
34. Tsilibaris V, Maenhaut-Michel G, Van Melderen L (2006) Biological roles of the Lon ATP-dependent protease. Res Microbiol 157(8):701–713
35. Barchinger SE, Ades SE (2013) Regulated proteolysis: control of the Escherichia coli σE-dependent cell envelope stress response. In: Dougan DA (ed) Regulated proteolysis in microorganisms. Springer, Subcell Biochem 66:129–160
36. Frees D, Brøndsted L, Ingmer H (2013) Bacterial proteases and virulence. In: Dougan DA (ed) Regulated proteolysis in microorganisms. Springer, Subcell Biochem 66:161–192
37. Molière N, Turgay K (2013) General and regulatory proteolysis in Bacillus subtilis. In: Dougan DA (ed) Regulated proteolysis in microorganisms. Springer, Subcell Biochem 66:73–103
38. Birghan C, Mundt E, Gorbalenya AE (2000) A non-canonical lon proteinase lacking the ATPase domain employs the ser-Lys catalytic dyad to exercise broad control over the life cycle of a double-stranded RNA virus. EMBO J 19(1):114–123
39. Botos I, Melnikov EE, Cherry S, Tropea JE et al (2004) The catalytic domain of Escherichia coli Lon protease has a unique fold and a Ser-Lys dyad in the active site. J Biol Chem 279(9):8140–8148
40. Polgar L (2005) The catalytic triad of serine peptidases. Cell Mol Life Sci 62(19–20): 2161–2172
41. Rotanova TV, Melnikov EE, Khalatova AG, Makhovskaya OV et al (2004) Classification of ATP-dependent proteases Lon and comparison of the active sites of their proteolytic domains. Eur J Biochem 271(23–24):4865–4871
42. Cha SS, An YJ, Lee CR, Lee HS et al (2010) Crystal structure of Lon protease: molecular architecture of gated entry to a sequestered degradation chamber. EMBO J 29(20):3520–3530
43. Stahlberg H, Kutejova E, Suda K, Wolpensinger B et al (1999) Mitochondrial Lon of Saccharomyces cerevisiae is a ring-shaped protease with seven flexible subunits. Proc Natl Acad Sci U S A 96(12):6787–6790
44. Vale RD (2000) AAA proteins. Lords of the ring. J Cell Biol 150(1):F13–F19
45. Rudyak SG, Brenowitz M, Shrader TE (2001) Mg2+-linked oligomerization modulates the catalytic activity of the Lon (La) protease from Mycobacterium smegmatis. Biochemistry 40(31):9317–9323
46. Park SC, Jia B, Yang JK, Van DL et al (2006) Oligomeric structure of the ATP-dependent protease La (Lon) of Escherichia coli. Mol Cells 21(1):129–134
47. Sousa MC, Trame CB, Tsuruta H, Wilbanks SM et al (2000) Crystal and solution structures of an HslUV protease-chaperone complex. Cell 103(4):633–643
48. Wang J, Hartling JA, Flanagan JM (1997) The structure of ClpP at 2.3 A resolution suggests a model for ATP-dependent proteolysis. Cell 91(4):447–456
49. Botos I, Melnikov EE, Cherry S, Kozlov S et al (2005) Atomic-resolution crystal structure of the proteolytic domain of Archaeoglobus fulgidus lon reveals the conformational variability in the active sites of lon proteases. J Mol Biol 351(1):144–157
50. Im YJ, Na Y, Kang GB, Rho SH et al (2004) The active site of a lon protease from Methanococcus jannaschii distinctly differs from the canonical catalytic Dyad of Lon proteases. J Biol Chem 279(51):53451–53457
51. Botos I, Melnikov EE, Cherry S, Khalatova AG (2004) Crystal structure of the AAA+alpha domain of E. coli Lon protease at 1.9A resolution. J Struct Biol 146(1–2):113–122

52. Li M, Rasulova F, Melnikov EE, Rotanova TV (2005) Crystal structure of the N-terminal domain of E. coli Lon protease. Protein Sci 14(11):2895–2900
53. Duman RE, Lowe J (2010) Crystal structures of Bacillus subtilis Lon protease. J Mol Biol 401(4):653–670
54. Gur E, Ottofuelling R, Dougan DA (2013) Machines of destruction: AAA+ proteases and the adaptors that control them. In: Dougan DA (ed) Regulated proteolysis in microorganisms. Springer, Subcell Biochem 66:3–33
55. Vineyard D, Patterson-Ward J, Lee I (2006) Single-turnover kinetic experiments confirm the existence of high- and low-affinity ATPase sites in Escherichia coli Lon protease. Biochemistry 45(14):4602–4610
56. Glynn SE, Martin A, Nager AR, Baker TA et al (2009) Structures of asymmetric ClpX hexamers reveal nucleotide-dependent motions in a AAA+protein-unfolding machine. Cell 139(4):744–756
57. Hersch GL, Burton RE, Bolon DN, Baker TA et al (2005) Asymmetric interactions of ATP with the AAA+ClpX6 unfoldase: allosteric control of a protein machine. Cell 121(7):1017–1027
58. Yakamavich JA, Baker TA, Sauer RT (2008) Asymmetric nucleotide transactions of the HslUV protease. J Mol Biol 380(5):946–957
59. Roudiak SG, Shrader TE (1998) Functional role of the N-terminal region of the Lon protease from Mycobacterium smegmatis. Biochemistry 37(32):11255–11263
60. Celis JE, Smith JD, Brenner S (1973) Correlation between genetic and translational maps of gene 23 in bacteriophage T4. Nat New Biol 241(109):130–132
61. Simon LD, Tomczak K, St John AC (1978) Bacteriophages inhibit degradation of abnormal proteins in E. coli. Nature 275(5679):424–428
62. Skorupski K, Tomaschewski J, Ruger W, Simon LD (1988) A bacteriophage T4 gene which functions to inhibit Escherichia coli Lon protease. J Bacteriol 170(7):3016–3024
63. Hilliard JJ, Maurizi MR, Simon LD (1998) Isolation and characterization of the phage T4 PinA protein, an inhibitor of the ATP-dependent lon protease of Escherichia coli. J Biol Chem 273(1):518–523
64. Hilliard JJ, Simon LD, Van Melderen L, Maurizi MR (1998) PinA inhibits ATP hydrolysis and energy-dependent protein degradation by Lon protease. J Biol Chem 273(1):524–527
65. Mogk A, Huber D, Bukau B (2011) Integrating protein homeostasis strategies in prokaryotes. Cold Spring Harb Perspect Biol 3(4)
66. Gur E, Sauer RT (2008) Recognition of misfolded proteins by Lon, a AAA(+) protease. Genes Dev 22(16):2267–2277
67. Ishii Y, Sonezaki S, Iwasaki Y, Miyata Y et al (2000) Regulatory role of C-terminal residues of SulA in its degradation by Lon protease in Escherichia coli. J Biochem 127(5):837–844
68. Gur E, Sauer RT (2009) Degrons in protein substrates program the speed and operating efficiency of the AAA + Lon proteolytic machine. Proc Natl Acad Sci U S A 106(44):18503–18508
69. Ishii Y, Amano F (2001) Regulation of SulA cleavage by Lon protease by the C-terminal amino acid of SulA, histidine. Biochem J 358(Pt 2):473–480
70. Monod J, Wyman J, Changeux JP (1965) On the nature of allosteric transitions: a plausible model. J Mol Biol 12:88–118

Chapter 3
FtsH Protease-Mediated Regulation of Various Cellular Functions

Takashi Okuno and Teru Ogura

Abstract FtsH, a member of the AAA (ATPases associated with a variety of cellular activities) family of proteins, is an ATP-dependent protease of ~71 kDa anchored to the inner membrane. It plays crucial roles in a variety of cellular processes. It is responsible for the degradation of both membrane and cytoplasmic substrate proteins. Substrate proteins are unfolded and translocated through the central pore of the ATPase domain into the proteolytic chamber, where the polypeptide chains are processively degraded into short peptides. FtsH is not only involved in the proteolytic elimination of unnecessary proteins, but also in the proteolytic regulation of a number of cellular functions. Its role in proteolytic regulation is achieved by one of two approaches, either the cellular levels of a regulatory protein are controlled by processive degradation of the entire protein, or the activity of a particular substrate protein is modified by processing. In the latter case, protein processing requires the presence of a stable domain within the substrate. Since FtsH does not have a robust unfolding activity, this stable domain is sufficient to abort processive degradation of the protein – resulting in release of a stable protein fragment.

Keywords FtsH • Processive degradation • Protein processing • Protein quality control • Regulation of lipid synthesis

T. Okuno
Department of Material and Biological Chemistry, Faculty of Science,
Yamagata University, Yamagata 990-8560, Japan
e-mail: okuno@sci.kj.yamagata-u.ac.jp

T. Ogura (✉)
Department of Molecular Cell Biology, Institute of Molecular
Embryology and Genetics, Kumamoto University,
Kumamoto 860-0811, Japan
e-mail: ogura@gpo.kumamoto-u.ac.jp

D.A. Dougan (ed.), *Regulated Proteolysis in Microorganisms*, Subcellular Biochemistry 66,
DOI 10.1007/978-94-007-5940-4_3,

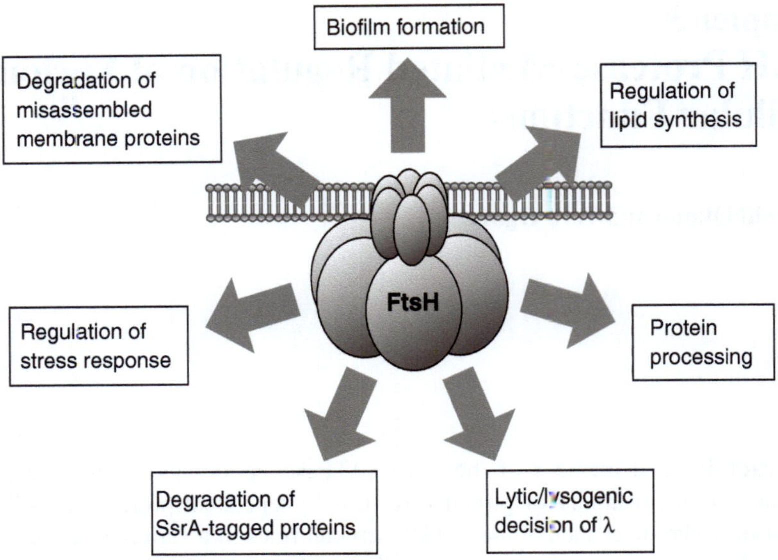

Fig. 3.1 Various cellular functions regulated by FtsH protease. FtsH, a membrane-bound AAA protease, is responsible for regulation of various cellular functions in *E. coli*. FtsH acts on both membrane and cytoplasmic substrate proteins. FtsH functions not only in the proteolytic elimination of unnecessary proteins but also controls the cellular levels of several regulatory proteins and the processing of specific substrate proteins (For further details refer to the main text)

Introduction

FtsH is an evolutionarily conserved protein that is present in all bacterial cells. It consists of transmembrane segment 1 (TM1), periplasmic domain, TM2, cytoplasmic ATPase and protease domains in this order from the N-terminus [1–6]. FtsH is a zinc-binding metalloprotease, which forms a homohexameric ring-shaped structure. It can degrade unstructured model substrates as well as structurally unstable substrate proteins, which can be easily unfolded. Unfolded polypeptide chains are translocated through the central pore of the ATPase domain into the proteolytic chamber. Proteolysis by FtsH is processive and most degradation products are short peptides of several to 20 amino acid residues in length [7]. In some cases, however, larger functional products containing at least one stable domain are released from FtsH as a result of incomplete processive proteolysis. This type of protein processing depends on domain stability and not sequence specificity. Eukaryotic homologs of FtsH have been identified in mitochondria and chloroplasts [8–11]. In mitochondria, there are two types of FtsH homologs, commonly referred to as *i*-AAA and *m*-AAA proteases depending on their topology in the mitochondrial inner membrane. Dysfunction of these proteins in humans has been related to a variety of diseases. In this chapter, however, we will mainly overview the cellular functions controlled by *Escherichia coli* FtsH (Fig. 3.1) and its regulatory mechanisms, and only briefly discuss some of the recent advances of other bacterial or mitochondrial homologs of FtsH.

Quality Control of Cytoplasmic Proteins

Regulation of the Heat Shock Response

Under normal cellular conditions, the heat shock transcriptional factor σ^{32} (which binds to the promoter region of heat shock genes) is rapidly degraded by FtsH. Upon shift to stress conditions, such as a temperature upshift, the cellular levels of σ^{32} increase ~15–20 fold in *E. coli*. This rapid increase in the cellular levels of σ^{32}, leads to an induction of stress response proteins, which is important for cell survival under these conditions. This increase in the level of σ^{32} is transient, which ensures the rapid and transient induction of stress response proteins. Sigma32 accumulates in *ftsH* mutant strains. The *ftsH* gene has been found to encode an inner-membrane anchored ATP-dependent AAA-type protease, which contributes to the efficient degradation of σ^{32} [12, 13]. Although the soluble cytoplasmic ATP-dependent proteases such as ClpAP, ClpXP and HslUV contribute to the degradation of σ^{32}, to some extent, the membrane-anchored FtsH recognizes and degrades σ^{32} preferentially. Currently, the precise amino acid sequence of σ^{32} responsible for recognition and degradation by FtsH has not yet been defined, however substitution of amino acid residues in region 2.1 and region C of σ^{32} affects FtsH-dependent stability of σ^{32} in the cell [14–16]. Since FtsH rapidly degrades destabilized σ^{32} *in vitro*, FtsH may only act on unstructured σ^{32}. Interestingly, to date, the efficient *in vivo* degradation of σ^{32} has not been reconstituted *in vitro* using purified FtsH. As such, it has been suggested that DnaK/DnaJ (DnaK/J) contributes to the stability of σ^{32} *in vivo*. Careful *in vitro* analysis has revealed that DnaJ binding to region 2.1 of σ^{32} destabilizes a distant region in close vicinity of the DnaK-binding site, and that DnaK destabilizes a region in the N-terminal domain [17]. If DnaK/J-induced destabilization of the N-terminal domain might facilitate degradation of σ^{32} by FtsH, it would be consistent with the fact that FtsH degrades σ^{32} from the N terminus to the C terminus [18]. So far, however, a DnaK/J-mediated stimulation of σ^{32} degradation by FtsH *in vitro* has not been reported. Indeed, the cooperative regulation of the FtsH activity by DnaK/J may be more complicated. A novel *in vitro* assay system, containing additional factors/components, needs to be developed to reveal the regulatory mechanism of FtsH-mediated degradation of σ^{32} by DnaK/J. It should also be noted that the transcriptional activity of σ^{32} is inhibited by binding of DnaK/J.

Degradation of SsrA-Tagged Proteins

SsrA RNA, also called tmRNA, is a specialized RNA that has properties of both a tRNA and an mRNA. When an mRNA lacks a stop codon, protein translation on the ribosome stalls, resulting in the production of an incompletely synthesized polypeptide. A short polypeptide "SsrA tag" is cotranslationally added to the C-terminus of the incomplete polypeptide in a reaction that is mediated by ribosome-bound SsrA

RNA. ATP-dependent proteases recognize and degrade these SsrA-tagged proteins to prevent accumulation of toxic incomplete polypeptides [18, 19]. The SsrA tag is attached to about 0.5% of newly synthesized polypeptides *in vivo*. Greater than 90% of SsrA-tagged polypeptides are digested by ClpXP protease [20] (for a recent review see [21]). The remaining 10% of the tagged polypeptides are removed from cells by ClpAP, Lon and FtsH. Therefore, FtsH partially but significantly contributes to degradation of SsrA-tagged polypeptides [22].

Quality Control of Membrane Proteins

FtsH is responsible for quality control of the inner-membrane environment. FtsH degrades membrane proteins, in an ATP-dependent manner, when they fail to form functional membrane protein complexes. For example, FtsH recognizes and degrades unassembled SecY, an integral membrane subunit of the protein translocation machinery (Sec translocase) in the inner membrane, and unassembled Fo subunit *a*, a membrane subunit of the ATP synthase [23, 24]. Similarly, when the Sec translocase becomes blocked with an inefficiently exported protein, the "jammed" SecY is degraded by FtsH [25]. The integral membrane protein YccA is also a proteolytic substrate of FtsH. Interestingly, YccA also modulates the FtsH-mediated degradation of membrane proteins. For example, YccA inhibits the FtsH-mediated degradation of blocked SecY. Nevertheless, the molecular mechanisms for the recognition of jammed SecY by FtsH and the inhibition of FtsH-mediated degradation by YccA remain elusive.

It has been proposed that FtsH cleaves polypeptide chains of substrate membrane proteins by extracting them from the inner membrane, releasing substrate-derived short peptides into the cytoplasm. To initiate degradation of membrane proteins, FtsH recognizes either N- or C-terminal cytoplasmic segments of a sufficient length (20 amino acid residues or more) [26, 27]. FtsH-mediated degradation of membrane proteins is processive, starting from one terminus, dislocating their transmembrane helices and periplasmic regions into the cytoplasmic side of the membrane, where the peptidase active site of FtsH is located. At present, the precise mechanism of this dislocation by FtsH has not been elucidated. Degradation by FtsH stops, when FtsH encounters the structurally stable domain [28]. Undigested fragments containing stable domains accumulate in the inner membrane [26].

It is conceivable that FtsH works cooperatively with an ATP-independent protease HtpX to remove substrate membrane proteins from the inner membrane [29]. HtpX is anchored to the inner membrane by the N-terminal transmembrane segment with overall topology similar to FtsH. The metal-binding protease active site of HtpX faces the cytoplasm. Although the physiological substrates of HtpX have not been identified yet, HtpX catalyzes the cleavage of casein and SecY polypeptide chains *in vitro* as well as the cleavage of overproduced SecY *in vivo*. HtpX is also responsible for endo-proteolytic cleavage within cytoplasmic regions of membrane proteins. A plausible scenario for the collaboration of FtsH and HtpX is that HtpX cleaves the cytoplasmic loops of substrate membrane proteins, generating a new cytoplasmic tail, which can be recognized by FtsH, resulting in the FtsH-mediated dislocation and degradation of the rest of the polypeptide chain [30].

Regulation of Lipid Synthesis

The composition and amount of lipid in the cell membrane is important for normal function. The balance of lipid composition in both inner and outer membranes of *E. coli* is tightly regulated. Since the biosynthesis of both phospholipids (PL) and lipopolysaccharides (LPS) are multistep pathways, which involve many different enzymes, the precise regulation (of amount and/or activity) of the enzymes that catalyze the committed steps in the pathway are critical for maintaining proper lipid composition. The cellular level of two key enzymes in biosynthesis of LPS is controlled by FtsH-mediated degradation. Loss of FtsH causes serious defects in the membrane function of *E. coli* and leads to cell death. The role of FtsH in the cell is not only restricted to proteolytic elimination of unnecessary proteins, but also to fine-tuning the cellular level of several critical proteins. Interestingly, the mitochondrial homolog of FtsH, Yme1, also participates in the regulation of lipid composition in the inner membrane of mitochondria as described below.

Synthesis of Lipid Molecules in E. coli

E. coli membranes are composed of two types of lipid molecules; PL and LPS. Both of which are synthesized by different pathways, and supplied to the membranes. The synthetic pathways and enzymes involved in lipid synthesis in *E. coli* are summarized in Fig. 3.2. For a detailed description of the biosynthesis of lipid molecules in *E. coli*, please refer to the following excellent reviews [31, 32]. The acyl donor, *R*-3-hydroxymyristoyl-ACP is an important branch point in the biosynthesis of both lipid molecules. In the synthesis of PL, *R*-3-hydroxymyristoyl-ACP is dehydrated by FabZ (*R*-3-hydroxy-acyl-ACP dehydrase), followed by elongation of short carbon units to produce long chain acyl-ACP species. Acyl-ACP is then transferred to lysophosphatidic acid (LPA) by either PlsX or PlsB, to produce phosphatidic acid (PA). Various types of PL are then synthesized from PA. In the synthesis of LPS, *R*-3-hydroxymyristoyl-ACP is attached to UDP-N-acetylglucosamine (UDP-GlcNAc) by LpxA to produce UDP-3-hydroxymyristoyl-GlcNAc. This reaction is the slowest step in LPS synthesis and the product of this reaction tends to return to the reactants. Therefore the next step in LPS synthesis, catalyzed by LpxC, is the rate-determining process [33]. Stimulation of LPS, but not PL, biosynthesis can exhaust the supply of the common intermediate, *R*-3-hydroxymyristoyl-ACP, causing an imbalance in biosynthesis of lipid molecules which prevents normal growth of *E. coli*. Therefore, for the correct maintenance of the PL:LPS ratios in the cell membrane, it is critical for the cell to balance the use of *R*-3-hydroxymyristoyl-ACP appropriately in both pathways.

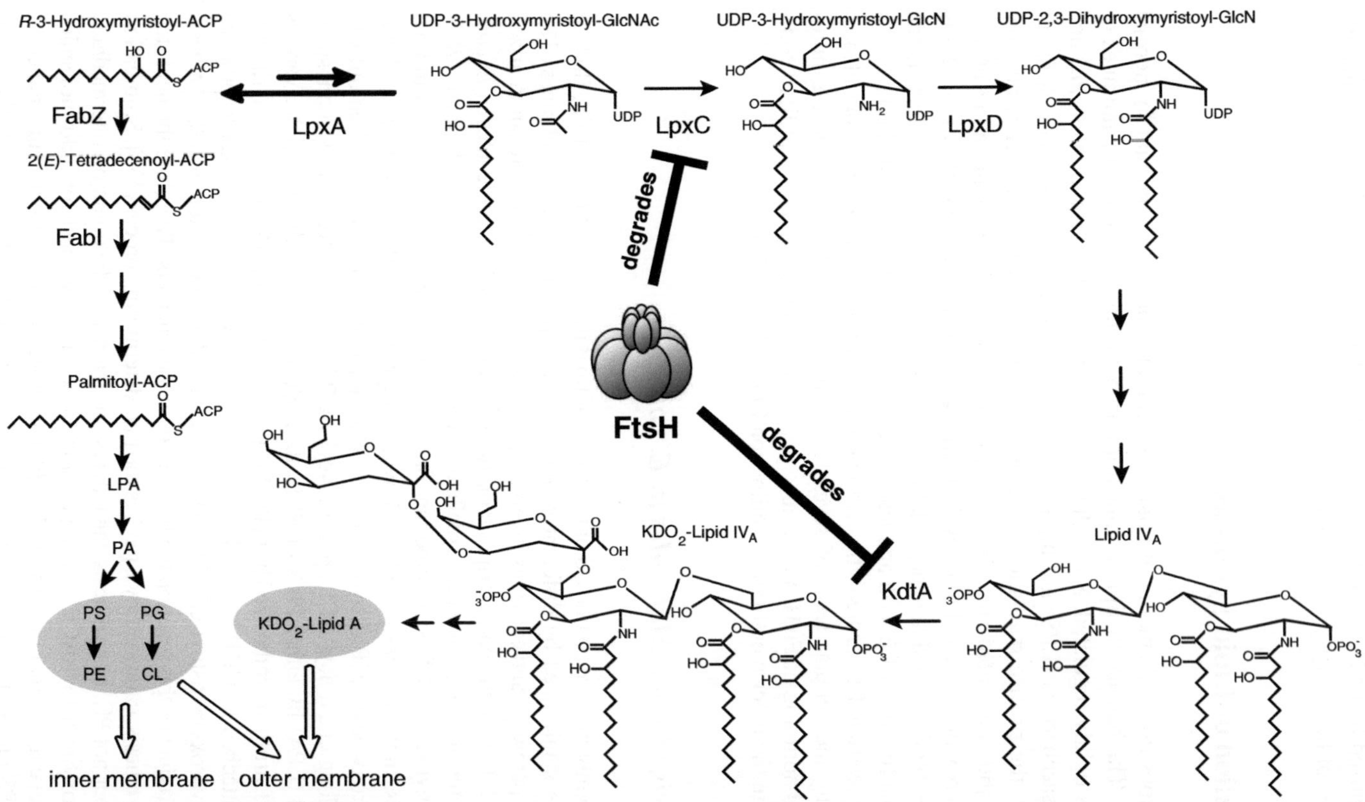

Fig. 3.2 Biosynthesis pathways of major membrane components in *E. coli* and regulation by FtsH. *Cartoon* illustrating the reaction steps in the biosynthesis of lipids that are regulated by FtsH. FtsH controls cellular levels of LpxC and KdtA by regulated degradation. *T-bars* represent negative control of the pathway, by the specific degradation of LpxC and KdtA. *ACP* acyl carrier protein, *GlcNAc* N-acetylglucosamine, *GlcN* glucosamine, *LPA* lysophosphatidic acid, *PA* phosphatidic acid, *PS* phosphatidylserine, *PG* phosphatidylglycerol, *PE* phosphatidylethanolamine, *CL* cardiolipin, *KDO* 3-deoxy-D-*manno*-octulosonic acids

Discovery of the Connection of FtsH to Lipid Synthesis

ftsH is an essential gene in *E. coli*. In the mid-1970s a thermosensitive mutant *E. coli* strain (*ftsH1*) was isolated [34]. Although this mutant strain was originally isolated as a cell division mutant, it was later (in the early 1990s) demonstrated to carry two mutations; *ftsH1* – responsible for lethality at high temperature and *ftsI372* – responsible for the cell division phenotype [35]. However, it wasn't until 1999, that the molecular basis of the lethal phenotype of the *ftsH1*(ts) mutation was clarified. The key finding to demonstrate the essential nature of the *ftsH* gene was the identification of a suppressor gene in another *ftsH* mutant, *tolZ21*. This mutant was isolated as a colicin tolerant strain, and it was found that *tolZ* was identical to *ftsH* [36]. The *tolZ21* mutation (H421Y) was located in a critical residue in the zinc-binding motif of FtsH essential for the metalloprotease activity, and thus it was expected that the mutant FtsH had no proteolytic activity. However, as the *tolZ21* mutant was viable, it suggests that, either the protease activity of FtsH is not required or that the *tolZ21* mutant carries a suppressor mutation. Precise genetic analysis revealed the presence of a suppressor mutation, and it was found that the suppressor mutation was an allele of *fabZ* [37]. Consistently, abnormal membrane structures accumulated in the periplasmic space of the *ftsH1* mutant at the non-permissive temperature. Biochemical analysis also showed an increase in the amount of LPS at non-permissive temperature [37]. Collectively, these data indicate that dysfunction of the FtsH protease causes alterations in lipid synthesis. Studies on two distinct *ftsH* mutant strains led to the discovery of a novel role of FtsH in the regulation of lipid synthesis.

Regulation of LpxC Levels by FtsH Protease

LpxC is a cytoplasmic enzyme responsible for catalyzing the committed step in LPS synthesis. It is composed of 205 amino acid residues and has a molecular mass of 33.9 kDa. A segment of the C-terminal sequence of LpxC (~20 amino acid residues) is required for degradation by FtsH [38, 39]. The C-terminus of LpxC, which resembles the SsrA tag, is rich in non-polar residues and mutation of which have been shown to stabilize LpxC [39]. Interestingly, although FtsH preferentially degrades LpxC, and hence regulates the amount of LpxC in enterobacteria [37], the FtsH-mediated degradation of LpxC is not conserved across all Gram-negative bacteria [40]. Accumulation of LpxC stimulates LPS biosynthesis, leading to a lethal imbalance in the PL:LPS ratio. Excess amounts of LPS in the cell result in the formation of abnormal membrane structures in the periplasmic space. The suppressor mutation *sfhC21* identified in the *tolZ21* mutant was shown to contain a point mutation in *fabZ*. This point mutation stimulates the activity of FabZ and compensates for the accumulation of LpxC, preventing the overproduction of LPS using *R*-3-hydroxymyristoyl-ACP for PL synthesis instead of LPS synthesis. Similarly, repression of LpxA or LpxD (two enzymes involved in LPS synthesis pathway, found before or after LpxC, in the pathway) also suppress the lethality of *ftsH1*. Indeed, to date, all identified suppressor mutations of *ftsH1* repress LPS synthesis.

Here an interesting question arises. Does the *sfhC21* mutation alone cause any defects in cell growth? Although it is reasonable to assume that the *sfhC21* mutation in *fabZ* causes an acceleration of PL synthesis and hence an increase in the PL:LPS ratio, this is not the case. In fact the *sfhC21* mutation in FabZ stabilizes LpxC, independent of the presence or absence of FtsH, and in fact the normal ration of PL and LPS is maintained in the strain that only carries the *sfhC21* mutation. Moreover, the stabilization of LpxC by the *sfhC21* mutation is substrate-specific, as the FtsH-mediated degradation of σ^{32} is not affected. The molecular mechanism, however, of this substrate-specific inhibition of FtsH-mediated degradation remains to be elucidated. Perhaps, alterations in acyl-ACP pools might modulate FtsH activity in a substrate-specific manner.

Degradation of KdtA

FtsH also regulates a late step in the LPS biosynthesis, the transfer of two 3-deoxy-D-manno-octulosonate (KDO) residues to lipid IV_A, which is catalyzed by KDO transferase (KdtA). KdtA is the sole enzyme that catalyzes transfer of two KDOs to lipid IV_A and hence is an essential glycosyltransferase in oligosaccharide biosynthesis [31]. KdtA is an inner-membrane protein, which is tethered to the membrane through an N-terminal transmembrane segment and its catalytic residues are presumed to face the cytoplasm. The *in vivo* half-life of KdtA is very short (~10 min) and the protease primarily responsible for its rapid degradation is FtsH [41]. The site or domain in KdtA to be recognized by FtsH has not been identified. As describe above, FtsH regulates the biosynthesis of LPS by controlling the amount of LpxC in the cell. Taken together, membrane-anchored FtsH protease plays important roles in the lipid biosynthesis by regulating the amount of two critical enzymes involved in the biosynthesis pathway of LPS.

Regulation of Lipid Composition in Mitochondria

As briefly mentioned above, homologs of FtsH are also found in the chloroplast and mitochondrion of eukaryotic cells. In mitochondria, there are two different FtsH-like proteases; referred to as *m*-AAA and *i*-AAA. In yeast, *i*-AAA is a homohexamer of Yme1 while *m*-AAA is a heterohexamer composed of two different proteins (Yta10 and Yta12). In contrast, humans contain a single *i*-AAA homohexamer (composed of YME1L) and two different *m*-AAA proteases (a homohexamer of AFG3L2 and a hetero-oligomer composed of AFG3L2 and paraplegin). All mitochondrial FtsH-like proteases are located in the inner membrane, *i*-AAA has one transmembrane segment near the N-terminus and its proteolytic active site is exposed to the intermembrane space, whereas *m*-AAA protease contains two transmembrane segments and its active site faces the matrix. Both *i*-AAA and *m*-AAA proteases are responsible for the proteolytic elimination of misfolded membrane proteins and protein processing in mitochondria. Dysfunction of *m*-AAA proteases in mitochondria causes neurodegenerative diseases [10, 42].

Although mitochondria have their own system for PL synthesis, several PLs, which constitute the inner and outer membranes of mitochondria, are supplied from the endoplasmic reticulum, as precursors. Cardiolipin (CL) and phosphatidylethanolamine (PE) are synthesized at the inner membrane, from phosphatidic acid and phosphatidylserine, respectively, and then transferred to the outer membrane through the intermembrane space. Homeostasis of CL and PE is regulated by two intermembrane space proteins, Ups1 and Ups2, respectively [43]. Although it is not yet clear how Ups1 and Ups2 regulate the concentration of CL and PE in the membrane [44]. Importantly, the level of Ups1 and Ups2 in the intermembrane space is regulated by rapid degradation by *i*-AAA protease [44]. Consistent with this, both Ups1 and Ups2 accumulate in mitochondria from the yeast *yme1* deletion strain. Overexpression of either Ups1 or Ups2 causes alterations in mitochondrial lipid composition, which leads to mitochondrial dysfunction [45]. Interestingly, the *i*-AAA-mediated degradation of Ups1 and Ups2 can be inhibited by binding of Mdm35 (a member of the twin Cx_9C protein family) [46].

Processing of Substrate Proteins

FtsH, in comparison to other AAA^+ proteases such as ClpXP, ClpAP, and HslUV has been demonstrated to have a "weak" unfolding activity [47, 48]. This distinguishing feature of FtsH plays an important role in its various *in vivo* functions. Although proteolysis by FtsH is processive, it is also abortive when FtsH encounters a tightly folded domain. Release of the stable polypeptide fragment, generated from abortive digestion by FtsH (and its homologs), is referred to as 'protein processing' and plays an important regulatory role in a number of cellular functions. In this section, 'protein processing' by FtsH and its homologs is summarised.

Processing of Colicins

Colicins are protein antibiotics that are released into the medium from *E. coli* cells carrying *colicin* genes, which kills other *E. coli* cells. Colicins released into the medium bind to receptors (BtuB, OmpF, FepA, etc.) on the outer membrane of *E. coli*, and are translocated, in cooperation with Tol and Ton translocators located in the inner membrane, into the periplasm. Then, colicins are imported into the cytoplasm, although the machineries for the translocation of colicins from the periplasm to the cytoplasm are not yet understood [49, 50]. Nuclease-type colicins must be imported into the cytoplasm of the target *E. coli* cell, where they disrupt DNA, tRNA and rRNA.

The *tolZ* mutant is tolerant to colicins E2, E3 and D. All of these colicins are nucleases, which, when translocated to the cytoplasm, act on either DNA (colicin E2), rRNA (colicin E3), or tRNA (colicin D). The *tolZ21* mutation, as mentioned previously, has been identified as a point mutation (H421Y) in the *ftsH* gene, which inactivates FtsH function [36]. Detailed analysis indicated that nuclease colicin toxicity is dependent on functional FtsH [51]. It has been shown that colicins D and E3, which are translocated by different machineries (BtuB/Tol and FepA/TonB, respectively), are processed by FtsH during their import into the cytoplasm [52]. Premature colicin D (75 kDa) is processed in an FtsH-dependent manner to yield a 12.4 kDa fragment containing the tRNase domain [52]. Production of the processed form of colicin E3 (15 kDa) was also found to be FtsH-dependent. Details of the processing of colicins by FtsH remain elusive [52]. Since FtsH leaves tightly folded domains of substrate proteins undegraded, it is possible that the processed form of colicin may result from abortive degradation by FtsH (see later).

Self-Processing of FtsH

The C-terminal seven amino acid residues of FtsH are removed auto-catalytically in a process that is, not only growth-phase dependent but also affected by mutations in *hflKC* [53]. Although the molecular mechanism of FtsH self-processing has not been elucidated, the processing site has been precisely determined. There is a clear preference for specific amino acid residues at the cleavage site. This is the sole example of FtsH-mediated site-specific cleavage. However, the biological significance of the self-processing of FtsH is unclear, since both the processed and full-length forms of FtsH are functionally indistinguishable. Although the self-processing of FtsH appears site-specific, the cleavage site may be determined by the combination of sequence preference and stability of the domain preceding the processing site. An example of the position-specific processing, whose cleavage site is simply determined by domain stability, has been reported for a mitochondrial FtsH homolog, *m*-AAA protease, and is discussed below (for a recent review see [54]). Such a possibility should be investigated for the self-processing of FtsH.

Molecular Mechanism of Substrate Processing

When FtsH encounters a stable domain within a substrate protein, processive degradation of polypeptide chain by FtsH is aborted, and the stable domain that cannot be unfolded for threading through the narrow pore of the FtsH ring, is released. SecY is an integral membrane protein, which contains ten transmembrane segments with both the N- and C-termini facing the cytoplasm. To better understand how FtsH processes integral membrane proteins, a number of SecY fusion proteins were generated. In one case, the eighth transmembrane helix (TM8) together with the following cytoplasmic region (30 amino acid residues) was fused to the C-terminal end of a periplasmic enzyme; alkaline phosphatase A (PhoA) to produce a model protein, PhoA-TM8-C_{30}. Although the reduced form of PhoA-TM8-C_{30} (lacking disulfide bonds in the PhoA domain) was completely degraded by FtsH, the oxidized form of PhoA-TM8-C_{30} (stabilized by disulfide bonds in the PhoA domain) was incompletely degraded and the stable PhoA domain fragment was released [26]. Collectively these data indicate that FtsH can initiate processive proteolysis from the C-terminus of the model substrate (PhoA-TM8-C_{30}), but cannot dislocate a stably folded domain (PhoA) from the periplasm to the cytoplasm.

Flavodoxin is a small flavin mononucleotide-containing protein, which plays an essential role in electron transfer pathways. Apo-flavodoxin, but not holo-flavodoxin, is degraded by FtsH *in vitro*. Interestingly, when apo-flavodoxin was attached to glutathione *S*-transferase (GST) or green fluorescent protein (GFP) such that FtsH can initiate processive proteolysis from the apo-flavodoxin moiety of the different model fusion substrates, FtsH was able to unfold and degrade the attached GST, but not the attached GFP. These data suggest that FtsH cannot unfold or degrade the thermally stable GFP, while it can unfold and degrade, to some extent, the GST moiety [55]. Therefore, the susceptibility of a domain to degradation by FtsH depends on thermal stability of that domain [56]. Interruption of processive proteolysis by a stably folded domain shown for these model substrates may be the molecular mechanism for FtsH-mediated processing of substrate proteins such as colicins and FtsH itself.

Processing of Substrate Proteins by Mitochondrial FtsH Homologs

The mitochondrial *m*-AAA and *i*-AAA proteases, participate in processing of several mitochondrial regulatory proteins by limited proteolysis as well as general quality control of mitochondrial proteins by complete digestion of damaged proteins [57, 58]. Yeast *m*-AAA protease is responsible for the processing of MrpL32, a component of the mitochondrial ribosome. MrpL32 is synthesized as a precursor in the cytoplasm, imported to the mitochondrial matrix, and processed to its mature form before it is assembled into mitochondrial ribosomes. The *m*-AAA protease cleaves the protein between the 71st and 72nd residue of the MrpL32 precursor to produce the mature form. Careful analysis demonstrated that processing of MrpL32 by the *m*-AAA protease depends on the folding of MrpL32 rather than on the specific recognition of the cleavage site [59]. The *m*-AAA protease initiates proteolysis from the N-terminus of MrpL32, which is halted by a tightly folded domain. Mammalian *m*-AAA proteases can also act as processing enzymes *in vivo* [60]. In murine mitochondria, the repertoire of *m*-AAA proteases is further expanded by the presence of an additional FtsH homolog AFG3L1 which forms various other hetero-oligomeric *m*-AAA complexes. The mitochondrial processing peptidase MPP generates an intermediate form of AFG3L2. This intermediate form is matured autocatalytically. AFG3L1 or AFG3L2 is also required for maturation of imported paraplegin after its cleavage by MPP. It is of great interest that mutations in different protease subunits are associated with distinct neuronal disorders in human; mutations in AFG3L2 are associated with spinocerebellar ataxia type 28 and spastic ataxia-neuropathy [41, 61], and those in paraplegin are associated with hereditary spastic paraplegia [62], respectively.

Mitochondrial *m*-AAA protease is also responsible for maturation of cytochrome *c* peroxidase (Ccp1) by rhomboid protease Pcp1 [63]. Premature Ccp1 is synthesized in the cytoplasm, translocated, and inserted into the inner membrane of mitochondria. The *m*-AAA protease mediates dislocation of the hydrophobic segment from the membrane in an ATP-dependent manner, making the processing site accessible for the rhomboid protease Pcp1. It should be noted that the maturation of Ccp1 depends only on the ATPase but not the proteolytic activity of the *m*-AAA protease.

Moreover, it has been shown that both *m*-AAA and *i*-AAA proteases are linked to the processing of the dynamin-like GTPase OPA1, a component of the mitochondrial fusion machinery [64, 65]. However, it is uncertain whether they act as processing enzymes or just assist processing by other proteases.

Biofilm Formation

Biofilm is an aggregate of microorganisms, in which cells stick to the surface of substrates and organisms. Development of the biofilm 'state' depends on the environments surrounding the bacteria, and is thought to be regulated by expression of multiple genes and operons [66, 67]. Because secretions, containing extracellular polysaccharide, wrap the aggregate of microorganisms, the microorganisms within the biofilm acquire physical strength and resistance to chemicals. Understanding the details of the biofilm 'life cycle' is important to develop procedures to control biofilm formation, which will be useful in medicine and food industry. At present, however, the precise processes that regulate the formation of biofilms are still unclear. Recently, the membrane anchored FtsH protease was implicated in the formation of biofilms. It was shown that a Δ*ftsH* mutant of *Lactobacillus plantarum* had a reduced capacity to form biofilms on abiotic surfaces [68]. Quantitative reverse transcription polymerase chain reaction (RT-PCR) studies revealed that expression of *ftsH* was upregulated during accretion of *Porphyromonas gingivalis* in heterotypic biofilms with *Streptococcus gordonii*. Contrary to the studies with *L. plantarum*, the Δ*ftsH* mutant of *P. gingivalis* formed more abundant biofilms with *S. gordonii* [69]. Taken together, it seems reasonable to assume that FtsH is involved in the biofilm formation in a variety of microorganisms. Yet, the roles of FtsH in the regulation of biofilm formation largely remain elusive.

Regulatory Proteins of FtsH

Following infection of an *E. coli* cell by a λ phage, the phage either enters the lytic or the lysogenic pathway. This choice is dependent on the physiological condition of the host cell and the decision is controlled by several key proteins encoded by the λ genome. For example, the amount of the transcription factor CII plays a crucial role in deciding which pathway to take. The cellular level of CII is primarily controlled by FtsH [70]. Indeed the degradation of CII by FtsH is very rapid (half-life ~2 min) under normal conditions, leading to the lytic pathway. However, under certain conditions, the degradation of CII by FtsH is inhibited, and CII accumulates in the cell, favouring the lysogenic pathway. The other key λ phage protein is λCIII, as it has been identified as an inhibitor of FtsH-mediated λCII degradation [71, 72]. This is because λCIII competes with λCII for binding to FtsH and is very slowly degraded by FtsH, CII is stabilized in the presence of CIII.

The proteolytic activity of FtsH is also modulated by two inner membrane proteins; HflK and HflC, which are homologs of the eukaryotic prohibitins. Both proteins have transmembrane segments near their N-termini and large periplasmic domains. Together they form a stable complex (HflKC). The HflKC complex binds to FtsH to form the FtsH holoenzyme, an exceptionally large complex (~1,000 kDa) with a proposed *in vivo* composition of $FtsH_6$•$HflK_6$•$HflC_6$ [73]. Binding of HflK or HflC to FtsH inhibits the degradation of CII [74, 75]. However, because the rate of σ^{32} degradation is not affected by the addition of HflKC, the HflKC complex does not simply decrease the proteolytic activity of FtsH. It has also been proposed that HflKC is involved in the regulation of proteolysis of membrane substrates [75, 76]. Currently however, the details of the structure of the FtsH holoenzyme and the selective regulation of the proteolytic activity of FtsH by HflKC remain to be elucidated.

Prohibitins, the eukaryotic homologs of HflK and HflC, are highly conserved membrane proteins that are required for normal cell growth and development. Prohibitins localize to the inner membrane of mitochondria and form large, multimeric ring-shaped complexes with a diameter of 20–25 nm. The function of prohibitins in mitochondria is related to the regulation of various mitochondrial functions such as respiration, stability of mitochondrial DNA, and maintenance of mitochondrial morphology [77–79]. Prohibitins associate with *m*-AAA protease and modulate its proteolytic activity. The loss of prohibitins in mitochondria stimulates the degradation of unassembled inner membrane proteins by *m*-AAA protease. On the other hand, it has also proposed that the prohibitin ring complex performs a scaffolding function to recruit *m*-AAA protease to a specific functional site in the inner membrane of mitochondria. The complexes of *m*-AAA proteases and prohibitins were also identified in plant mitochondria [80].

Conclusions and Perspectives

Since the discovery of FtsH protease, a number of substrate proteins have been identified. FtsH recognizes a wide range of substrate proteins, and thus is involved in the regulation of a variety of cellular processes. In some substrates, an unstructured tail of ~20 amino acid residues, located at either the N- or C-terminus is recognized by FtsH to initiate processive proteolysis. Since FtsH lacks a robust unfolding activity, it is primarily responsible for (a) the selective degradation of structurally unstable proteins or (b) processing of specific protein substrates as a result of encountering a stable domain, which aborts processive proteolysis by FtsH. This processing role has been more extensively studied in eukaryotic FtsH homologs present in mitochondria. Indeed, recent studies revealed that both substrate selectivity and weak unfolding ability of FtsH are crucial for its regulatory roles in diverse cellular activities. To date, however, the precise mechanisms of substrate recognition, unfolding, and processive degradation by FtsH are still largely unclear. Further investigations will be of importance to understand the molecular mechanism of FtsH, which executes the regulation of a variety of cellular processes.

Acknowledgements We are grateful to Yoshinori Akiyama for stimulating discussion and invaluable comments and to David Dougan for generous editorial suggestions to the manuscript. We also thank to Chiyome Ichinose for secretarial assistant. This work was supported in part by grants from the Ministry of Education, Culture, Science, Sports and Technology of Japan.

References

1. Bieniossek C, Niederhauser B, Baumann UM (2009) The crystal structure of apo-FtsH reveals domain movements necessary for substrate unfolding and translocation. Proc Natl Acad Sci U S A 106(51):21579–21584
2. Krzywda S, Brzozowski AM, Verma C, Karata K (2002) The crystal structure of the AAA domain of the ATP-dependent protease FtsH of *Escherichia coli* at 1.5 A resolution. Structure 10(8):1073–1083
3. Langklotz S, Baumann U, Narberhaus F (2012) Structure and function of the bacterial AAA protease FtsH. Biochim Biophys Acta 1823(1):40–48
4. Ogura T, Okuno T, Suno R, Akiyama Y (2013) FtsH protease. In: Rawlings ND, Salvesen G (eds) Handbook of Proteolytic Enzymes, 3rd edn. Elsevier, pp 685–692
5. Okuno T, Ogura T (2008) FtsH protease, a eubacterial membrane-bound AAA protease. In: Kutejova E (ed) ATP-Dependent Proteases, Research Signport, pp 87–114
6. Suno R, Niwa H, Tsuchiya D, Zhang X et al (2006) Structure of the whole cytosolic region of ATP-dependent protease FtsH. Mol Cell 22(5):575–585
7. Asahara Y, Atsuta K, Motohashi K, Taguchi H et al (2000) FtsH recognizes proteins with unfolded structure and hydrolyzes the carboxyl side of hydrophobic residues. J Biochem 127(5):931–937
8. Gerdes F, Tatsuta T, Langer T (2012) Mitochondrial AAA proteases – towards a molecular understanding of membrane-bound proteolytic machines. Biochim Biophys Acta 1823(1):49–55
9. Nixon PJ, Michoux F, Yu J, Boehm M et al (2010) Recent advances in understanding the assembly and repair of photosystem II. Ann Bot 106(1):1–16
10. Rugarli EI, Langer T (2006) Translating m-AAA protease function in mitochondria to hereditary spastic paraplegia. Trends Mol Med 12(6):262–269
11. Sakamoto W (2006) Protein degradation machineries in plastids. Annu Rev Plant Biol 57:599–621
12. Herman C, Thevenet D, D'Ari R, Bouloc P (1995) Degradation of sigma 32, the heat shock regulator in *Escherichia coli*, is governed by HflB. Proc Natl Acad Sci U S A 92(8):3516–3520
13. Tomoyasu T, Gamer J, Bukau B, Kanemori M et al (1995) *Escherichia coli* FtsH is a membrane-bound, ATP-dependent protease which degrades the heat-shock transcription factor sigma 32. EMBO J 14(11):2551–2560

14. Obrist M, Langklotz S, Milek S, Fuhrer F et al (2009) Region C of the *Escherichia coli* heat shock sigma factor RpoH (sigma 32) contains a turnover element for proteolysis by the FtsH protease. FEMS Microbiol Lett 290(2):199–208
15. Obrist M, Milek S, Klauck E, Hengge R (2007) Region 2.1 of the *Escherichia coli* heat-shock sigma factor RpoH (sigma32) is necessary but not sufficient for degradation by the FtsH protease. Microbiology 153(8):2560–2571
16. Obrist M, Narberhaus F (2005) Identification of a turnover element in region 2.1 of *Escherichia coli* sigma32 by a bacterial one-hybrid approach. J Bacteriol 187(11):3807–3813
17. Rodriguez F, Arsene-Ploetze F, Rist W, Rudiger S et al (2008) Molecular basis for regulation of the heat shock transcription factor sigma32 by the DnaK and DnaJ chaperones. Mol Cell 32(3):347–358
18. Okuno T, Yamada-Inagawa T, Karata K, Yamanaka K et al (2004) Spectrometric analysis of degradation of a physiological substrate sigma32 by *Escherichia coli* AAA protease FtsH. J Struct Biol 146(1–2):148–154
19. Karzai AW, Roche ED, Sauer RT (2000) The SsrA-SmpB system for protein tagging, directed degradation and ribosome rescue. Nat Struct Biol 7(6):449–455
20. Lies M, Maurizi MR (2008) Turnover of endogenous SsrA-tagged proteins mediated by ATP-dependent proteases in *Escherichia coli*. J Biol Chem 283(34):22918–22929
21. Gur E, Ottofuelling R, Dougan DA (2013) Machines of destruction – AAA+ proteases and the adaptors that control them. In: Dougan DA (ed) Regulated proteolysis in microorganisms. Springer, Subcell Biochem 66:3–33
22. Herman C, Thevenet D, Bouloc P, Walker GC et al (1998) Degradation of carboxy-terminal-tagged cytoplasmic proteins by the *Escherichia coli* protease HflB (FtsH). Genes Dev 12(9):1348–1355
23. Akiyama Y, Kihara A, Ito K (1996) Subunit a of proton ATPase F0 sector is a substrate of the FtsH protease in *Escherichia coli*. FEBS Lett 399(1–2):26–28
24. Akiyama Y, Kihara A, Tokuda H, Ito K (1996) FtsH (HflB) is an ATP-dependent protease selectively acting on SecY and some other membrane proteins. J Biol Chem 271(49): 31196–31201
25. van Stelten J, Silva F, Belin D, Silhavy TJ (2009) Effects of antibiotics and a proto-oncogene homolog on destruction of protein translocator SecY. Science 325(5941):753–756
26. Chiba S, Akiyama Y, Ito K (2002) Membrane protein degradation by FtsH can be initiated from either end. J Bacteriol 184(17):4775–4782
27. Chiba S, Akiyama Y, Mori H, Matsuo E et al (2000) Length recognition at the N-terminal tail for the initiation of FtsH-mediated proteolysis. EMBO Rep 1(1):47–52
28. Kihara A, Akiyama Y, Ito K (1999) Dislocation of membrane proteins in FtsH-mediated proteolysis. EMBO J 18(11):2970–2981
29. Shimohata N, Chiba S, Saikawa N, Ito K et al (2002) The Cpx stress response system of *Escherichia coli* senses plasma membrane proteins and controls HtpX, a membrane protease with a cytosolic active site. Genes Cells 7(7):653–662
30. Akiyama Y (2009) Quality control of cytoplasmic membrane proteins in *Escherichia coli*. J Biochem 146(4):449–454
31. Raetz CR, Reynolds CM, Trent MS, Bishop RE (2007) Lipid A modification systems in gram-negative bacteria. Annu Rev Biochem 76:295–329
32. Zhang YM, Rock CO (2008) Membrane lipid homeostasis in bacteria. Nat Rev Microbiol 6(3):222–233
33. Anderson MS, Bull HG, Galloway SM, Kelly TM et al (1993) UDP-N-acetylglucosamine acyltransferase of *Escherichia coli*. The first step of endotoxin biosynthesis is thermodynamically unfavorable. J Biol Chem 268(26):19858–19865
34. Santos D, De Almeida DF (1975) Isolation and characterization of a new temperature-sensitive cell division mutant of *Escherichia coli* K-12. J Bacteriol 124(3):1502–1507
35. Begg KJ, Tomoyasu T, Donachie WD, Khattar M et al (1992) *Escherichia coli* mutant Y16 is a double mutant carrying thermosensitive ftsH and ftsI mutations. J Bacteriol 174(7):2416–2417

36. Qu JN, Makino SI, Adachi H, Koyama Y et al (1996) The tolZ gene of *Escherichia coli* is identified as the ftsH gene. J Bacteriol 178(12):3457–3461
37. Ogura T, Inoue K, Tatsuta T, Suzaki T et al (1999) Balanced biosynthesis of major membrane components through regulated degradation of the committed enzyme of lipid A biosynthesis by the AAA protease FtsH (HflB) in *Escherichia coli.* Mol Microbiol 31(3):833–844
38. Fuhrer F, Langklotz S, Narberhaus F (2006) The C-terminal end of LpxC is required for degradation by the FtsH protease. Mol Microbiol 59(3):1025–1036
39. Fuhrer F, Muller A, Baumann H, Langklotz S et al (2007) Sequence and length recognition of the C-terminal turnover element of LpxC, a soluble substrate of the membrane-bound FtsH protease. J Mol Biol 372(2):485–496
40. Langklotz S, Schakermann M, Narberhaus F (2011) Control of lipopolysaccharide biosynthesis by FtsH-mediated proteolysis of LpxC is conserved in enterobacteria but not in all gram-negative bacteria. J Bacteriol 193(5):1090–1097
41. Katz C, Ron EZ (2008) Dual role of FtsH in regulating lipopolysaccharide biosynthesis in *Escherichia coli.* J Bacteriol 190(21):7117–7122
42. Pierson TM, Adams D, Bonn F, Martinelli P et al (2011) Whole-exome sequencing identifies homozygous AFG3L2 mutations in a spastic ataxia-neuropathy syndrome linked to mitochondrial m-AAA proteases. PLoS Genet 7(10):e1002325
43. Tamura Y, Endo T, Iijima M, Sesaki H (2009) Ups1p and Ups2p antagonistically regulate cardiolipin metabolism in mitochondria. J Cell Biol 185(6):1029–1045
44. Potting C, Wilmes C, Engmann T, Osman C et al (2010) Regulation of mitochondrial phospholipids by Ups1/PRELI-like proteins depends on proteolysis and Mdm35. EMBO J 29(17):2888–2898
45. Osman C, Haag M, Potting C, Rodenfels J et al (2009) The genetic interactome of prohibitins: coordinated control of cardiolipin and phosphatidylethanolamine by conserved regulators in mitochondria. J Cell Biol 184(4):583–596
46. Tamura Y, Iijima M, Sesaki H (2010) Mdm35p imports Ups proteins into the mitochondrial intermembrane space by functional complex formation. EMBO J 29(17):2875–2887
47. Herman C, Prakash S, Lu CZ, Matouschek A et al (2003) Lack of a robust unfoldase activity confers a unique level of substrate specificity to the universal AAA protease FtsH. Mol Cell 11(3):659–669
48. Koodathingal P, Jaffe NE, Kraut DA, Prakash S et al (2009) ATP-dependent proteases differ substantially in their ability to unfold globular proteins. J Biol Chem 284(28):18674–18684
49. Cascales E, Buchanan SK, Duche D, Kleanthous C et al (2007) Colicin biology. Microbiol Mol Biol Rev 71(1):158–229
50. Kleanthous C (2010) Swimming against the tide: progress and challenges in our understanding of colicin translocation. Nat Rev Microbiol 8(12):843–848
51. Walker D, Mosbahi K, Vankemmelbeke M, James R et al (2007) The role of electrostatics in colicin nuclease domain translocation into bacterial cells. J Biol Chem 282(43):31389–31397
52. Chauleau M, Mora L, Serba J, de Zamaroczy M (2011) FtsH-dependent processing of RNase colicins D and E3 means that only the cytotoxic domains are imported into the cytoplasm. J Biol Chem 286(33):29397–29407
53. Akiyama Y (1999) Self-processing of FtsH and its implication for the cleavage specificity of this protease. Biochemistry 38(36):11693–11699
54. Voos W, Ward LA, Truscott KN (2013) The role of AAA+ proteases in mitochondrial protein biogensis, homeostasis and activity control. In: Dougan DA (ed) Regulated proteolysis in microorganisms. Springer, Subcell Biochem 66:223–263
55. Okuno T, Yamanaka K, Ogura T (2006) An AAA protease FtsH can initiate proteolysis from internal sites of a model substrate, apo-flavodoxin. Genes Cells 11(3):261–268
56. Ayuso-Tejedor S, Nishikori S, Okuno T, Ogura T et al (2010) FtsH cleavage of non-native conformations of proteins. J Struct Biol 171(2):117–124
57. Koppen M, Langer T (2007) Protein degradation within mitochondria: versatile activities of AAA proteases and other peptidases. Crit Rev Biochem Mol Biol 42(3):221–242
58. Martinelli P, Rugarli EI (2010) Emerging roles of mitochondrial proteases in neurodegeneration. Biochim Biophys Acta 1797(1):1–10

59. Bonn F, Tatsuta T, Petrungaro C, Riemer J et al (2011) Presequence-dependent folding ensures MrpL32 processing by the m-AAA protease in mitochondria. EMBO J 30(13):2545–2556
60. Koppen M, Bonn F, Ehses S, Langer T (2009) Autocatalytic processing of m-AAA protease subunits in mitochondria. Mol Biol Cell 20(19):4216–4224
61. Di Bella D, Lazzaro F, Brusco A, Plumari M et al (2010) Mutations in the mitochondrial protease gene AFG3L2 cause dominant hereditary ataxia SCA28. Nat Genet 42(4):313–321
62. Casari G, De Fusco M, Ciarmatori S, Zeviani M et al (1998) Spastic paraplegia and OXPHOS impairment caused by mutations in paraplegin, a nuclear-encoded mitochondrial metalloprotease. Cell 93(6):973–983
63. Tatsuta T, Augustin S, Nolden M, Friedrichs B (2007) m-AAA protease-driven membrane dislocation allows intramembrane cleavage by rhomboid in mitochondria. EMBO J 26(2):325–335
64. Ehses S, Raschke I, Mancuso G, Bernacchia A et al (2009) Regulation of OPA1 processing and mitochondrial fusion by m-AAA protease isoenzymes and OMA1. J Cell Biol 187(7):1023–1036
65. Ishihara N, Fujita Y, Oka T, Mihara K (2006) Regulation of mitochondrial morphology through proteolytic cleavage of OPA1. EMBO J 25(13):2966–2977
66. Lopez D, Vlamakis H, Kolter R (2010) Biofilms. Cold Spring Harb Perspect Biol 2(7):a000398
67. McDougald D, Rice SA, Barraud N, Steinberg PD et al (2012) Should we stay or should we go: mechanisms and ecological consequences for biofilm dispersal. Nat Rev Microbiol 10(1):39–50
68. Bove P, Capozzi V, Garofalo C, Rieu A et al (2012) Inactivation of the *ftsH* gene of *Lactobacillus plantarum* WCFS1: effects on growth, stress tolerance, cell surface properties and biofilm formation. Microbiol Res 167(4):187–193
69. Simionato MR, Tucker CM, Kuboniwa M, Lamont G et al (2006) *Porphyromonas gingivalis* genes involved in community development with *Streptococcus gordonii*. Infect Immun 74(11):6419–6428
70. Herman C, Ogura T, Tomoyasu T, Hiraga S et al (1993) Cell growth and lambda phage development controlled by the same essential *Escherichia coli* gene, ftsH/hflB. Proc Natl Acad Sci U S A 90(22):10861–10865
71. Herman C, Thevenet D, D'Ari R, Bouloc P (1997) The HflB protease of *Escherichia coli* degrades its inhibitor lambda cIII. J Bacteriol 179(2):358–363
72. Kobiler O, Rokney A, Oppenheim AB (2007) Phage lambda CIII: a protease inhibitor regulating the lysis-lysogeny decision. PLoS One 2(4):e363
73. Saikawa N, Akiyama Y, Ito K (2004) FtsH exists as an exceptionally large complex containing HflKC in the plasma membrane of *Escherichia coli*. J Struct Biol 146(1–2):123–129
74. Bandyopadhyay K, Parua PK, Datta AB, Parrack P (2010) *Escherichia coli* HflK and HflC can individually inhibit the HflB (FtsH)-mediated proteolysis of lambdaCII in vitro. Arch Biochem Biophys 501(2):239–243
75. Kihara A, Akiyama Y, Ito K (1997) Host regulation of lysogenic decision in bacteriophage lambda: transmembrane modulation of FtsH (HflB), the cII degrading protease, by HflKC (HflA). Proc Natl Acad Sci U S A 94(11):5544–5549
76. Kihara A, Akiyama Y, Ito K (1998) Different pathways for protein degradation by the FtsH/HflKC membrane-embedded protease complex: an implication from the interference by a mutant form of a new substrate protein, YccA. J Mol Biol 279(1):175–188
77. Artal-Sanz M, Tavernarakis N (2009) Prohibitin and mitochondrial biology. Trends Endocrinol Metab 20(8):394–401
78. Merkwirth C, Langer T (2009) Prohibitin function within mitochondria: essential roles for cell proliferation and cristae morphogenesis. Biochim Biophys Acta 1793(1):27–32
79. Osman C, Merkwirth C, Langer T (2009) Prohibitins and the functional compartmentalization of mitochondrial membranes. J Cell Sci 122(Pt 21):3823–3830
80. Piechota J, Kolodziejczak M, Juszczak I, Sakamoto W et al (2010) Identification and characterization of high molecular weight complexes formed by matrix AAA proteases and prohibitins in mitochondria of *Arabidopsis thaliana*. J Biol Chem 285(17):12512–12521

Part II
Regulatory Proteolysis in Bacteria

Chapter 4
General and Regulatory Proteolysis in *Bacillus subtilis*

Noël Molière and Kürşad Turgay

Abstract The soil-dwelling bacterium *Bacillus subtilis* is widely used as a model organism to study the Gram-positive branch of Bacteria. A variety of different developmental pathways, such as endospore formation, genetic competence, motility, swarming and biofilm formation, have been studied in this organism. These processes are intricately connected and regulated by networks containing e.g. alternative sigma factors, two-component systems and other regulators. Importantly, in some of these regulatory networks the activity of important regulatory factors is controlled by proteases. Furthermore, together with chaperones, the same proteases constitute the cellular protein quality control (PQC) network, which plays a crucial role in protein homeostasis and stress tolerance of this organism. In this review, we will present the current knowledge on regulatory and general proteolysis in *B. subtilis* and discuss its involvement in developmental pathways and cellular stress management.

Introduction

The soil bacterium *Bacillus subtilis* encounters quickly changing and often unfavorable conditions in its natural habitat. During evolution these environmental cues might have been important for the establishment of a wide variety of elaborate stress response and developmental pathways, which *B. subtilis* cells exploit in order to adapt to their environment. For example, *B. subtilis* cells can grow normally even above 50 °C and they are quite resistant to osmotic stress. Furthermore, when the cells enter stationary phase the bacterium can differentiate into many different cell types, such as endospores, cells competent for DNA uptake and biofilm forming cells. At the heart of these pathways are complex signal transduction systems, which

N. Molière (✉) • K. Turgay
Institut für Mikrobiologie, Leibniz Universität Hannover,
Schneiderberg 50, 30167 Hannover, Germany
e-mail: moliere@ifmb.uni-hannover.de

D.A. Dougan (ed.), *Regulated Proteolysis in Microorganisms*, Subcellular Biochemistry 66,
DOI 10.1007/978-94-007-5940-4_4,

integrate different environmental cues to modify gene expression in order to respond to the external conditions [1].

Interestingly, controlled protein degradation is intricately involved in these regulatory pathways. Here the stability of transcription factors or other cellular regulators, which e.g. directly affect signal transduction pathways, are specifically modulated by proteolysis. At first glance, the control of cellular regulatory proteins by proteolysis might be considered rather inefficient and wasteful, given the amount of energy required for the synthesis of such a protein. However, using proteolysis in regulatory modules has certain advantages and can thus be beneficial to the survival of cells. The first advantage of protein degradation is its irreversibility. By irreversibly removing one component of a pathway, equilibriums can be shifted and bistable states can be stabilized. Secondly, proteolytic systems can exhibit very fast response times. For example, if a protein is constitutively degraded and then stabilized in response to a signal, this results in a fast, switch-like behavior [2–4].

At the same time protein degradation is also utilized as part of the cellular protein quality control (PQC) system. Proteases, in conjunction with chaperone systems, degrade a large variety of unfolded, misfolded or damaged proteins, which would otherwise be detrimental to cellular function. The cellular PQC system becomes especially important during heat, oxidative or salt stress, which are potentially damaging conditions for proteins but also other cellular components [5–7].

Protease Systems

Hsp100/Clp and AAA+(ATPases Associated with Various Cellular Activities) Proteases

Here, we will focus on the role of Hsp100/Clp and related AAA+protease complexes in regulatory and general proteolysis. These bipartite ATP fueled protease complexes consist of a hexameric ring of Hsp100/Clp proteins, which are ATP driven unfoldases of the AAA+ family associated with a barrel-like structure formed by the higher oligomeric peptidase complex. The active peptidase sites are not accessible to substrates outside of the compartmentalized protease, unless the substrate proteins are recognized, unfolded and transferred by the associated hexameric ATPases into the proteolytic chamber [8–10].

In *B. subtilis* the ATPase components are the Hsp100/Clp proteins ClpC, ClpE, or ClpX, which associate with the double-heptameric ClpP peptidase complex, ClpY which associates with the ClpQ peptidase, or the closely related AAA+protease complexes with a similar architecture; LonA, LonB and the membrane associated FtsH [4]. Of these seven different ATP driven proteases, ClpCP and ClpXP are intricately involved in cellular signaling processes and developmental programs such as sporulation, competence development, motility, biofilm formation and in stress response pathways. Interestingly, the same protease complexes serve as an integral part of the PQC network through general proteolysis of unfolded or misfolded proteins. The important role of these proteases in *B. subtilis* is supported by the relatively severe and pleiotropic phenotypes of *clp* mutants [4].

The synthesis of ClpC, ClpP and ClpE is controlled by the repressor CtsR (see section "Heat Shock Response") and strongly induced by heat stress [11–14]. ClpE synthesis is very tightly regulated and it is present in only very low amounts in non-stressed cells [15]. To date, the functions assigned to ClpE suggest a role in protein disaggregation and possible subsequent degradation of aggregated proteins as well as modulation of the heat shock response [16].

ClpC is present in non-stressed cells and plays an important regulatory role during normal logarithmic growth. During heat shock ClpCP is produced in high amounts and can degrade or disaggregate protein aggregates [17–20]. An interesting and unusual feature of ClpC is its absolute requirement of adaptor proteins for its activity. Adaptor proteins modulate the substrate specificity of the Clp/Hsp100 ATPases, usually by tethering the substrate to the ATPase [4]. However, for ClpC it could be demonstrated that in the absence of an adaptor protein, ClpC is monomeric and the ATPase is inactive. The adaptors have been shown to bind to the N-terminal domain (NTD) and linker domain of ClpC and trigger its oligomerization into a hexameric ring-shaped complex with ATPase activity [20–23].

Currently, three adaptor proteins for ClpC have been identified: MecA targets the competence master regulator ComK for degradation by ClpCP [24], its paralog YpbH permits degradation of unfolded and aggregated model proteins *in vitro* [20, 25]. MscB has recently been identified as an adaptor protein, which targets the class III heat shock repressor CtsR to ClpCP for degradation [21]. The identification of different adaptor proteins, which can target various substrates for degradation by ClpCP might explain the concurrent involvement of ClpC in many distinct regulatory and general processes [4]. There is some evidence that more unknown ClpC adaptors might exist. For example, *in vitro* degradation experiments with the known *in vivo* ClpCP substrates MurAA [26] and SpoIIAB [27] were unsuccessful, possibly because the correct adaptor was missing from the reaction.

ClpX is quite abundant under non-stress conditions and its stress-induced expression is regulated by an unknown mechanism. Mutants in *clpX* have a pleiotropic phenotype of slow growth on solid media, an apparent long lag phase during growth in liquid medium, and a defect in competence, which is mostly due to over-production of the ClpXP substrate Spx [28, 29]. Spx is an important transcriptional regulator of oxidative stress in *B. subtilis* (see section "Oxidative Stress Response and Spx"). The degradation of Spx is modulated by the adaptor protein YjbH and its recently identified anti-adaptor protein YirB [30–32]. ClpX is also involved in the degradation of SsrA-tagged proteins and in the degradation of unfolded and aggregated proteins [33].

LonA and LonB are hexameric proteases, in which the ATPase domain and the protease domain are present on a single polypeptide chain [34]. The ClpYQ (CodWX) complex is homologous to the HslUV complex in *Escherichia coli* [35, 36] and structurally resembles the Clp protease complexes, although the protease component ClpQ exhibits sixfold and not sevenfold symmetry [37, 38]. FtsH is a zinc metalloprotease, which is embedded in the cytoplasmic face of the membrane [39]. The deletion of *ftsH* displays a pleiotropic phenotype in *B. subtilis* [39, 40].

Regulated Intramembrane Proteolysis

We will also briefly discuss regulated intramembrane proteolysis (RIP), which is a process that both prokaryotes and eukaryotes use for signaling across cell membranes. Here a signaling module consists of a cytosolic transcription factor that is kept inactive either as a preprotein with a trans-membrane domain or by a membrane-localized inhibiting factor. The trans-membrane activation of the regulatory protein is initiated and completed by the consecutive action of two membrane-localized proteases on this preprotein (or inhibiting factor). Upon an external signal the site-1 protease initiates the signaling cascade by processing of the preprotein (or inhibitor). This first proteolytic event makes its trans-membrane portion accessible to the site-2 protease (also named intramembrane cleaving protease (I-CLiP), resulting in cleavage of the intramembrane domain thereby causing release and activation of the cytosolic transcription factor [41–43]. In *B. subtilis* RIP is involved, not only in the envelope stress response, but also during sporulation in signal transduction between the mother cell and forespore [43–45] (see section "Membrane Stress and Regulated Intramembrane Proteolysis (RIP) of the Extracytoplasmic Function (ECF) Sigma Factors").

Regulatory Proteolysis

Competence Development

Competence is the ability of a subpopulation of stationary phase *B. subtilis* cells to take up DNA from the environment and incorporate it into their genome by homologous recombination. This process is controlled by the master regulator ComK, a transcriptional activator, which controls genes required for DNA binding, processing and transport into the cell. During logarithmic growth, ComK is inhibited [46, 47] and targeted by MecA for degradation by ClpCP [24]. This process results in a very low steady state concentration of ComK and ensures a very tight control of competence during exponential growth. At high cell density, however, a quorum sensing pathway activates the expression of the anti adaptor ComS, which binds to MecA with high affinity and thus competes with the binding of ComK to MecA [24, 48] (Fig. 4.1a). Subsequently, the released ComK activates its transcription from the *comK* promotor, leading to positive auto-regulation and bi-stable competence gene expression in the subpopulation of cells that reach a threshold level of ComK by stabilization due to the proteolytic ComS-anti adaptor switch (Fig. 4.1a) [49–51].

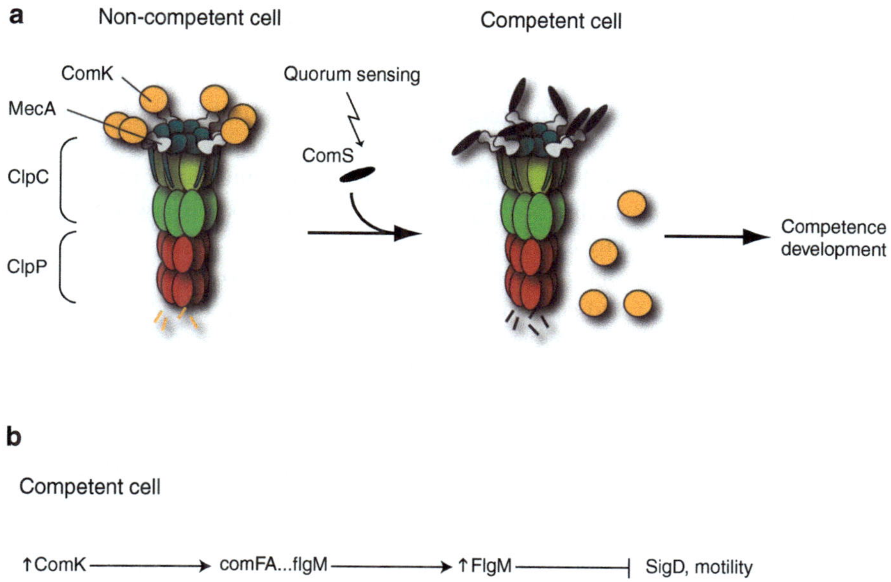

Fig. 4.1 (**a**) **Competence regulation by ClpCP/MecA**. In non-competent cells, ComK, the master regulator of competence development, is inhibited by MecA and targeted for degradation by ClpCP/MecA. This results in a low concentration of ComK. At high cell densities, a quorum sensing signal activates a cascade culminating in the expression of the peptide ComS. ComS is also a target of the ClpCP/MecA complex and competes with ComK for binding to MecA. Due to the high concentration of ComS, this peptide is preferentially degraded. In turn, ComK accumulates and activates the *comK* promotor. This positive autoregulatory loop results in high levels of ComK and competence development. (**b**) **Motility regulation by ClpCP/MecA**. The gene encoding FlgM (the flagellar anti-sigma factor) is located downstream of the competence genes *comFA*, separated by a weak terminator. Under conditions of high ComK levels, transcriptional read-through from *comFA* into *flgM* results in accumulation of FlgM and inhibition of σ^D-dependent motility gene expression

Swimming Motility

Swimming motility is a post-exponential process in *B. subtilis*, during which single flagellated cells swim through liquid medium by a stochastic run-and-tumble path that is guided by chemotaxis. In contrast, swarming motility describes a coordinated movement of tight bundles of cells over semi-solid surfaces, which requires flagellar rotation and the presence of surfactants [52]. Both processes require the expression of motility genes, which are organized in a hierarchy of two classes of genes: the first class of early flagellar genes is encoded in the 27 kb *fla/che* operon, which is transcribed by RNA polymerase (RNAP) in complex with the housekeeping sigma factor (σ^A). Examples of early flagellar genes are those encoding the ring complexes and the hook basal body complex. σ^D is an alternative sigma factor, which is encoded close to the 3′-end of the *fla/che* operon. The second class of late flagellar genes consist of separate transcriptional units, which all exhibit a σ^D-dependent promoter.

These genes include those encoding Hag, the structural subunit of the flagellum, and FliD, the cap of the flagellum, as well as their cognate export chaperones. The promoter binding activity of σ^D is inhibited by the anti-sigma factor FlgM [53, 54] possibly by direct protein-protein interaction as was demonstrated for FlgM and σ^{28} [55]. In *Salmonella*, FlgM is exported by the flagellar type III secretion system after completion of the hook basal body, releasing the flagellar sigma factor and enabling efficient production of flagellin [56, 57]. In *B. subtilis* it is not known, whether FlgM is exported or inactivated by a different mechanism.

Already shortly after the discovery of the *clp* genes in *B. subtilis*, a motility defect of the *clpP*, *clpX* and *clpC* mutants was described [58]. In the case of *clpC*, this phenotype was examined in detail, and was shown to be dependent on *comK*. A ComK-controlled competence gene *comFA* happens to be located directly in front of the operon containing *flgM*, separated by a weak terminator. Consequently, when high levels of ComK are present in the cell (i.e. in a *clpC* mutant or in competent cells), *comFA* is activated, resulting in read-through transcription of *flgM* (Fig. 4.1b). The excess production of FlgM anti-sigma factor inhibits σ^D and leads to a block of σ^D-dependent gene expression and motility (Fig. 4.1b) [59].

The biological implication of this mechanism may be to ensure mutually exclusive development of motility and competence. Competent cells are physiologically very different from exponentially growing cells and presumably it would be a waste of energy for them to express flagella. According to this model, the protease ClpCP together with the adaptor protein MecA would then promote motility gene expression during exponential phase by maintaining low ComK levels through proteolysis.

Interestingly, one study reported a *comK*-independent effect of *clpC* on swimming motility [60]. This suggests that other ClpCP substrates may also be involved in the regulation of motility. One candidate is the response regulator DegU, which is degraded by ClpCP/MecA [61] and represses the promotor of the *fla/che* operon in its phosphorylated form [62]. The specific recognition and degradation of only phosphorylated DegU [61] by ClpCP could be a very interesting regulatory mechanism and should be analysed in more detail.

However, the effect of the DegS-DegU two component system appears to be complex and is controversially discussed, because DegU has also been described as an activator of the *fla/che* promotor both in its phosphorylated and unphosphorylated forms [63–65]. The activating effect of DegU also seems to be stronger in strains, in which the flagellar activator SwrA is expressed. The presence of SwrA leads to hyper-flagellation required for swarming motility and SwrA is not expressed in laboratory strains due to a frameshift mutation in the *swrA* gene [65–67]. Whether the *comK*-independent effect of *clpC* on motility is due to *degU* awaits further characterization.

Recently, an interesting connection between FlgM and the ClpCP substrate DegU was discovered. Kearns and co-workers found that phosphorylated DegU (DegU ~ P) acts as a positive transcriptional regulator of *flgM* [68]. By degrading DegU ~ P, ClpCP would then facilitate the expression of the σ^D-regulon by yet another mechanism that acts on *flgM* transcription.

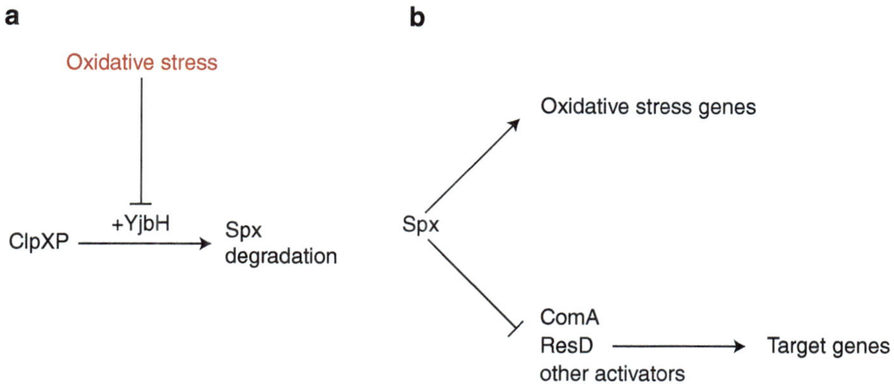

Fig. 4.2 Oxidative stress response mediated by Spx. (**a**) During normal growth, Spx is rapidly degraded by ClpXP, assisted by the adaptor protein YjbH. In response to oxidative stress, Spx is stabilized. (**b**) Spx acts as a transcriptional regulator, which positively regulates the expression of oxidative stress response genes and negatively regulates gene expression governed by transcriptional activators such as ComA and ResD

Stress Responses

Oxidative Stress Response and Spx

The oxidative stress response in *B. subtilis* and other Gram-positive bacteria is governed by the transcriptional regulator Spx. The *spx* gene (*yjbD*) was originally identified as a suppressor of *clpP* and *clpX*. Subsequently, Spx was shown to be a substrate of the ClpXP protease that is constantly degraded during normal growth (Fig. 4.2). Importantly, upon oxidative stress, introduced by diamide (an agent that introduces the formation of cytosolic disulfide bonds), paraquat or H_2O_2 treatment, the degradation of Spx is inhibited (Fig. 4.2). In *clpX* or *clpP* mutants, Spx accumulates to high levels under non-stress conditions, which has a detrimental effect on growth and leads to frequent suppressor mutations [28, 29]. Spx proteolysis is enhanced by the ClpX adaptor protein YjbH [30, 32], which itself might be regulated by oxidative stress [30]. Interestingly, it was proposed that ClpX can be inactivated by oxidation of a conserved zinc cluster, which is part of its N-terminal domain [69]. Taken together, the inactivation of ClpX and YjbH would result in rapid stabilization of Spx after redox stress. Recently, a novel anti-adaptor YirB was discovered [31]. The expression of this protein stabilizes Spx by inhibition of YjbH-mediated degradation of Spx by ClpXP, both *in vivo* and *in vitro* [31]. The expression of Spx is also transcriptionally induced during stress by the inactivation of the repressors PerR and YodB, which normally bind and repress one of the five known *spx* promoters [70, 71]. Furthermore, *spx* transcription is activated by heat shock and other stresses [72].

Spx acts as a negative regulator of a set of genes, including competence genes. According to the interference model, which has been directly demonstrated for the response regulators ResD and ComA [73], Spx inhibits the transcriptional activation of genes by binding to the C-terminal domain (CTD) of the alpha subunit of the RNA polymerase (RNAP). Thereby the transcription of genes, which require transcriptional activators that bind to the CTD of the alpha subunit are inhibited (Fig. 4.2). The formation of this inhibitory Spx-alpha CTD complex has also been investigated at the molecular level by X-ray crystallography [74].

Interestingly, analysis of the Spx dependent transcriptome, suggested that Spx also positively influences the transcription of a large number of genes [75]. Recent experiments suggest that this positive influence could be achieved directly by Spx in complex with the CTD of the RNAP alpha subunit by enhancing the binding of RNAP to certain promoters [76, 77]. The genes within the Spx regulon encode enzymes with a role in processing or detoxification of reactive oxygen and nitrogen species, i.e. thioredoxin, thioredoxin reductase and superoxide dismutase (Fig. 4.2) [28, 29, 75].

Similar to the proteolytic switch involving ClpCP/MecA and ComK, ClpXP-mediated Spx proteolysis may serve as a very tight control mechanism of Spx under non-stress conditions, combined with a rapid response time following oxidative stress.

Heat Shock Response

In contrast to *E. coli*, where the heat shock response is controlled by the alternative sigma factor RpoH (σ^{32}), four different mechanisms of the heat shock response have been described for *B. subtilis* [78]. Class I heat shock genes, including the *dnaK* and *groEL/groES* operons, contain operator sites known as CIRCE-elements in their 5′-untranslated regions, which are bound and repressed by HrcA, which is activated by the chaperonin GroEL [79]. Promoters of class II heat shock genes are bound by the alternative sigma factor σ^{B}, and are induced not only by heat shock, but also by other kinds of stress [80]. The class III heat shock genes are controlled by the transcriptional repressor CtsR and include the *clpC* operon (containing *ctsR*, *mcsA*, *mcsB* and *clpC*) and the *clpE* and *clpP* genes [12–14, 81]. Heat shock genes, which are controlled by an unknown mechanism, have been grouped together in class IV. These genes include *htpG* (encoding Hsp90), *clpX* and the *lon* genes [78]. The class V heat shock genes are activated by extracellular protein folding stress as well as heat shock and are controlled by the two-component system CssS/CssR [82, 83].

CtsR

In this section, we will focus on the class III heat shock genes and the regulation of its repressor CtsR by proteolysis. CtsR is degraded by ClpCP, mediated by its adaptor protein McsB during heat stress, but not at normal growth temperatures [14, 21]. McsB is not only an adaptor protein, but also a protein kinase [84]. McsB was first characterized as a tyrosine kinase [84] and later shown to be the first protein arginine kinase [85]. McsB autophosphorylates arginine residues in the presence of McsA [86] and can phosphorylate substrate proteins, among them CtsR [85]. The kinase activity of McsB is inhibited by ClpC and counter-acted by the arginine phosphatase YwlE [84, 87]. The general impact and physiological significance of arginine phosphorylation awaits further characterization. Importantly, as an adaptor protein McsB only binds and activates ClpC and targets CtsR for degradation in the presence of McsA when it is activated and autophosphorylated [21, 88]. Interestingly, phosphorylated McsB displays a high affinity for ClpC *in vitro* and thus can compete for binding to ClpC with the other adaptor protein MecA [21]. In addition, phosphorylated McsB can interfere with the DNA binding of CtsR [84, 85], leading to de-repression of CtsR-controlled genes. Based on these results a titration model for the activation of the class III heat shock genes was suggested, in which heat is sensed by an increase of unfolded proteins targeted by MecA to ClpCP [21, 84]. However, recently it was demonstrated that CtsR directly senses a temperature shift, resulting in its inactivation and subsequent release from DNA [88]. Neither MscB arginine phosphorylation nor CtsR degradation were required for this process, suggesting that proteolysis serves primarily to remove inactive CtsR molecules from the cell. This finding is also corroborated by the observation that CtsR controls heat shock genes in organisms that lack a McsB homologue. The current model for the activation of class III heat shock genes is as follows: CtsR changes its conformation at high temperatures, causing it to dissociate from its target class III heat shock genes resulting in their transcription. At the same time, ClpC is removed from the inhibitory McsA-McsB-ClpC complex, possibly involving MecA targeting unfolded proteins to ClpC, which results in the activation of the MscB arginine kinase activity. McsB phosphorylates itself and CtsR in the presence of McsA. Phosphorylated, heat-inactivated CtsR cannot rebind its operator sites and is also targeted for degradation by the proteolytic McsA-McsB-ClpCP complex, thus stabilizing the class III heat shock "ON" state (Fig. 4.3a).

Effect of Oxidative Stress on McsA

The CtsR regulon is not only de-repressed by heat shock, but also by oxidative stress [89, 90]. Recently, the mechanism of CtsR inactivation was shown to be very different from heat inactivation [87]. First of all, the activation of the class III heat shock genes by oxidizing agents is markedly slower than activation by heat stress, suggesting an indirect mechanism. Furthermore, McsB, but not its kinase activity, is required for this process in contrast to heat activation, which depends only on CtsR [88] and CtsR is not degraded during oxidative stress. McsA, the co-adaptor and modulator of CtsR, contains two clusters of conserved cysteine residues, which could constitute sensor domains to oxidative stress. Indeed, both clusters (one of which contains a zinc ion in its reduced state) were oxidized by disulfide stress *in vivo* [87]. Furthermore, McsA is also irreversibly modified by oxidative stress, resulting in a faster migrating form on SDS-PAGE gels. On the basis of N-terminal and C-terminal tagging experiments, this result was interpreted as proteolytic cleavage of McsA. Interestingly, the presence of the smaller band depended on the presence of ClpC and ClpP, suggesting a direct or indirect involvement of ClpCP in the putative cleavage reaction. Importantly, modification of McsA by oxidative stress led to dissociation of the McsA-McsB complex. McsB is able to release CtsR from DNA also in the absence

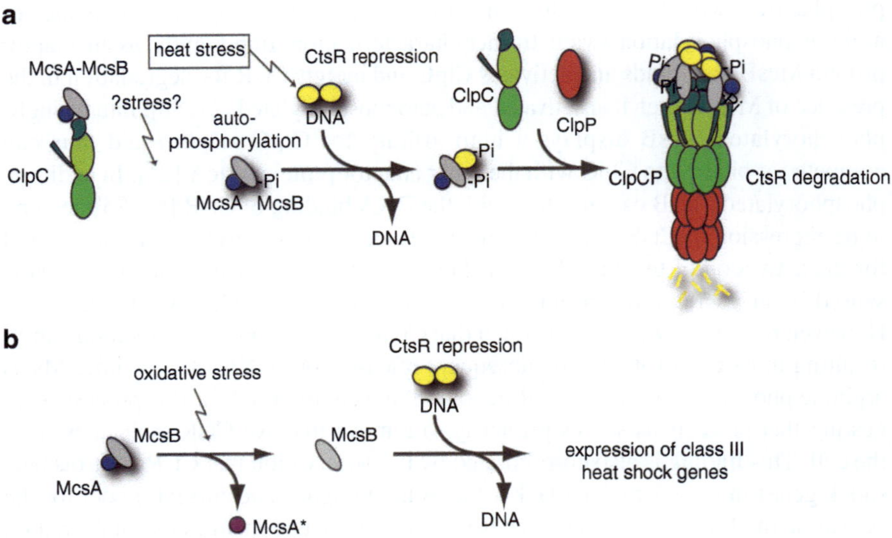

Fig. 4.3 Regulation of class III heat shock genes. (**a**) **CtsR regulation by heat.** In non-stressed cells, CtsR represses transcription of class III heat shock genes by binding to CIRCE operator sites. The kinase activity of MscB is inhibited by ClpC. Heat stress is directly sensed by CtsR, which is released from the DNA. This leads to derepression of class III heat shock genes. At the same time, the interaction of ClpC with McsB is disrupted by stress (possibly by unfolded proteins targeted for degradation by ClpCP/MecA) and McsB auto-phosphorylates in the presence of McsA. Phosphorylated McsB further contributes to release of CtsR from DNA and targets CtsR for degradation by ClpCP. (**b**) **CtsR regulation by oxidative stress**. McsA is oxidized and presumably processed by redox stress and dissociates from McsB. Monomeric McsB, which is not phosphorylated under these conditions, releases CtsR from DNA. Class III heat shock genes are derepressed

of McsA [84] and McsA acts as an inhibitor of McsB with regards to the ability to remove CtsR from DNA [14]. Thus, the results presented by Gerth and colleagues led to the following model [87]. In response to oxidative stress, McsA is proteolytically cleaved following modification of its cysteine residues, resulting in its dissociation from McsB (Fig. 4.3b). In turn, McsB is able to bind to CtsR and remove the protein from its operator DNA, causing derepression of the CtsR regulon [87] (Fig. 4.3b). Consequently, the heat shock "sensing" repressor, CtsR is also involved in responding to oxidative stress, via the McsA-McsB system.

Membrane Stress and Regulated Intramembrane Proteolysis (RIP) of the Extracytoplasmic Function (ECF) Sigma Factors

Regulated intramembrane proteolysis (RIP) is used by both prokaryotes and eukaryotes for signaling across cell membranes. RIP has been studied in great detail in *E. coli* using the activation of the extracytoplasmic function (ECF) sigma factor σ^E as a paradigm. ECF sigma factors constitute a special group of sigma factors (Group

IV of the σ^{70} subfamily) that are involved in the signaling of extracytoplasmic conditions and are characterized by a two-domain architecture [44]. There are seven ECF sigma factors in *B. subtilis* (σ^M, σ^V, σ^W, σ^X, σ^Y, σ^Z, and YlaC), whereas *Streptomyces coelicolor* contains about 50. The best characterized ECF sigma factor in *B. subtilis* is σ^W. This sigma factor is activated by alkaline stress, phage infection, salt stress, antimicrobial peptides and antibiotics such as vancomycin resulting in the expression of genes involved in the response to envelope stress and antibiotics [44]. In analogy to the *E. coli* σ^E/RseA system, *sigW*, the gene encoding σ^W is cotranscribed with *rsiW*, encoding a membrane bound, σ^W anti-sigma factor [91] (see also [92]). The cytoplasmic N-terminal domain of RsiW binds to, and inhibits the activity of, σ^W [91] (Fig. 4.4a). However, in contrast to the cascade in *E. coli*, site-1 cleavage is not executed by any of the DegS homologs HtrA, HtrB or HtrC [93]. Instead, PrsW (YpdC) is the site-1 protease of RsiW [94, 95] (Fig. 4.4a). PrsW is homologous to the eukaryotic type II CAAX prenyl endopeptidase family [94] and removes 40 amino acids from the extracellular domain of RsiW, cleaving between alanine 168 and serine 169 [96]. Additional processing by an unknown protease is required to facilitate cleavage by the site-2 protease and in a reconstituted *E. coli* system the tail-specific protease (Tsp) is able to fulfill this task [96] (Fig. 4.4a). The site-2 protease of RsiW was identified as RasP (YluC) [91] a zinc metalloprotease with an extracytoplasmic PDZ-domain and four transmembrane helices (Fig. 4.4a). This protease belongs to the site-2 protease (S2P) family of I-CLiPs [96] and processes RsiW after site-1 cleavage close to two highly conserved alanine residues [97]. This process requires the presence of the ABC transporter EcsAB, but the mechanism of this regulation remains unknown [98]. After site-2 cleavage and release from the cytoplasmic face of the membrane, the RsiW fragment remains bound to σ^W in the cytosol, until it is degraded by ClpXP [97] (Fig. 4.4a). The degradation of this fragment, by ClpXP or ClpEP but not cleavage by site-2 proteolysis, requires a stretch of three alanine residues at the new C-terminus of RsiW [97].

Two publications suggest that the substrate specificity of RasP is much broader than previously anticipated. Surprisingly, the cell-division protein FtsL is cleaved by RasP. Importantly, this cleavage is required for normal cell division, demonstrated by a short cell phenotype of a strain, in which FtsL could not be processed [99]. Furthermore, the cell division protein DivIC was shown to protect FtsL from proteolysis [100].

Recently, Akiyama and colleagues have demonstrated that RasP can also act as a signal peptide peptidase (SPP) that cleaves signal peptides (SP) in the membrane, after they have been removed from preproteins by the action of signal peptidase [101]. Likewise, the *E. coli* I-CLiP protease RseP, was also shown to act as the site-2 protease of pro-σ^E. Consistent with these findings, deletion of *rseP* inhibited SP cleavage *in vivo* while in contrast disruption of *sppA* (previously proposed to be the *E. coli* SPP) had no effect on SP cleavage [101].

It is fascinating to see how a protease, which was originally thought to have only one or two regulatory substrates, is suddenly found to be involved in such a broad fundamental process as signal peptide cleavage during protein secretion. These

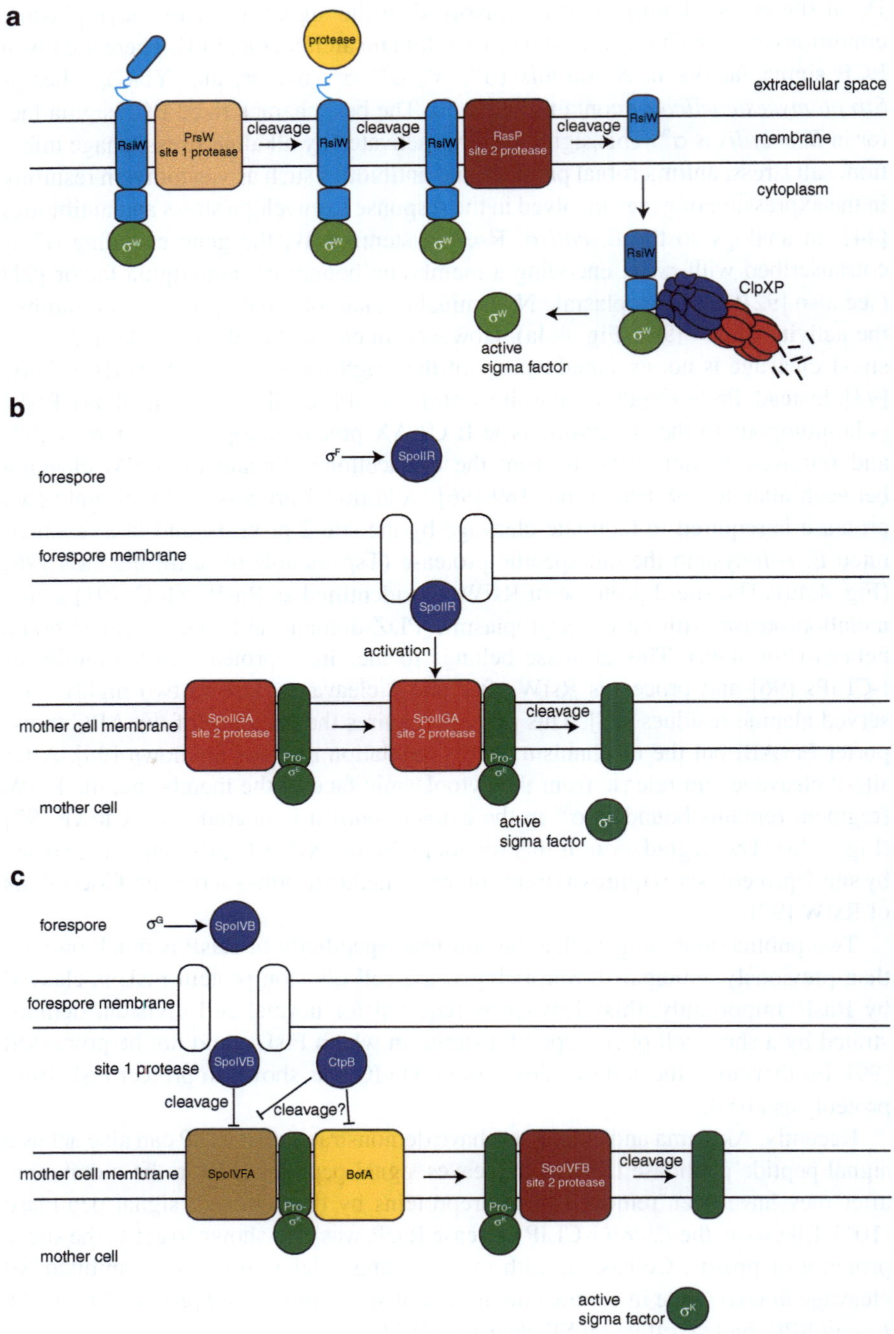

Fig. 4.4 Regulated intramembrane proteolysis (RIP) in *B. subtilis*. (a) RIP in the extracytoplasmic stress response. The extracytoplasmic stress response is controlled by the alternative sigma factor σ^W. In non-stressed cells, σ^W is bound and inhibited by the cytoplasmic domain of the membrane-embedded anti-sigma factor RsiW. Upon exposure to stress, the extracytoplasmic domain of

experiments suggest that many membrane localized proteases including FtsH and HtrA (DegP) may be involved in both regulation and PQC, just like the cytosolic protease systems.

Sporulation

Sporulation is a developmental pathway that is initiated in stationary phase under nutrient-limiting conditions and by which cells form durable and highly resistant endospores. These spores are able to preserve their genetic information during adverse environmental conditions and can later germinate, when nutrients are more abundant. During the early stage of sporulation, a *B. subtilis* cell divides asymmetrically into a smaller forespore and a larger mother cell. Subsequently, the forespore is engulfed by the mother cell through membrane fusion and eventually the mother cell lyses, releasing the developed spore. The importance of proteolysis in the sporulation of *B. subtilis* cells, is supported by the observation that FtsH, ClpP, ClpC and ClpX are all required for sporulation [27, 40, 58, 102, 103].

The early steps of sporulation (i.e. the formation of the asymmetric septum) are regulated by phosphorylation of the response regulator Spo0A, which is controlled by a phosphorelay system involving five different histidine kinases and several phosphatases. A self-reinforcing cycle involving Spo0A, the alternative sigma factor σ^H and the repressor AbrB, contributes to the production of high levels of phosphorylated Spo0A and entry into the sporulation pathway. Briefly, in its phosphorylated form, Spo0A represses the *abrB* gene, leading to derepression of the *spo0H* gene, encoding the alternative sigma factor σ^H, which directs transcription of *spo0A*. This leads to higher levels of Spo0A and activation of a positive autoregulatory loop triggering the differentiation into a spore-forming cell. Spo0A and σ^H activate several down-stream genes, i.e. the *spoIIA* locus, encoding σ^F (see below) and initiates the formation of the asymmetric division septum. After the

RsiW is processed by the site-1 protease PrsW and by a second unknown protease. This cleavage makes the transmembrane domain of RsiW susceptible to cleavage by the site-2 protease RasP. Cleavage by RasP releases σ^W from the membrane. The RsiW fragment still bound to σ^W is degraded by ClpXP in the cytoplasm. Subsequently, σ^W binds to RNAP and activates the transcription of its target genes. (**b**) RIP in σ^E regulation. σ^E is produced in the cytoplasm of the mother cell during sporulation as a proprotein (pro-σ^E) with a transmembrane domain and inserted into the mother cell membrane. SpoIIR is expressed in the forespore under the control of σ^F and secreted into the intermembrane space between forespore and mother cell. SpoIIR activates the site-2 protease SpoIIGA. In turn, SpoIIGA cleaves pro-σ^E, resulting in the release of σ^E from the membrane, which then mediates the expression of early mother cell specific genes. (**c**) RIP in σ^K regulation. Pro-σ^K is produced in the mother cell cytoplasm and inserted into the membrane, where it forms a complex with BofA and SpoIVFA, which protect pro-σ^K from proteolysis by the site-2 protease SpoIVFB. The late forespore-specific sigma factor σ^G directs expression of the site-1 protease SpoIVB, which is secreted into the intermembrane space and proteolytically processes SpoIFA, resulting in the destabilization of the pro-σ^K-BofA-SpoIVFA complex. In turn, SpoIVFB cleaves pro-σ^K and σ^K is released from the membrane to activate the expression of late mother cell specific genes

formation of the division septum, four compartment-specific alternative sigma factors (responsible for the expression of different classes of sporulation genes) are activated in a controlled temporal and spatial order. First, σ^F is activated in the forespore, next σ^E is activated in the mother cell. Following spore engulfment, σ^G replaces σ^F in the spore and finally, σ^K directs gene expression in the late mother cell [45, 104].

FtsH

Interestingly, one of the Spo0A phosphatases, Spo0E, was identified as an FtsH substrate [105], explaining the earlier observation that Spo0A activity is reduced in *ftsH* mutant cells [40]. FtsH localization to the asymmetric division septum is also consistent with its role in sporulation [106]. Furthermore, the FtsH protease is regulated by the small peptide SpoVM, which is essential for sporulation. SpoVM, not only binds to and inhibits FtsH but also serves as a substrate of the protease [107, 108].

LonA and LonB

The LonA protease was first identified in a screen for factors that down-regulate σ^G activity. Although the mechanism of σ^G inhibition by LonA was not investigated in detail, it was suggested that LonA is directly responsible for the degradation of σ^G [109]. LonA and LonB were also reported to down-regulate σ^H levels posttranslationally at low pH during sporulation, although this had no effect on σ^H directed gene expression [103]. These data are consistent with the model that σ^H is targeted to LonA and LonB at low pH, and that LonA and LonB also influence a downstream process, which is required for σ^H–mediated gene expression. LonB seems to have a special role in sporulation, since the *lonB* gene features a σ^F–dependent promotor and is specifically expressed under sporulation conditions. Neither a *lonB* mutant nor a strain over-expressing LonB affected sporulation, but interestingly expressing *lonA* from the sporulation dependent *lonB*-promoter lead to a sporulation defect and the down-regulation of σ^G–dependent gene expression [110]. LonA localizes to the nucleoid of cells during normal growth and to the forespore compartment during sporulation. However, LonB localized to the forespore membrane early in sporulation and to the whole forespore during later stages of the process [111].

ClpCP

In the pre-divisional cell, σ^F is already transcribed (activated by Spo0A), but is kept inactive by its anti-sigma factor SpoIIAB. SpoIIAB is antagonized by its anti-anti-sigma factor SpoIIAA. According to the partner-switch model, SpoIIAB

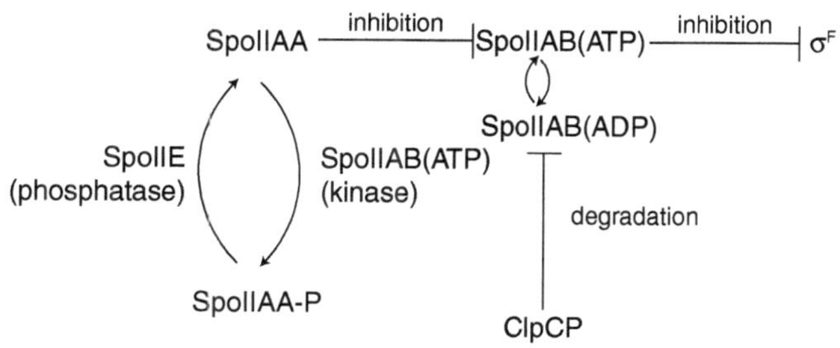

Fig. 4.5 Role of ClpC in σ^F regulation. The anti-sigma factor of σ^F, SpoIIAB can exist in an ATP-bound and an ADP-bound form. In the ATP-bound form, SpoIIAB binds to and inhibits σ^F. In the ADP-bound form, SpoIIAB is bound and inhibited by the anti-anti sigma factor SpoIIAA. SpoIIAB also acts as a kinase, which phosphorylates and inactivates SpoIIAA. SpoIIAA is dephosphorylated by the phosphatase SpoIIE. ClpCP degrades SpoIIAB, thus stabilizing the free active form of σ^F

can exist either in a complex with σ^F in its ATP-bound form, with SpoIIAA in the ADP-bound form or as a monomer bound to ADP. In the ATP-SpoIIAB-σ^F complex, SpoIIAB can act as a protein serine kinase that phosphorylates and inactivates SpoIIAA. The kinase activity of SpoIIAB is counteracted by the phosphatase SpoIIE. In the forespore after formation of the division septum, SpoIIE is able to dephosphorylate SpoIIAA, causing the formation of the inhibitory SpoIIAA-SpoIIAB-ADP complex and the release of σ^F, which can then interact with RNAP and activate its cognate promoters [45, 104] (Fig. 4.5). Interestingly, SpoIIAB is degraded by ClpCP but only in its monomeric form [27] (Fig. 4.5). This degradation is assumed to stabilize the free form of σ^F by shifting the equilibrium between the complexes described above. Targeting of SpoIIAB to ClpCP depends on a tripeptide motif (LCN) located at the C-terminus of SpoIIAB [27, 112]. Importantly, the proteolytic tag (LCN) on SpoIIAB is required for both normal σ^F-dependent gene expression and sporulation [27]. Given the degradation of SpoIIAB cannot be reconstituted *in vitro* it is plausible that an unknown adaptor protein or targeting mechanism may be responsible for the recognition of the LCN tag and targeting of SpoIIAB to ClpCP for degradation. However, SpoIIAB is also unstable *in vivo* when overexpressed during exponential growth, suggesting that the putative adaptor protein is unlikely to be expressed only during sporulation.

Interestingly the genes for SpoIIE, ClpC and ClpP are located close to the origin of replication of the *B. subtilis* chromosome and therefore are expressed together early in forespore development supporting the proteolytic part of the molecular mechanism resulting in forespore specific activation of σ^F [113]. This model is also supported by the localization of ClpC during sporulation, as determined from GFP-fusion experiments [114]. While ClpX-GFP and ClpP-GFP form foci close to the polar septum of the mother cell, ClpC-GFP exhibits a dynamic localization. The polar foci that are normally observed for ClpC-GFP during growth, delocalize early during sporulation. During engulfment of the prespore, ClpC-GFP forms a helical pattern in the mother cell with a number of ClpC foci co-localizing with the engulfing membrane. Later, ClpC forms foci at the distal prespore membrane, which subsequently delocalize. Using GFP fusions to truncated ClpC and ClpX, a quantitative measure of ClpC and ClpX in the forespore and mother cell was achieved, which displayed a diffuse localization pattern avoiding the complication of foci formation. Interestingly, the truncated GFP-ClpC protein localized preferentially to the forespore, whereas GFP-ClpX localized to the mother cell. This localization could be explained by the presence of a putative σ^F–dependent promoter in the *clpC* operon between *ctsR* and *mcsA*. Finally, the GFP-LCN fusion protein is degraded exclusively in the forespore, while the GFP-AAV fusion protein, which is targeted to ClpXP, is only degraded in the mother cell. This distribution of the various components could contribute to preferential degradation of SpoIIAB in the forespore and to σ^F-activity. The biological implication of the preferred location of ClpX and the complex dynamic distribution of ClpC during sporulation remains to be elucidated [114].

ClpXP

ClpX mutants have a sporulation defect and are compromised in σ^H-dependent gene expression [103, 115]. Later, it was reported that ClpX influences RNAP holoenzyme composition, triggering preferred incorporation of σ^H into RNAP during stationary phase [115]. The sporulation defect of the *clpP* and *clpX* mutants is also partially suppressed by mutating *spx* [116]. Spx protein accumulates in *clpP* and *clpX* mutant cells. By binding to the RNAP alpha CTD, Spx interferes with the binding of other activators, such as ComA to RNAP [73]. The ComP-ComA two-component system is required for the expression of the *srf* operon [117], which is required for sporulation [73] and for production of the RapC and RapF phosphatases involved in dephosphorylation of Spo0A. Taken together, these data suggest that *clpX* positively regulates sporulation by at least two different mechanisms, one of which is dependent on *spx*. These mechanisms are still not very well understood and await further investigation.

RIP in Sporulation

Communication between the forespore and mother cell compartments during two different stages of sporulation is mediated by regulated intramembrane proteolysis. Both the early mother cell-specific sigma factor σ^E and the late mother cell specific sigma factor σ^K are synthesized as preproteins with an N-terminal transmembrane domain that is inserted into the membrane adjacent to the forespore [118–120]. In both cases, a signal to activate the sigma factors is initiated by the respective early or late forespore specific sigma factors, ensuring a strict temporal order of gene expression by communication across compartments [45, 104].

SigmaE (σ^E) SigmaE is synthesized as an inactive precursor protein with an N-terminal transmembrane domain [118]. Interestingly, the proprotein (pro-σ^E) is produced in both the mother cell and the forespore membrane, but is selectively removed from the forespore membrane [121, 122]. Conversion of pro-σ^E to the active form only occurs after formation of the polar septum [123]. Intramembrane proteolytic cleavage of the prosequence requires the membrane protein SpoIIGA [124–126] (Fig. 4.4b). Only relatively recently, it was biochemically demonstrated that SpoIIGA is a novel type of aspartic protease that cleaves pro-σ^E [127, 128]. The activation of SpoIIGA requires SpoIIR, which is produced in the forespore under σ^F control [129–131]. SpoIIR was shown to interact with SpoIIGA, but the mechanism by which SpoIIR activates SpoIIGA is still not understood. Processing of SpoIIGA by SpoIIR or another site-1-protease has not been observed. Instead Kroos and colleagues have proposed the formation of inactive oligomers by SpoIIGA, which are dissociated by binding of SpoIIR [128]. In summary, SpoIIR is synthesized in the forespore under the control of σ^F and activates the membrane localized SpoIIGA, which in turn proteolytically activates the membrane associated pro-σ^E to form the transcription factor σ^E.

SigmaK σ^K is proteolytically processed by the intramembrane zinc metalloprotease SpoIVFB, which resembles eukaryotic site-2 proteases [132, 133]. SpoIVFB forms a complex with the proteins SpoIVFA and BofA [120, 134] (Fig. 4.4c). BofA inhibits the proteolytic activity of SpoIVFB toward pro-σ^K and SpoIVFA indirectly inhibits SpoIVFB by recruiting BofA to the complex [134]. Activated by σ^G, the serine protease SpoIVB is produced in the forespore and presumably secreted into the intermembrane space between forespore and mother cell, where it cleaves SpoIVFA [135, 136] (Fig. 4.4c).

According to one model [136], based on the heterologous expression of the signaling complex in *E. coli,* SpoIVB induced cleavage of SpoIVFA destabilizes the heterotrimeric complex, which leads to BofA degradation by another secreted serine protease, CtpB [137] and activation of SpoIVFB. Based on *in vitro* experiments in detergent micelles and membrane vesicles, Campo and Rudner suggest another model, in which the SpoIVFB-SpoIVFA-BofA complex remains intact even after cleavage of SpoIVFA by SpoIVB and argue that this cleavage leads to a conformational change promoting activation of SpoIVFB [138]. Furthermore, these authors demonstrated that CtpB processes SpoIVFA, but not BofA. CtpB is also cleaved by SpoIVB, but this processing appeared to have no effect on the activation of CtpB *in vivo* [139]. Contrary to initial assumptions, CtpB is mainly produced in the forespore under the control of σ^G, just like SpoIVB [139]. In summary, both models propose that SpoIVB and CtpB jointly contribute to the proteolytic processing and activation of the SpoIVFA, which leads to the activation of the site-2 protease SpoIVFB and the processing of pro-σ^K into its active form. Subsequently, σ^K is released from the membrane to recruit RNAP to its cognate promoters (Fig. 4.4c).

This is the second example where a developmental signal from the forespore is conveyed via different membrane localized proteases over two membranes, resulting in the transmembrane proteolytical activation of an alternative sigma factor.

Biofilm Formation

Biofilms are complex communities of bacteria growing on solid surfaces and air liquid interfaces held together by an extracellular matrix. The bacteria in this multicellular heterogenous biofilm population undertake a drastic lifestyle change from the better studied often motile and planktonic state, to this special sessile and stationary-phase developmental state [140, 141].

Laboratory strains of *B. subtilis* are unable to form robust biofilms due to mutations acquired during cultivation under laboratory conditions. However, biofilm formation has been investigated in this model organism, using "wild" *B. subtilis* isolates [142, 143]. It could be demonstrated that *B. subtilis* biofilms consist of long chains of individual cells that are held together laterally by an extracellular matrix composed of exopolysaccharides (EPS) and fibers of the amyloid-like protein TasA [144, 145]. The regulation of biofilm development is intricately connected with several other developmental processes e.g. sporulation, motility and competence development [146–149].

The operons encoding the EPS (*epsA-O*) and TasA (*tapA-sipW-tasA*) are controlled by the transcriptional repressor SinR, which is antagonized by the anti-repressor SinI. SinI in turn is activated by Spo0A, a global regulator of sporulation and other stationary phase processes, leading to expression of EPS and TasA and the formation of biofilm [149, 150]. By investigation of cell chaining during biofilm formation it could be demonstrated that SinR is involved in a bistable regulatory feedback loop with another transcriptional regulator, SlrR [151, 152] (Fig. 4.6). In this double negative feedback loop, SinR represses the *slrR* gene and SlrR protein forms a heterodimeric complex with SinR that switches the DNA binding specificity of SinR. Whereas SinR on its own represses *slrR*, *epsA-0* and *tapA-sipW-tasA*, these genes are derepressed in the presence of the SinR-SlrR complex. Interestingly, the SinR-SlrR dimer represses a completely different set of genes, among them the ORFs encoding autolysins (enzymes, which degrade peptidoglycan during cell separation) and flagellin [152]. This results in the production of more SlrR, titration of more SinR and progression into a biofilm-promoting SlrR "HIGH" state, in which biofilm genes are expressed while motility and cell separation genes are repressed (Fig. 4.6). During exponential growth, this pathway is activated stochastically, resulting in a small fraction of cells forming chains. However, when the culture is committed to biofilm formation, the entire population is forced into the SlrR "HIGH" state by the expression of SinI through Spo0A [152]. Biofilms are not infinitely stable, but exhibit a tendency to disassemble after several days of growth [153], which is accompanied by a decrease in cellular SlrR levels [154]. This observation prompted an investigation into how the switch from the SlrR "HIGH" state back to the SlrR "LOW" state is achieved [154]. Interestingly, SlrR bears a LexA-like autocleavage sequence in the linker between its N- and C-terminal domains. Notably, cleavage into a smaller SlrR species was observed *in vivo*. SlrR was also found to be unstable *in vivo* and this degradation depended on the presence of the autocleavage region. Furthermore, mutation of *clpC* partially stabilized SlrR in a late biofilm-forming culture and a *clpC* mutant exhibited severe cell chaining during exponential growth, which was suppressed by an *slrR* mutant. These data imply that SlrR is removed by proteolysis during the late stages of biofilm formation to return the cell to the SlrR "LOW" state [154].

General Proteolysis

Degradation of Unfolded and Aggregated Proteins

B. subtilis must adapt to extreme and quickly changing temperatures in its natural environment. Consequently, this organism is very well suited to grow at temperatures well over 50 °C [155]. Growth at such high temperatures requires an intricate protein quality control (PQC) system to cope with an increased amount of unfolded and misfolded proteins, which tend to form toxic aggregates. In all organisms, PQC systems consist of chaperones and proteases [5, 156]. Chaperones such as DnaK and GroEL either bind to unfolded substrate proteins to prevent their aggregation, actively refold the substrates or provide a sequestered environment for protein folding [157]. In cooperation with the chaperones, proteases recognize unfolded, misfolded and aggregated proteins to degrade them in a controlled manner [6]. Finally, unfoldases, such as *E. coli* ClpB or the yeast homolog Hsp104, together with their cognate Hsp70 chaperone systems, actively solubilize and refold protein aggregates [158–160]. ClpB is a Clp/Hsp100 ATPase, which lacks the ability to form a complex with ClpP and acts as a protein disaggregation enzyme together with DnaK in conjunction with small heat shock proteins in *E. coli* [161–163].

Although many of the chaperone systems are shared between *B. subtilis* and *E. coli* (DnaK/DnaJ/GrpE, GroEL/GroES, trigger factor, HtpG), *B. subtilis* lacks homologues of the proteins involved in protein disaggregation in *E. coli* (i.e. ClpB or the small heat shock proteins IbpA and IbpB). This raised the question, whether aggregates can be dissolved and refolded by other chaperones in *B. subtilis* or whether aggregates are irreversibly removed by proteases. The relatively severe heat shock phenotypes of *clpC*, *clpX* and *clpP* mutants in contrast to the comparably mild effect of a *dnaK* mutant in *B. subtilis* seems to suggest that general proteolysis is very important for PQC in this organism. However, the direct contribution of ClpC and ClpP to PQC is difficult to assess, because these proteins are also involved in regulation of the stress response. Likewise, in *E. coli* DnaK plays an important function in the heat shock regulation of σ^{32} [164–169], which could partly explain the severe heat shock phenotype of the *E. coli dnaK* mutant. In *B. subtilis*, GroEL/GroES regulates the class I heat shock repressor HrcA [79], but DnaK is not known to be involved in heat shock regulation.

However, there is substantial evidence that Clp proteases are directly involved in the degradation of unfolded or aggregated proteins [17]. ClpC, ClpE, ClpX, and ClpP localize to inclusion bodies formed by expression of puromycyl peptides or the aggregating protein PorA [16, 19, 170, 171]. Also, the turnover of puromycyl peptides decreased in all four mutant strains [16, 19]. Finally, ClpC in the presence of the adaptor protein MecA was shown to degrade aggregated model proteins

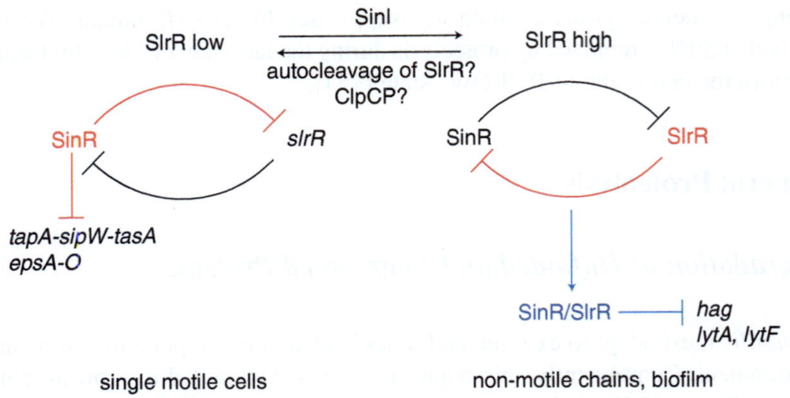

Fig. 4.6 The SlrR biofilm/motility switch. During exponential growth, SinR represses the biofilm operons *tapA-sipW-tasA* and *epsA-O* in most cells and motility genes and autolysins are expressed, resulting in single motile cells. The *slrR* gene is also repressed by SinR, which leads to a "low" steady state level of SlrR. The anti-repressor SinI is expressed in response to Spo0A activation in biofilm-forming cells or stochastically in exponential phase in a small number of cells. SinI inhibits SinR, which causes derepression of the *slrR* gene. SlrR forms a heterodimer with SinR. This heterodimer acts as a transcriptional repressor with different target specificity compared to SinR alone. The autolysin genes and *hag* encoding flagellin are directly repressed by SinR/SlrR, while the biofilm operons *tapA-sipW-tasA* and *epsA-O* and *slrR* are de-repressed. This double negative feedback loop results in a state, in which a high level of SlrR is expressed and cells are present as non-motile chains expressing biofilm genes. To return from the SlrR high state to the SlrR low state, LexA-like auto-cleavage of SlrR and ClpC are required, but the mechanism of this process is currently unknown

in vitro. Interestingly, the ClpC-MecA complex also displayed disaggregation activity in the absence of ClpP *in vitro* similar to *E. coli* ClpB [20]. This result suggests that ClpC could substitute for the ClpB-mediated disaggregation activity *in vivo*, although the ClpC-mediated disaggregation was slower than the ClpCP-mediated degradation *in vitro*. Whether ClpC function is always associated with ClpP *in vivo* remains to be elucidated.

The ability to survive an otherwise lethal heat shock after preconditioning at a sublethal temperature (a process known as thermotolerance) is directly connected to the presence of ClpB (in *E. coli*), or Hsp104 (in yeast) [158–163]. Interestingly, although *B. subtilis* lack homologs of both ClpB and the small heat shock proteins, *B. subtilis* cells exhibit thermotolerance. Future research will be aimed at elucidating the involvement and interplay of chaperones and proteases in PQC, stress response and thermotolerance.

Trans-translation and SsrA-Tagging in B. subtilis

Trans-translation is a process by which bacteria rescue stalled ribosomes. Ribosome stalling can occur for a number of reasons, i.e. as a consequence of damaged mRNAs lacking a stop codon, during the translation of proteins containing numerous rare codons or as a regulatory process. This translational block is relieved by an unusual small RNA, *ssrA* (also known as tmRNA), which structurally resembles both a tRNA and a mRNA [172]. The *ssrA* RNA binds to stalled ribosomes with the help of the protein SmpB and triggers translation to resume. Alanine, which is conjugated to the tRNA acceptor stem of *ssrA* is first incorporated into the polypeptide chain. Subsequently, the ribosome deciphers the sequence encoded in the mRNA-like domain of *ssrA* and adds a short peptide tag of 15 amino acids to the C-terminus of the polypeptide. This peptide sequence, known as the SsrA-tag, is recognized by proteases and targets the tagged protein for rapid degradation [173]. Proteolysis of SsrA-tagged proteins is believed to be important for survival of the bacterial cell, because truncated proteins arising from translational stalling are often misfolded and tend to aggregate [172]. In contrast to *E. coli*, where different proteases can degrade SsrA-tagged proteins, ClpXP is the only protease identified with this activity in *B. subtilis* [33]. Although LonA, LonB, and ClpYQ have yet to be tested [33], ClpCP, ClpEP and FtsH all lack the ability to degrade SsrA-tagged proteins. Although the sequence of the *B. subtilis* SsrA tag (AGKTNSFNQNVALAA) and the *E. coli* SsrA tag (ANDENYALAA) are quite divergent, the four C-terminal residues are identical. These C-terminal alanine residues are critical for degradation by ClpXP while the upstream sequence of the *E. coli* SsrA tag is required for ClpA and SspB binding [33] and consistently, *B. subtilis* lacks the adaptor protein SspB, which enhances the targeting of SsrA-tagged proteins to ClpXP in *E. coli* (see also [174]). In *B. subtilis* the abundance of SsrA-tagged proteins increased at 50 °C and the nature of the tagged proteins varied considerably between 37 and 50 °C [175]. Only eight SsrA-tagged substrates have been identified to date [175]. At 37 °C, three tagged proteins were identified, two of unknown function, YqaP and YtoQ, along with TreP (trehalose permease, a component of the phosphotransfer system). At 50 °C, five tagged proteins were identified, YloN (unknown function), PerR (repressor of oxidative stress genes), TufA (elongation factor Tu), FolA (dehydroneopterin aldolase) and GsiB (a σ^B-dependent general stress protein of unknown function) [175].

The *B. subtilis ssrA* gene is dispensable under non-stress conditions, but becomes essential for growth at high temperature, high concentration of ethanol or cadmium chloride [176], and low temperature [177]. The *ssrA* gene is induced by heat by an unknown mechanism [176] and resides in an operon with *secG* (component of the SecYEG translocon channel), *yvaK* (carboxyl esterase), *rnr* (RNase R) and *smpB*, which is subject to complex regulation and features five promotors, including a weak σ^B-dependent and a heat-inducible σ^A-dependent promoter [177]. These data indicate that trans-translation and/or SsrA-tagging is important during stress, when translational stalling may occur more often or may be more damaging to the cell. One result reported by Muto and colleagues, suggests that trans-translation and not the degradation of the SsrA-tagged proteins is important for growth at high temperature [176]. In this experiment, the two C-terminal alanine residues of the SsrA tag were replaced by aspartic acid residues and the strain (*ssrA*^{DD}) was grown at high temperature. In contrast to cells, in which *ssrA* was completely removed or the alanylation site was mutated, the *ssrA*^{DD} strain grew as well as the wild type at high temperature [176]. These data suggest that potentially aggregated SsrA-tagged substrates can be dealt with by other chaperones and proteases, which are highly abundant during heat stress, and that the principle task of the trans-translation system is the reconstitution of functional ribosomes.

Localization of Clp Proteases During Regulatory and General Proteolysis

Recently, the cellular localization of Clp proteases fused to GFP was investigated by three independent studies [111, 114, 171]. Interestingly, ClpP and its Clp ATPases were not localized uniformly throughout the cells, but clustered in 1–2 foci close to the cell poles. Furthermore, ClpP colocalized with its ATPases in CFP/YFP double labeling experiments [111, 114, 171]. The foci were still present in long filamentous cells induced by depletion of FtsZ, implying that localization depends on occlusion from the nucleoid. During cell division, ClpP foci localized to the division site, suggesting an unknown activity of ClpP in this process. ClpP clusters were also monitored within a growing microcolony. Under these conditions, the foci were highly dynamic. Clusters appeared and disappeared quickly at different locations in the cell, whereas individual clusters did not move very far. After heat shock, more foci of ClpP, ClpX and ClpC appeared, and ClpE formed visible clusters that were not observed during non-stress conditions. Also, the intensity of the clusters increased after heat stress. Strikingly, the adaptor protein McsB and the ClpCP substrate CtsR localized to the same locations as ClpCP after heat shock. This pattern already occurred at 30 °C in an *ywlE* mutant, in which McsB is hyperphosphorylated, and was unchanged in a trapped complex with ClpC. Although colocalization was not directly demonstrated, these results suggest that regulatory proteolysis of CtsR during heat shock is confined to specific locations in the cell. Furthermore, ClpP and all three ATPases colocalized with PorA-aggregates at the cell poles, demonstrating that general proteolysis is also localized close to the poles [171].

Antibiotics and Proteolysis

Recently, ClpP was identified as a target for acyldepsipeptide (ADEP) antibiotics, in both Gram-negative and Gram-positive bacteria. ADEP has an unusual mode of action, because it activates and dysregulates the protease systems, which results in subsequent cell death [178]. It could be demonstrated that ADEP binds specifically to a pocket on ClpP, which is necessary for the interaction with the associated Hsp100/Clp ATPase [179]. The binding of ADEP thereby interferes with the protease complex formation and the recognition of cellular substrates is prevented [180]. In addition, ADEP binding induces a conformational change, which opens the ClpP complex resulting in the recognition of larger unfolded proteins for proteolysis [179, 180]. Thereby substrate recognition, normally controlled by the ATPase is prevented, and unfolded proteins i.e. nascent polypeptide chains emanating from the ribosome are recognised and degraded by ClpP [180]. Recently the cell division protein FtsZ was identified as a new substrate for the activated ADEP-ClpP complex in *B. subtilis*, explaining the observed inhibition of cell division mediated by ADEP [181].

Another compound specifically designed to inhibit ClpP [182] was shown to be effective e.g. against ClpP of apicoplasts in *Plasmodium falciparum* [183].

A third compound with antibiotic activity against *Mycobacterium tuberculosis*, cyclomarin, was identified and it could be demonstrated that cyclomarin targets ClpC in this organism [184].

The recent discovery of antibiotics, which target regulatory and general proteolysis in bacteria, highlights the importance of the described protease systems for bacterial physiology.

Concluding Remarks

Although tremendous progress has been made in the last couple of years in the field of proteolysis in *B. subtilis*, many processes involving proteases are still poorly understood. Novel substrates involved in regulatory pathways continue to be discovered and the exact role of proteolysis in protein quality control remains to be elucidated. These are only a few examples of the exciting questions in the field of proteolysis in *B. subtilis* to be addressed in the future.

Acknowledgements The work in the Laboratory of KT is supported by the Deutsche Forschungsgemeinschaft.

References

1. Lopez D, Vlamakis H, Kolter R (2009) Generation of multiple cell types in Bacillus subtilis. FEMS Microbiol Rev 33(1):152–163
2. Gottesman S (2003) Proteolysis in bacterial regulatory circuits. Annu Rev Cell Dev Biol 19:565–587
3. Jenal U, Hengge-Aronis R (2003) Regulation by proteolysis in bacterial cells. Curr Opin Microbiol 6(2):163–172
4. Kirstein J, Moliere N, Dougan DA, Turgay K (2009) Adapting the machine: adaptor proteins for Hsp100/Clp and AAA+proteases. Nat Rev Microbiol 7(8):589–599
5. Hartl FU, Hayer-Hartl M (2002) Molecular chaperones in the cytosol: from nascent chain to folded protein. Science 295(5561):1852–1858
6. Wickner S, Maurizi MR, Gottesman S (1999) Posttranslational quality control: folding, refolding, and degrading proteins. Science 286(5446):1888–1893
7. Turgay K (2010) Role of proteolysis and chaperones in stress response and regulation. In: Storz G, Hengge R (eds) Bacterial stress responses, 2nd edn. ASM Press, American Society for Microbiology, Washington, DC, pp 75–90
8. Kress W, Maglica Z, Weber-Ban E (2009) Clp chaperone-proteases: structure and function. Res Microbiol 160(9):618–628

9. Sauer RT, Baker TA (2011) AAA+proteases: ATP-fueled machines of protein destruction. Annu Rev Biochem 80:587–612
10. Sauer RT, Bolon DN, Burton BM, Burton RE et al (2004) Sculpting the proteome with AAA(+) proteases and disassembly machines. Cell 119(1):9–18
11. Derre I, Rapoport G, Devine K, Rose M et al (1999) ClpE, a novel type of HSP100 ATPase, is part of the CtsR heat shock regulon of Bacillus subtilis. Mol Microbiol 32(3):581–593
12. Derre I, Rapoport G, Msadek T (1999) CtsR, a novel regulator of stress and heat shock response, controls clp and molecular chaperone gene expression in gram-positive bacteria. Mol Microbiol 31(1):117–131
13. Krüger E, Hecker M (1998) The first gene of the Bacillus subtilis clpC operon, ctsR, encodes a negative regulator of its own operon and other class III heat shock genes. J Bacteriol 180(24):6681–6688
14. Krüger E, Zühlke D, Witt E, Ludwig H et al (2001) Clp-mediated proteolysis in Gram-positive bacteria is autoregulated by the stability of a repressor. EMBO J 20(4):852–863
15. Gerth U, Kirstein J, Mostertz J, Waldminghaus T et al (2004) Fine-tuning in regulation of Clp protein content in Bacillus subtilis. J Bacteriol 186(1):179–191
16. Miethke M, Hecker M, Gerth U (2006) Involvement of Bacillus subtilis ClpE in CtsR degradation and protein quality control. J Bacteriol 188(13):4610–4619
17. Kock H, Gerth U, Hecker M (2004) The ClpP peptidase is the major determinant of bulk protein turnover in Bacillus subtilis. J Bacteriol 186(17):5856–5864
18. Krüger E, Völker U, Hecker M (1994) Stress induction of *clpC* in *Bacillus subtilis* and its involvement in stress tolerance. J Bacteriol 176:3360–3367
19. Krüger E, Witt E, Ohlmeier S, Hanschke R et al (2000) The clp proteases of Bacillus subtilis are directly involved in degradation of misfolded proteins. J Bacteriol 182(11):3259–3265
20. Schlothauer T, Mogk A, Dougan DA, Bukau B et al (2003) MecA, an adaptor protein necessary for ClpC chaperone activity. Proc Natl Acad Sci U S A 100(5):2306–2311
21. Kirstein J, Dougan DA, Gerth U, Hecker M et al (2007) The tyrosine kinase McsB is a regulated adaptor protein for ClpCP. EMBO J 26(8):2061–2070
22. Kirstein J, Schlothauer T, Dougan DA, Lilie H et al (2006) Adaptor protein controlled oligomerization activates the AAA+protein ClpC. EMBO J 25(7):1481–1491
23. Wang F, Mei Z, Qi Y, Yan C et al (2011) Structure and mechanism of the hexameric MecA-ClpC molecular machine. Nature 471(7338):331–335
24. Turgay K, Hahn J, Burghoorn J, Dubnau D (1998) Competence in Bacillus subtilis is controlled by regulated proteolysis of a transcription factor. EMBO J 17(22):6730–6738
25. Persuh M, Mandic-Mulec I, Dubnau D (2002) A MecA paralog, YpbH, binds ClpC, affecting both competence and sporulation. J Bacteriol 184(8):2310–2313
26. Kock H, Gerth U, Hecker M (2004) MurAA, catalysing the first committed step in peptidoglycan biosynthesis, is a target of Clp-dependent proteolysis in Bacillus subtilis. Mol Microbiol 51(4):1087–1102
27. Pan Q, Garsin DA, Losick R (2001) Self-reinforcing activation of a cell-specific transcription factor by proteolysis of an anti-sigma factor in B. subtilis. Mol Cell 8(4):873–883
28. Zuber P (2004) Spx-RNA polymerase interaction and global transcriptional control during oxidative stress. J Bacteriol 186(7):1911–1918
29. Zuber P (2009) Management of oxidative stress in Bacillus. Annu Rev Microbiol 63:575–597
30. Garg SK, Kommineni S, Henslee L, Zhang Y et al (2009) The YjbH protein of Bacillus subtilis enhances ClpXP-catalyzed proteolysis of Spx. J Bacteriol 191(4):1268–1277
31. Kommineni S, Garg SK, Chan CM, Zuber P (2011) YjbH-enhanced proteolysis of Spx by ClpXP in Bacillus subtilis is inhibited by the small protein YirB (YuzO). J Bacteriol 193(9):2133–2140
32. Larsson JT, Rogstam A, von Wachenfeldt C (2007) YjbH is a novel negative effector of the disulphide stress regulator, Spx, in Bacillus subtilis. Mol Microbiol 66(3):669–684
33. Wiegert T, Schumann W (2001) SsrA-mediated tagging in Bacillus subtilis. J Bacteriol 183(13):3885–3889

34. Duman RE, Lowe J (2010) Crystal structures of Bacillus subtilis Lon protease. J Mol Biol 401(4):653–670
35. Kang MS, Kim SR, Kwack P, Lim BK et al (2003) Molecular architecture of the ATP-dependent CodWX protease having an N-terminal serine active site. EMBO J 22(12):2893–2902
36. Kang MS, Lim BK, Seong IS, Seol JH et al (2001) The ATP-dependent CodWX (HslVU) protease in Bacillus subtilis is an N-terminal serine protease. EMBO J 20(4):734–742
37. Bochtler M, Hartmann C, Song HK, Bourenkov GP et al (2000) The structures of HslU and the ATP-dependent protease HslU-HslV. Nature 403(6771):800–805
38. Sousa MC, Trame CB, Tsuruta H, Wilbanks SM et al (2000) Crystal and solution structures of an HslUV protease-chaperone complex. Cell 103(4):633–643
39. Ito K, Akiyama Y (2005) Cellular functions, mechanism of action, and regulation of FtsH protease. Annu Rev Microbiol 59:211–231
40. Deuerling E, Mogk A, Richter C, Purucker M et al (1997) The ftsH gene of Bacillus subtilis is involved in major cellular processes such as sporulation, stress adaptation and secretion. Mol Microbiol 23(5):921–933
41. Wolfe MS (2009) Intramembrane proteolysis. Chem Rev 109(4):1599–1612
42. Urban S (2009) Making the cut: central roles of intramembrane proteolysis in pathogenic microorganisms. Nat Rev Microbiol 7(6):411–423
43. Brown MS, Ye J, Rawson RB, Goldstein JL (2000) Regulated intramembrane proteolysis: a control mechanism conserved from bacteria to humans. Cell 100(4):391–398
44. Heinrich J, Wiegert T (2009) Regulated intramembrane proteolysis in the control of extracytoplasmic function sigma factors. Res Microbiol 160(9):696–703
45. Rudner DZ, Losick R (2001) Morphological coupling in development: lessons from prokaryotes. Dev Cell 1(6):733–742
46. Kong L, Dubnau D (1994) Regulation of competence-specific gene expression by Mec-mediated protein-protein interaction in Bacillus subtilis. Proc Natl Acad Sci U S A 91(13):5793–5797
47. Turgay K, Hamoen LW, Venema G, Dubnau D (1997) Biochemical characterization of a molecular switch involving the heat shock protein ClpC, which controls the activity of ComK, the competence transcription factor of Bacillus subtilis. Genes Dev 11(1):119–128
48. Prepiak P, Dubnau D (2007) A peptide signal for adapter protein-mediated degradation by the AAA+ protease ClpCP. Mol Cell 26(5):639–647
49. Maamar H, Dubnau D (2005) Bistability in the Bacillus subtilis K-state (competence) system requires a positive feedback loop. Mol Microbiol 56(3):615–624
50. Smits WK, Eschevins CC, Susanna KA, Bron S et al (2005) Stripping Bacillus: ComK autostimulation is responsible for the bistable response in competence development. Mol Microbiol 56(3):604–614
51. Dubnau D, Losick R (2006) Bistability in bacteria. Mol Microbiol 61(3):564–572
52. Kearns DB (2010) A field guide to bacterial swarming motility. Nat Rev Microbiol 8(9):634–644
53. Caramori T, Barilla D, Nessi C, Sacchi L et al (1996) Role of FlgM in sigma D-dependent gene expression in Bacillus subtilis. J Bacteriol 178(11):3113–3118
54. Fredrick K, Helmann JD (1996) FlgM is a primary regulator of sigmaD activity, and its absence restores motility to a sinR mutant. J Bacteriol 178(23):7010–7013
55. Sorenson MK, Ray SS, Darst SA (2004) Crystal structure of the flagellar sigma/anti-sigma complex sigma(28)/FlgM reveals an intact sigma factor in an inactive conformation. Mol Cell 14(1):127–138
56. Chevance FF, Hughes KT (2008) Coordinating assembly of a bacterial macromolecular machine. Nat Rev Microbiol 6(6):455–465
57. Aldridge P, Hughes KT (2002) Regulation of flagellar assembly. Curr Opin Microbiol 5(2):160–165
58. Msadek T, Dartois V, Kunst F, Herbaud ML et al (1998) ClpP of Bacillus subtilis is required for competence development, motility, degradative enzyme synthesis, growth at high temperature and sporulation. Mol Microbiol 27(5):899–914
59. Liu J, Zuber P (1998) A molecular switch controlling competence and motility: competence regulatory factors ComS, MecA, and ComK control sigmaD-dependent gene expression in Bacillus subtilis. J Bacteriol 180(16):4243–4251

60. Rashid MH, Tamakoshi A, Sekiguchi J (1996) Effects of mecA and mecB (clpC) mutations on expression of sigD, which encodes an alternative sigma factor, and autolysin operons and on flagellin synthesis in Bacillus subtilis. J Bacteriol 178(16):4861–4869
61. Ogura M, Tsukahara K (2010) Autoregulation of the Bacillus subtilis response regulator gene degU is coupled with the proteolysis of DegU-P by ClpCP. Mol Microbiol 75(5):1244–1259
62. Amati G, Bisicchia P, Galizzi A (2004) DegU-P represses expression of the motility fla-che operon in Bacillus subtilis. J Bacteriol 186(18):6003–6014
63. Verhamme DT, Kiley TB, Stanley-Wall NR (2007) DegU co-ordinates multicellular behaviour exhibited by Bacillus subtilis. Mol Microbiol 65(2):554–568
64. Kobayashi K (2007) Gradual activation of the response regulator DegU controls serial expression of genes for flagellum formation and biofilm formation in Bacillus subtilis. Mol Microbiol 66(2):395–409
65. Tsukahara K, Ogura M (2008) Promoter selectivity of the Bacillus subtilis response regulator DegU, a positive regulator of the fla/che operon and sacB. BMC Microbiol 8:8
66. Kearns DB, Chu F, Rudner R, Losick R (2004) Genes governing swarming in Bacillus subtilis and evidence for a phase variation mechanism controlling surface motility. Mol Microbiol 52(2):357–369
67. Calvio C, Celandroni F, Ghelardi E, Amati G et al (2005) Swarming differentiation and swimming motility in Bacillus subtilis are controlled by swrA, a newly identified dicistronic operon. J Bacteriol 187(15):5356–5366
68. Hsueh YH, Cozy LM, Sham LT, Calvo RA et al (2011) DegU-phosphate activates expression of the anti-sigma factor FlgM in Bacillus subtilis. Mol Microbiol 81(4):1092–1108
69. Zhang Y, Zuber P (2007) Requirement of the zinc-binding domain of ClpX for Spx proteolysis in Bacillus subtilis and effects of disulfide stress on ClpXP activity. J Bacteriol 189(21):7669–7680
70. Leelakriangsak M, Kobayashi K, Zuber P (2007) Dual negative control of spx transcription initiation from the P3 promoter by repressors PerR and YodB in Bacillus subtilis. J Bacteriol 189(5):1736–1744
71. Leelakriangsak M, Zuber P (2007) Transcription from the P3 promoter of the Bacillus subtilis spx gene is induced in response to disulfide stress. J Bacteriol 189(5):1727–1735
72. Petersohn A, Brigulla M, Haas S, Hoheisel JD et al (2001) Global analysis of the general stress response of Bacillus subtilis. J Bacteriol 183(19):5617–5631
73. Nakano S, Nakano MM, Zhang Y, Leelakriangsak M et al (2003) A regulatory protein that interferes with activator-stimulated transcription in bacteria. Proc Natl Acad Sci U S A 100(7):4233–4238
74. Newberry KJ, Nakano S, Zuber P, Brennan RG (2005) Crystal structure of the Bacillus subtilis anti-alpha, global transcriptional regulator, Spx, in complex with the alpha C-terminal domain of RNA polymerase. Proc Natl Acad Sci U S A 102(44):15839–15844
75. Nakano S, Kuster-Schock E, Grossman AD, Zuber P (2003) Spx-dependent global transcriptional control is induced by thiol-specific oxidative stress in Bacillus subtilis. Proc Natl Acad Sci U S A 100(23):13603–13608
76. Reyes DY, Zuber P (2008) Activation of transcription initiation by Spx: formation of transcription complex and identification of a Cis-acting element required for transcriptional activation. Mol Microbiol 69(3):765–779
77. Nakano MM, Lin A, Zuber CS, Newberry KJ et al (2010) Promoter recognition by a complex of Spx and the C-terminal domain of the RNA polymerase alpha subunit. PLoS One 5(1):e8664
78. Hecker M, Schumann W, Volker U (1996) Heat-shock and general stress response in Bacillus subtilis. Mol Microbiol 19(3):417–428
79. Mogk A, Homuth G, Scholz C, Kim L et al (1997) The GroE chaperonin machine is a major modulator of the CIRCE heat shock regulon of Bacillus subtilis. EMBO J 16(15):4579–4590
80. Hecker M, Pane-Farre J, Volker U (2007) SigB-dependent general stress response in Bacillus subtilis and related gram-positive bacteria. Annu Rev Microbiol 61:215–236
81. Derre I, Rapoport G, Msadek T (2000) The CtsR regulator of stress response is active as a dimer and specifically degraded in vivo at 37 degrees C. Mol Microbiol 38(2):335–347

82. Hyyrylainen HL, Bolhuis A, Darmon E, Muukkonen L et al (2001) A novel two-component regulatory system in Bacillus subtilis for the survival of severe secretion stress. Mol Microbiol 41(5):1159–1172
83. Darmon E, Noone D, Masson A, Bron S et al (2002) A novel class of heat and secretion stress-responsive genes is controlled by the autoregulated CssRS two-component system of Bacillus subtilis. J Bacteriol 184(20):5661–5671
84. Kirstein J, Zuhlke D, Gerth U, Turgay K et al (2005) A tyrosine kinase and its activator control the activity of the CtsR heat shock repressor in B. subtilis. EMBO J 24(19):3435–3445
85. Fuhrmann J, Schmidt A, Spiess S, Lehner A et al (2009) McsB is a protein arginine kinase that phosphorylates and inhibits the heat-shock regulator CtsR. Science 324(5932): 1323–1327
86. Elsholz AK, Hempel K, Michalik S, Gronau K et al (2011) Activity control of the ClpC adaptor McsB in Bacillus subtilis. J Bacteriol 193(15):3887–3893
87. Elsholz AK, Hempel K, Pother DC, Becher D et al (2011) CtsR inactivation during thiol-specific stress in low GC, Gram+ bacteria. Mol Microbiol 79(3):772–785
88. Elsholz AK, Michalik S, Zuhlke D, Hecker M et al (2010) CtsR, the Gram-positive master regulator of protein quality control, feels the heat. EMBO J 29(21):3621–3629
89. Leichert LI, Scharf C, Hecker M (2003) Global characterization of disulfide stress in Bacillus subtilis. J Bacteriol 185(6):1967–1975
90. Mostertz J, Scharf C, Hecker M, Homuth G (2004) Transcriptome and proteome analysis of Bacillus subtilis gene expression in response to superoxide and peroxide stress. Microbiology 150(Pt 2):497–512
91. Schöbel S, Zellmeier S, Schumann W, Wiegert T (2004) The Bacillus subtilis sigmaW anti-sigma factor RsiW is degraded by intramembrane proteolysis through YluC. Mol Microbiol 52(4):1091–1105
92. Barchinger SE, Ades SE (2013) Regulated proteolysis: control of the Escherichia coli σE-dependent cell envelope stress response. In: Dougan DA (ed) Regulated proteolysis in microorganisms. Springer, Subcell Biochem 66:129–160
93. Noone D, Howell A, Devine KM (2000) Expression of ykdA, encoding a Bacillus subtilis homologue of HtrA, is heat shock inducible and negatively autoregulated. J Bacteriol 182(6):1592–1599
94. Ellermeier CD, Losick R (2006) Evidence for a novel protease governing regulated intramembrane proteolysis and resistance to antimicrobial peptides in Bacillus subtilis. Genes Dev 20(14):1911–1922
95. Heinrich J, Wiegert T (2006) YpdC determines site-1 degradation in regulated intramembrane proteolysis of the RsiW anti-sigma factor of Bacillus subtilis. Mol Microbiol 62(2):566–579
96. Heinrich J, Hein K, Wiegert T (2009) Two proteolytic modules are involved in regulated intramembrane proteolysis of Bacillus subtilis RsiW. Mol Microbiol 74(6):1412–1426
97. Zellmeier S, Schumann W, Wiegert T (2006) Involvement of Clp protease activity in modulating the Bacillus subtilis sigmaw stress response. Mol Microbiol 61(6):1569–1582
98. Heinrich J, Lunden T, Kontinen VP, Wiegert T (2008) The Bacillus subtilis ABC transporter EcsAB influences intramembrane proteolysis through RasP. Microbiology 154(Pt 7):1989–1997
99. Bramkamp M, Weston L, Daniel RA, Errington J (2006) Regulated intramembrane proteolysis of FtsL protein and the control of cell division in Bacillus subtilis. Mol Microbiol 62(2):580–591
100. Wadenpohl I, Bramkamp M (2010) DivIC stabilizes FtsL against RasP cleavage. J Bacteriol 192(19):5260–5263
101. Saito A, Hizukuri Y, Matsuo E, Chiba S et al (2011) Post-liberation cleavage of signal peptides is catalyzed by the site-2 protease (S2P) in bacteria. Proc Natl Acad Sci U S A 108(33):13740–13745
102. Gerth U, Kruger E, Derre I, Msadek T et al (1998) Stress induction of the Bacillus subtilis clpP gene encoding a homologue of the proteolytic component of the Clp protease and the involvement of ClpP and ClpX in stress tolerance. Mol Microbiol 28(4):787–802

103. Liu J, Cosby WM, Zuber P (1999) Role of lon and ClpX in the post-translational regulation of a sigma subunit of RNA polymerase required for cellular differentiation in Bacillus subtilis. Mol Microbiol 33(2):415–428
104. Errington J (2003) Regulation of endospore formation in Bacillus subtilis. Nat Rev Microbiol 1(2):117–126
105. Le AT, Schumann W (2009) The Spo0E phosphatase of Bacillus subtilis is a substrate of the FtsH metalloprotease. Microbiology 155(Pt 4):1122–1132
106. Wehrl W, Niederweis M, Schumann W (2000) The FtsH protein accumulates at the septum of Bacillus subtilis during cell division and sporulation. J Bacteriol 182(13):3870–3873
107. Cutting S, Anderson M, Lysenko E, Page A et al (1997) SpoVM, a small protein essential to development in Bacillus subtilis, interacts with the ATP-dependent protease FtsH. J Bacteriol 179(17):5534–5542
108. Prajapati RS, Ogura T, Cutting SM (2000) Structural and functional studies on an FtsH inhibitor from Bacillus subtilis. Biochim Biophys Acta 1475(3):353–359
109. Schmidt R, Decatur AL, Rather PN, Moran CP Jr et al (1994) Bacillus subtilis lon protease prevents inappropriate transcription of genes under the control of the sporulation transcription factor sigma G. J Bacteriol 176(21):6528–6537
110. Serrano M, Hovel S, Moran CP Jr, Henriques AO et al (2001) Forespore-specific transcription of the lonB gene during sporulation in Bacillus subtilis. J Bacteriol 183(10):2995–3003
111. Simmons LA, Grossman AD, Walker GC (2008) Clp and Lon proteases occupy distinct subcellular positions in Bacillus subtilis. J Bacteriol 190(20):6758–6768
112. Pan Q, Losick R (2003) Unique degradation signal for ClpCP in Bacillus subtilis. J Bacteriol 185(17):5275–5278
113. Dworkin J, Losick R (2001) Differential gene expression governed by chromosomal spatial asymmetry. Cell 107(3):339–346
114. Kain J, He GG, Losick R (2008) Polar localization and compartmentalization of ClpP proteases during growth and sporulation in Bacillus subtilis. J Bacteriol 190(20):6749–6757
115. Liu J, Zuber P (2000) The ClpX protein of Bacillus subtilis indirectly influences RNA polymerase holoenzyme composition and directly stimulates sigma- dependent transcription. Mol Microbiol 37(4):885–897
116. Nakano MM, Hajarizadeh F, Zhu Y, Zuber P (2001) Loss-of-function mutations in yjbD result in ClpX- and ClpP-independent competence development of Bacillus subtilis. Mol Microbiol 42(2):383–394
117. Kobayashi K, Ogura M, Yamaguchi H, Yoshida K et al (2001) Comprehensive DNA microarray analysis of Bacillus subtilis two-component regulatory systems. J Bacteriol 183(24):7365–7370
118. LaBell TL, Trempy JE, Haldenwang WG (1987) Sporulation-specific sigma factor sigma 29 of Bacillus subtilis is synthesized from a precursor protein, P31. Proc Natl Acad Sci U S A 84(7):1784–1788
119. Zhang B, Hofmeister A, Kroos L (1998) The prosequence of pro-sigmaK promotes membrane association and inhibits RNA polymerase core binding. J Bacteriol 180(9):2434–2441
120. Cutting S, Oke V, Driks A, Losick R et al (1990) A forespore checkpoint for mother cell gene expression during development in B. subtilis. Cell 62(2):239–250
121. Fujita M, Losick R (2002) An investigation into the compartmentalization of the sporulation transcription factor sigmaE in Bacillus subtilis. Mol Microbiol 43(1):27–38
122. Ju J, Luo T, Haldenwang WG (1998) Forespore expression and processing of the SigE transcription factor in wild-type and mutant Bacillus subtilis. J Bacteriol 180(7):1673–1681
123. Pogliano K, Hofmeister AE, Losick R (1997) Disappearance of the sigma E transcription factor from the forespore and the SpoIIE phosphatase from the mother cell contributes to establishment of cell-specific gene expression during sporulation in Bacillus subtilis. J Bacteriol 179(10):3331–3341
124. Jonas RM, Weaver EA, Kenney TJ, Moran CP Jr et al (1988) The Bacillus subtilis spoIIG operon encodes both sigma E and a gene necessary for sigma E activation. J Bacteriol 170(2):507–511

125. Peters HK 3rd, Haldenwang WG (1994) Isolation of a Bacillus subtilis spoIIGA allele that suppresses processing-negative mutations in the Pro-sigma E gene (sigE). J Bacteriol 176(24):7763–7766
126. Stragier P, Bonamy C, Karmazyn-Campelli C (1988) Processing of a sporulation sigma factor in Bacillus subtilis: how morphological structure could control gene expression. Cell 52(5):697–704
127. Imamura D, Kuwana R, Kroos L, Feig M et al (2011) Substrate specificity of SpoIIGA, a signal-transducing aspartic protease in Bacilli. J Biochem 149(6):665–671
128. Imamura D, Zhou R, Feig M, Kroos L (2008) Evidence that the Bacillus subtilis SpoIIGA protein is a novel type of signal-transducing aspartic protease. J Biol Chem 283(22):15287–15299
129. Hofmeister AE, Londono-Vallejo A, Harry E, Stragier P et al (1995) Extracellular signal protein triggering the proteolytic activation of a developmental transcription factor in B. subtilis. Cell 83(2):219–226
130. Karow ML, Glaser P, Piggot PJ (1995) Identification of a gene, spoIIR, that links the activation of sigma E to the transcriptional activity of sigma F during sporulation in Bacillus subtilis. Proc Natl Acad Sci U S A 92(6):2012–2016
131. Londono-Vallejo JA, Stragier P (1995) Cell-cell signaling pathway activating a developmental transcription factor in Bacillus subtilis. Genes Dev 9(4):503–508
132. Rudner DZ, Fawcett P, Losick R (1999) A family of membrane-embedded metalloproteases involved in regulated proteolysis of membrane-associated transcription factors. Proc Natl Acad Sci U S A 96(26):14765–14770
133. Yu YT, Kroos L (2000) Evidence that SpoIVFB is a novel type of membrane metalloprotease governing intercompartmental communication during Bacillus subtilis sporulation. J Bacteriol 182(11):3305–3309
134. Rudner DZ, Losick R (2002) A sporulation membrane protein tethers the pro-sigmaK processing enzyme to its inhibitor and dictates its subcellular localization. Genes Dev 16(8):1007–1018
135. Cutting S, Driks A, Schmidt R, Kunkel B et al (1991) Forespore-specific transcription of a gene in the signal transduction pathway that governs Pro-sigma K processing in Bacillus subtilis. Genes Dev 5(3):456–466
136. Zhou R, Kroos L (2005) Serine proteases from two cell types target different components of a complex that governs regulated intramembrane proteolysis of pro-sigmaK during Bacillus subtilis development. Mol Microbiol 58(3):835–846
137. Pan Q, Losick R, Rudner DZ (2003) A second PDZ-containing serine protease contributes to activation of the sporulation transcription factor sigmaK in Bacillus subtilis. J Bacteriol 185(20):6051–6056
138. Campo N, Rudner DZ (2006) A branched pathway governing the activation of a developmental transcription factor by regulated intramembrane proteolysis. Mol Cell 23(1):25–35
139. Campo N, Rudner DZ (2007) SpoIVB and CtpB are both forespore signals in the activation of the sporulation transcription factor sigmaK in Bacillus subtilis. J Bacteriol 189(16): 6021–6027
140. Lopez D, Vlamakis H, Kolter R (2010) Biofilms. Cold Spring Harb Perspect Biol 2(7):a000398
141. Kolter R, Greenberg EP (2006) Microbial sciences: the superficial life of microbes. Nature 441(7091):300–302
142. Hamon MA, Lazazzera BA (2001) The sporulation transcription factor Spo0A is required for biofilm development in Bacillus subtilis. Mol Microbiol 42(5):1199–1209
143. Branda SS, Gonzalez-Pastor JE, Ben-Yehuda S, Losick R et al (2001) Fruiting body formation by Bacillus subtilis. Proc Natl Acad Sci U S A 98(20):11621–11626
144. Branda SS, Chu F, Kearns DB, Losick R et al (2006) A major protein component of the Bacillus subtilis biofilm matrix. Mol Microbiol 59(4):1229–1238
145. Romero D, Aguilar C, Losick R, Kolter R (2010) Amyloid fibers provide structural integrity to Bacillus subtilis biofilms. Proc Natl Acad Sci U S A 107(5):2230–2234
146. Stanley NR, Britton RA, Grossman AD, Lazazzera BA (2003) Identification of catabolite repression as a physiological regulator of biofilm formation by Bacillus subtilis by use of DNA microarrays. J Bacteriol 185(6):1951–1957

147. Branda SS, Gonzalez-Pastor JE, Dervyn E, Ehrlich SD et al (2004) Genes involved in formation of structured multicellular communities by Bacillus subtilis. J Bacteriol 186(12):3970–3979
148. Hamon MA, Stanley NR, Britton RA, Grossman AD et al (2004) Identification of AbrB-regulated genes involved in biofilm formation by Bacillus subtilis. Mol Microbiol 52(3):847–860
149. Kearns DB, Chu F, Branda SS, Kolter R et al (2005) A master regulator for biofilm formation by Bacillus subtilis. Mol Microbiol 55(3):739–749
150. Chu F, Kearns DB, Branda SS, Kolter R et al (2006) Targets of the master regulator of biofilm formation in Bacillus subtilis. Mol Microbiol 59(4):1216–1228
151. Chu F, Kearns DB, McLoon A, Chai Y et al (2008) A novel regulatory protein governing biofilm formation in Bacillus subtilis. Mol Microbiol 68(5):1117–1127
152. Chai Y, Norman T, Kolter R, Losick R (2010) An epigenetic switch governing daughter cell separation in Bacillus subtilis. Genes Dev 24(8):754–765
153. Kolodkin-Gal I, Romero D, Cao S, Clardy J et al (2010) D-amino acids trigger biofilm disassembly. Science 328(5978):627–629
154. Chai Y, Kolter R, Losick R (2010) Reversal of an epigenetic switch governing cell chaining in Bacillus subtilis by protein instability. Mol Microbiol 78(1):218–229
155. Völker U, Engelmann S, Maul B, Riethdorf S et al (1994) Analysis of the induction of general stress proteins of Bacillus subtilis. Microbiology 140(Pt 4):741–752
156. Hartl FU, Hayer-Hartl M (2009) Converging concepts of protein folding in vitro and in vivo. Nat Struct Mol Biol 16(6):574–581
157. Bukau B, Horwich AL (1998) The Hsp70 and Hsp60 chaperone machines. Cell 92(3):351–366
158. Glover JR, Lindquist S (1998) Hsp104, Hsp70, and Hsp40: a novel chaperone system that rescues previously aggregated proteins. Cell 94(1):73–82
159. Goloubinoff P, Mogk A, Zvi AP, Tomoyasu T et al (1999) Sequential mechanism of solubilization and refolding of stable protein aggregates by a bichaperone network. Proc Natl Acad Sci U S A 96(24):13732–13737
160. Mogk A, Tomoyasu T, Goloubinoff P, Rudiger S et al (1999) Identification of thermolabile Escherichia coli proteins: prevention and reversion of aggregation by DnaK and ClpB. EMBO J 18(24):6934–6949
161. Mogk A, Deuerling E, Vorderwulbecke S, Vierling E et al (2003) Small heat shock proteins, ClpB and the DnaK system form a functional triade in reversing protein aggregation. Mol Microbiol 50(2):585–595
162. Mogk A, Schlieker C, Friedrich KL, Schonfeld HJ et al (2003) Refolding of substrates bound to small Hsps relies on a disaggregation reaction mediated most efficiently by ClpB/DnaK. J Biol Chem 278(33):31033–31042
163. Weibezahn J, Tessarz P, Schlieker C, Zahn R et al (2004) Thermotolerance requires refolding of aggregated proteins by substrate translocation through the central pore of ClpB. Cell 119(5):653–665
164. Tilly K, McKittrick N, Zylicz M, Georgopoulos C (1983) The dnaK protein modulates the heat-shock response of Escherichia coli. Cell 34(2):641–646
165. Straus D, Walter W, Gross CA (1990) DnaK, DnaJ, and GrpE heat shock proteins negatively regulate heat shock gene expression by controlling the synthesis and stability of sigma 32. Genes Dev 4(12A):2202–2209
166. Gamer J, Bujard H, Bukau B (1992) Physical interaction between heat shock proteins DnaK, DnaJ, GrpE and the bacterial heat shock transcription factor σ^{32}. Cell 69:833–842
167. Liberek K, Galitski TP, Zylicz M, Georgopoulos C (1992) The DnaK chaperone modulates the heat shock response of Escherichia coli by binding to the sigma 32 transcription factor. Proc Natl Acad Sci U S A 89(8):3516–3520
168. Gamer J, Multhaupt G, Tomoyasu T, McCarty JS et al (1996) A cycle of binding and release of the DnaK, DnaJ and GrpE chaperones regulates activity of the *Escherichia coli* heat shock transcription factor σ^{32}. EMBO J 15:607–617
169. Rodriguez F, Arsene-Ploetze F, Rist W, Rudiger S et al (2008) Molecular basis for regulation of the heat shock transcription factor sigma32 by the DnaK and DnaJ chaperones. Mol Cell 32(3):347–358

170. Jürgen B, Hanschke R, Sarvas M, Hecker M et al (2001) Proteome and transcriptome based analysis of Bacillus subtilis cells overproducing an insoluble heterologous protein. Appl Microbiol Biotechnol 55(3):326–332
171. Kirstein J, Strahl H, Moliere N, Hamoen LW et al (2008) Localization of general and regulatory proteolysis in Bacillus subtilis cells. Mol Microbiol 70(3):682–694
172. Moore SD, Sauer RT (2007) The tmRNA system for translational surveillance and ribosome rescue. Annu Rev Biochem 76:101–124
173. Keiler KC, Waller PR, Sauer RT (1996) Role of a peptide tagging system in degradation of proteins synthesized from damaged messenger RNA. Science 271(5251):990–993
174. Gur E, Ottofuelling R, Dougan DA (2012) Machines of destruction – AAA+proteases and the adaptors that control them. In: Dougan DA (ed) Regulated proteolysis in microorganisms. Springer, Dordrecht, pp xxx–xxx
175. Fujihara A, Tomatsu H, Inagaki S, Tadaki T et al (2002) Detection of tmRNA-mediated trans-translation products in Bacillus subtilis. Genes Cells 7(3):343–350
176. Muto A, Fujihara A, Ito KI, Matsuno J et al (2000) Requirement of transfer-messenger RNA for the growth of Bacillus subtilis under stresses. Genes Cells 5(8):627–635
177. Shin JH, Price CW (2007) The SsrA-SmpB ribosome rescue system is important for growth of Bacillus subtilis at low and high temperatures. J Bacteriol 189(10):3729–3737
178. Brötz-Oesterhelt H, Beyer D, Kroll HP, Endermann R et al (2005) Dysregulation of bacterial proteolytic machinery by a new class of antibiotics. Nat Med 11(10):1082–1087
179. Lee BG, Park EY, Lee KE, Jeon H et al (2010) Structures of ClpP in complex with acyldepsipeptide antibiotics reveal its activation mechanism. Nat Struct Mol Biol 17(4):471–478
180. Kirstein J, Hoffmann A, Lilie H, Schmidt R et al (2009) The antibiotic ADEP reprogrammes ClpP, switching it from a regulated to an uncontrolled protease. EMBO Mol Med 1(1):37–49
181. Sass P, Josten M, Famulla K, Schiffer G et al (2011) Antibiotic acyldepsipeptides activate ClpP peptidase to degrade the cell division protein FtsZ. Proc Natl Acad Sci U S A 108(42):17474–17479
182. Bottcher T, Sieber SA (2008) Beta-lactones as specific inhibitors of ClpP attenuate the production of extracellular virulence factors of Staphylococcus aureus. J Am Chem Soc 130(44):14400–14401
183. Rathore S, Sinha D, Asad M, Bottcher T et al (2010) A cyanobacterial serine protease of Plasmodium falciparum is targeted to the apicoplast and plays an important role in its growth and development. Mol Microbiol 77(4):873–890
184. Schmitt EK, Riwanto M, Sambandamurthy V, Roggo S et al (2011) The natural product cyclomarin kills Mycobacterium tuberculosis by targeting the ClpC1 subunit of the caseinolytic protease. Angew Chem Int Ed Engl 50(26):5889–5891

Chapter 5
Proteolytic Regulation of Stress Response Pathways in *Escherichia coli*

Dimce Micevski and David A. Dougan

Abstract Maintaining correct cellular function is a fundamental biological process for all forms of life. A critical aspect of this process is the maintenance of protein homeostasis (proteostasis) in the cell, which is largely performed by a group of proteins, referred to as the protein quality control (PQC) network. This network of proteins, comprised of chaperones and proteases, is critical for maintaining proteostasis not only during favourable growth conditions, but also in response to stress. Indeed proteases play a crucial role in the clearance of unwanted proteins that accumulate during stress, but more importantly, in the activation of various different stress response pathways. In bacteria, the cells response to stress is usually orchestrated by a specific transcription factor (sigma factor). In *Escherichia coli* there are seven different sigma factors, each of which responds to a particular stress, resulting in the rapid expression of a specific set of genes. The cellular concentration of each transcription factor is tightly controlled, at the level of transcription, translation and protein stability. Here we will focus on the proteolytic regulation of two sigma factors (σ^{32} and σ^{S}), which control the heat and general stress response pathways, respectively. This review will also briefly discuss the role proteolytic systems play in the clearance of unwanted proteins that accumulate during stress.

Introduction

Like many living organisms, bacteria are constantly challenged with changing environmental conditions. In order to survive these changes, bacteria have developed a number of different cellular strategies. In cases where the stress is short-lived (e.g.

D. Micevski • D.A. Dougan (✉)
Department of Biochemistry, La Trobe Institute for Molecular Science (LIMS),
La Trobe University, Melbourne 3086, Australia
e-mail: d.dougan@latrobe.edu.au

D.A. Dougan (ed.), *Regulated Proteolysis in Microorganisms*, Subcellular Biochemistry 66, DOI 10.1007/978-94-007-5940-4_5, © Springer Science+Business Media Dordrecht 2013 heat-shock) they have developed sophisticated networks or programs to combat the effects of the stress, while in cases where the stress may be prolonged (e.g. when nutrients are depleted) they can enter a "hibernation"-like state, waiting for the return of better conditions. In fact, bacteria have developed several distinct pathways, each of which is tailored to a particular type of stress. In most cases, the response is controlled by a master regulator (or sigma factor), which in turn activates the expression of a particular set of genes (or regulon) that restore cellular homeostasis. In *Escherichia coli*, there are seven different sigma factors (σ^{70}, σ^{54}, σ^{38}, σ^{32}, σ^{28}, σ^{24} and σ^{18}), all of which compete for binding to the RNA polymerase (RNAP) core enzyme, for the transcription of a specific set of genes. As such, the cellular levels of these master regulators (and their affinity to RNAP) are crucial for the activation and/or maintenance of these different stress responses. Given this, it is not surprising that the active cellular concentration of these regulatory proteins is tightly controlled, not only at the transcriptional and translational levels, but also at the post-translational level through protein degradation. Hence proteases play a key role in the regulation of stress response pathways. This review will focus primarily on the general stress response and the heat shock response in *E. coli*. For a detailed description of the extracytoplasmic or extracellular stress response please refer to the accompanying review by Brachinger and Ades [1].

General Stress Response

As the name suggests, the general stress response is a common cellular response that is activated by a range of different conditions, from nutrient starvation and moderate temperature downshifts [2] to high osmolarity [3] and pH downshifts. It is characterised by a number of distinct morphological and physiological changes [4], which protects the cell from assault by these different stresses. As such, the general stress response acts as a pre-emptive measure to prevent subsequent cellular damage. This response occurs through the activation of a common set of genes that are up regulated by an alternative sigma factor subunit of RNAP, commonly referred to as the stationary-phase sigma factor (σ^{38}), also known as σ^{S} [4–6]. The following section will describe some of these pathways, focusing in particular on the role of proteases in controlling the general stress response in *E. coli*.

The Master Regulator of the General Stress Response, σ^{S}

SigmaS (σ^{S}) was first discovered as the master regulator of stationary-phase [4]. It is an inducible subunit of RNAP, which is related to the constitutively expressed vegetative or housekeeping sigma-factor, σ^{70}, and as such competes for binding to the RNAP core enzyme [7] (Fig. 5.1a). Under normal cellular conditions (i.e. in rapidly growing cells) the levels of σ^{S} are low [8]. However, during stationary phase

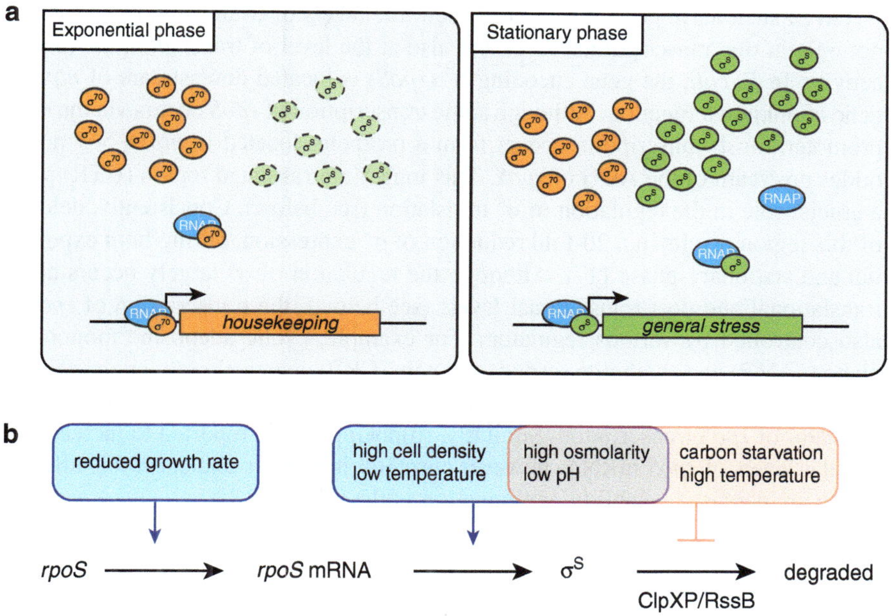

Fig. 5.1 In *E. coli*, the levels of SigmaS are controlled not only at the transcriptional and translational level, but also through protein degradation. (**a**) During exponential phase transcription is largely restricted to housekeeping genes. This is due, partly to the greater abundance of σ^{70}, but also to its higher affinity for RNAP core enzyme, in comparison to most alternative sigma factors including σ^S. During stationary phase, despite its weak affinity for RNAP (relative to σ^{70}), the rapid increase in the level of σ^S, permits competitive binding to RNAP and hence, transcription switches to general stress genes. (**b**) Different environmental stresses modulate the cellular levels of σ^S, by targeting different processes; (i) reduced growth rate stimulates transcription, (ii) high cell density or low temperature stimulate *rpoS* translation, (iii) carbon starvation or high temperature inhibits σ^S degradation and (iv) high osmolarity or pH, both stimulate *rpoS* translation and inhibit σ^S degradation

and in response to a number of different stresses, such as anaerobiosis [9], oxidative stress [10], and osmotic stress [5] the cellular levels of σ^S rise rapidly (Fig. 5.1a, b). As a result of this increase in the cellular concentration of σ^S, RNA polymerase is directed to a specific set of promoters resulting in the expression of downstream σ^S-dependent genes, which control the metabolic state of a cell (Fig. 5.1) [11, 12]. Indeed, σ^S has been implicated in the regulation (either directly or indirectly) of approximately 500 genes, which equates to approximately 10 % of the *E. coli* genome [12–14]. These genes are expressed not only during the transition into stationary-phase, but also in response to a number of different stresses [12], i.e. under conditions of nutrient limitation, in which the cells switch from optimal growth to a "maintenance" state. Similarly, σ^S also controls the expression of genes that mediate programmed cell death, in which the sacrifice of a small population of cells under extreme stress provides a supply of nutrients to other cells permitting their survival [15]. In addition to these survival mechanisms, σ^S also controls virulence genes in pathogenic enteric bacteria [reviewed by 16].

Given such an important role in the cell, the levels of σ^S are tightly controlled, not only at the transcriptional level, but also at the level of translation and protein activity. In *E. coli*, the gene encoding σ^S (*rpoS*) is located downstream of *nlpD*, a gene of unknown function. Although some transcription of *rpoS* occurs via the *nlpD* promoter, most transcription occurs from a promoter located in *nlpD*, 567 nucleotides upstream of the AUG of *rpoS*. This long 5′ untranslated region (UTR) plays a crucial role in the regulation of σ^S translation (see below). Consistently, deletion of this region results in a 20-fold reduction of σ^S expression, during both exponential and stationary-phase [17]. Although the regulation of σ^S largely occurs at the translational and post-translational levels (see below), the transcription of *rpoS* is also controlled by various regulators. For example, cyclic adenosine monophosphate (cAMP) and catabolite response protein (CRP) negatively regulate *rpoS* transcription [4], while in contrast the two-component system (BarA/UvrY) is a positive regulator of *rpoS* transcription. Similarly, (p)ppGpp is also reported to increase the cellular levels of *rpoS* mRNA, however currently it remains unclear if this effect is due to an increase in stability or elongation of the mRNA [17].

Translational Control of σ^S

At the translational level, the expression of σ^S is stimulated by a variety of different conditions, including hyperosmotic shift [3, 18], low temperature [2], and acid pH [19]. This activation, of *rpoS* translation, is regulated by the structural rearrangement of the *rpoS* mRNA [19], which is mediated by the RNA-chaperone, Hfq [20] and several regulatory small RNAs [21–24]. Specifically, the long 5′ UTR of *rpoS* mRNA, is proposed to form an intra-molecular stem loop structure, which under normal conditions occludes the ribosome-binding site (RBS) and hence limits σ^S translation [18]. The translation of σ^S, can however be stimulated by various non-coding small RNAs (sRNAs), such as DsrA and RprA, in the presence of the RNA chaperone, Hfq [25–28]. Hfq is a small RNA-binding protein that not only stabilises the sRNAs, but also enhances RNA-RNA interactions [29–32]. *E. coli* strains bearing mutations in *hfq* are sensitive to multiple stresses and hence exhibit a similar phenotype to *rpoS* mutant strains [31]. Consistently, Hfq plays a role in the translation of *rpoS* [20], however the mode of action by which Hfq functions is currently unclear. Despite this, a number of models have currently been proposed. The first, suggests that Hfq acts on *rpoS* mRNA directly by stabilising the secondary structure of the *rpoS* mRNA [20, 33–36]. Binding of Hfq is thought to shift the equilibrium of the *rpoS* mRNA secondary structure, from a less active form, where translation is inefficient, to an active form that permits easy access to the ribosome. An alternate model suggests that Hfq does not affect the secondary structure of *rpoS* mRNA. This model describes Hfq as a 'platform' for binding of other regulatory molecules, which are involved in the translational control of *rpoS*. Consistent with this model, Hfq interacts directly with several small regulatory RNAs (DsrA, RprA, ArcZ and OxyS), which have been shown to regulate *rpoS* translation [2, 36]. Moreover, Hfq is able to stimulate base pairing, between the sRNA and the target mRNA to promote

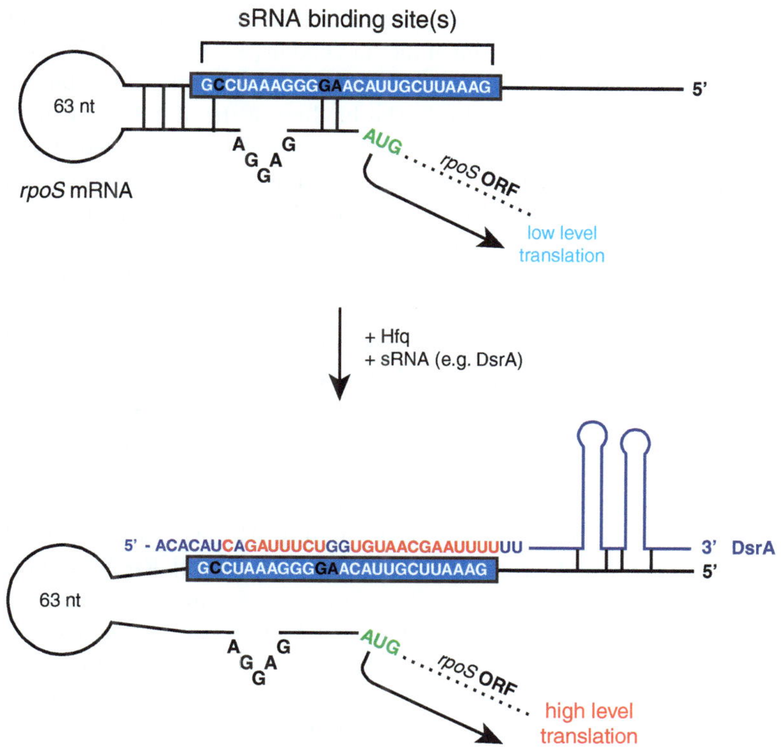

Fig. 5.2 Activation of σ^S translation by sRNAs (e.g. DsrA). Under non-stressed conditions, σ^S translation is low, due largely to a stem loop structure in the *rpoS* mRNA, which occludes the RBS and initiating AUG. In the presence of the RNA chaperone, sRNAs such as DsrA hybridise to the mRNA exposing the RBS and the initiating AUG

a specific response, which either inhibits or enhances translational initiation [37]. To date, a total of four sRNAs have been identified, which function to regulate σ^S translation. Three of which (DsrA, RprA, and ArcZ) positively regulate *rpoS* translation [38–40], while a single sRNA, OxyS, has a negative effect on *rpoS* translation [36, 41]. Each of these sRNAs is expressed in response to different stress conditions [24, 33, 42].

Although four different sRNAs have been shown to effect *rpoS* translation, a common model can be drawn from a single example and hence this section will focus on the most extensively studied – DsrA. DsrA (downstream from RcsA) was originally discovered in a study that examined capsule regulation in *E. coli* [43] and later shown to be required for translation of *rpoS* at low temperature [42, 44]. Biochemical analysis of DsrA has revealed that the mechanism that DsrA employs to promote translation of *rpoS* mRNA is via an interaction with the *rpoS* mRNA, which is facilitated by Hfq [45]. It is an 87 nucleotide RNA which folds into a stem loop structure and contains a small single-stranded region that is complementary to an element within the 5′ UTR of *rpoS* mRNA (Fig. 5.2). *In vivo* studies have shown that DsrA hybridises to the predicted *rpoS* mRNA duplex segment (self inhibitory stem) at a position that lies upstream of the start codon [39, 40, 46, 47]. This hybridisation induces a structural change in *rpoS* mRNA that permits accessibility to the RBS present on the other strand [45]. The binding of the sRNA is facilitated by formation of a ternary complex with Hfq, which results in the activation of translation. In order for sRNAs, such as DsrA to activate *rpoS* translation, Hfq requires a $(AAN)_4$ repeat element located at the 5′ UTR of *rpoS* [26, 48]. Interestingly, both *rpoS* mRNA and DsrA appear to bind to the same binding site within the proximal RNA-binding domain of Hfq [49]. As such, the current model suggests that Hfq enhances the interaction of DsrA and the *rpoS* mRNA by (a) increasing the local concentration of both RNAs and (b) unwinding the inhibitory stem of the *rpoS* mRNA [49, 50]. Recently however, the role of DsrA and its involvement in stimulating *rpoS* translation has also expanded to include stabilisation of *rpoS* mRNA by base pairing to these sRNAs, potentially preventing the RNase E-dependent degradation of the target mRNA [51].

Consistent with the findings for DsrA; RprA and ArcZ also regulate translation of σ^S by base pairing to the 5′ UTR of *rpoS* [40, 47]. All three of these sRNAs are expressed in response to different stress conditions [24, 33, 42]. RprA (RpoS regulator), a 105 nucleotide RNA, was identified during a screening of a multi-copy suppressor library that increased the translation of *rpoS*-lacZ (translational fusion) in the absence of *dsrA* [33]. RprA in contrast to DsrA has been found to stimulate σ^S synthesis in response to cell envelope stress, and a modest effect has been observed in response to osmotic stress [33, 40]. ArcZ functions to positively regulate σ^S translation [21, 24]. Processing of ArcZ from a 121 nucleotide RNA to a stable 56 nucleotide species is required for the formation of a strong Hfq-dependent ternary complex with the 5′ UTR of *rpoS* mRNA [21, 47, 52]. Although the sequences of these regulatory RNAs differ, the mechanistic details appear to be conserved [47, 53]. Common to all of the sRNAs, they each interact with Hfq and activate translation by opening the stem-loop structure of the *rpoS* 5′ UTR, allowing access to the RBS [39, 40, 53]. Apart from binding to *rpoS* mRNA to promote translation, it's also postulated that hybridisation of sRNAs to target mRNA promotes stabilisation of the target mRNA, in turn protecting it from degradation [51]. In contrast to the other sRNAs, OxyS is a negative regulator of *rpoS* translation. Encoded by the *oxyS* gene, this regulatory sRNA is induced upon exposure to hydrogen peroxide (oxidative stress) [36, 41]. Consistent with the positively regulating sRNAs, OxyS also associates with Hfq [36], however the mechanism of action of OxyS, is currently poorly understood. Nonetheless, based on secondary structure predictions, OxyS seems to share structural similarities with DsrA [19]. However, in contrast to DsrA, a linker region in OxyS seems to align with the RBS of the *rpoS* mRNA, suggesting that repression of σ^S translation may occur through the occlusion of the RBS by OxyS base pairing [32]. However, evidence for a direct interaction with *rpoS* mRNA is currently lacking. On the other hand, co-immunoprecipitation experiments, which confirm an interaction with Hfq, suggest an alternative model for the regulation of *rpoS* translation by OxyS [36]. This model proposes that the negative regulation of *rpoS*, by OxyS, is achieved through competitive binding to the Hfq-sRNAs binding site. For

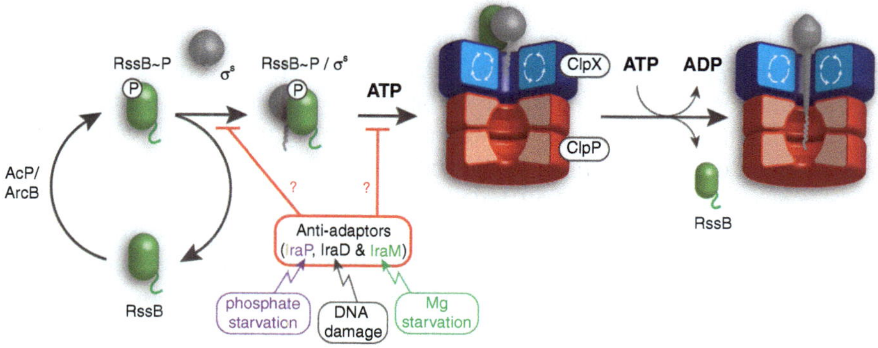

Fig. 5.3 Degradation of σ^S by ClpXP requires the adaptor protein, RssB. The adaptor protein, RssB (*green*) is phosphorylated by either the phosphordonor, acetyl phosphate (AcP) or the response regulator, ArcB. The phosphorylated form of RssB binds to σ^S (*grey*) with high affinity. Binding of RssB to σ^S, triggers a conformational change in σ^S, exposing a ClpX-recognition motif. The RssB/σ^S complex, then docks to ClpX by an unknown mechanism, and delivers the substrate to the ClpXP protease for ATP-dependent degradation. The degradation of σ^S can be inhibited by specific anti-adaptors (i.e. IraP, IraD and IraM), which are induced by phosphate starvation, DNA damage or Mg starvation, respectively

instance, competitive binding to Hfq may hinder the binding of positive regulatory sRNAs, which in turn inhibits *rpoS* translation [36].

Regulated Degradation of σ^S

Although the transcriptional and translational regulation of σ^S is rather extensive, this only accounts for a fraction of the overall regulation. The turnover of σ^S plays a major role in controlling the cellular levels of σ^S during the various growth phases. As mentioned previously, the relative amounts of σ^S present during exponential growth are extremely low, which is largely a result of degradation by the energy-dependent AAA+protease, ClpXP [54], and is dependent on the two-component response regulator RssB [55–58] (see Fig. 5.3). In the absence of RssB, the ClpXP protease is unable to recognise σ^S, and hence RssB is required for the successful removal of σ^S from the cell. Importantly, RssB itself is not degraded in the process of substrate delivery and therefore is able to perform numerous cycles of substrate binding and delivery [58]. As a consequence, the limiting factor for σ^S degradation is RssB, which is present within the cell at very low levels (approximately 1 molecule of RssB for every 25 molecules of σ^S) [59]. Independent of the levels of RssB, the interaction with σ^S is modulated by phosphorylation [55, 58, 60, 61]. Although the phosphate donor (acetyl phosphate, AcP) and the two-component system ArcA/B have been shown to trigger RssB phosphorylation [61], a dedicated phosphatase or histidine kinase has yet to be identified. Regardless of how RssB is phosphorylated, the mechanistic effect of phosphorylation on RssB and the relevance it plays with respect to the regulation of σ^S remains unclear and currently several models exist to describe the contribution of phosphorylation.

Moreover, it has been recently shown that the stability of σ^S is also controlled by another group of proteins termed Ira (Inhibitor of RssB activity). Currently, three Ira proteins have been identified: IraP, IraD and IraM, which stabilise σ^s in response to phosphate starvation, DNA damage and magnesium starvation, respectively [62, 63]. However, to date little is known about the mechanism by which these anti-adaptors inhibit the RssB-mediated delivery of σ^S to ClpXP. The following section will focus on our current understanding of the components involved in the regulated turnover of σ^S.

The Protease – ClpXP

In the cytosol of *E. coli* there are five ATP-dependent proteases (ClpXP, ClpAP, HslUV, Lon and FtsH). Each protease is composed of two components; a peptidase component and an AAA+ (ATPase associated with a variety of cellular activities) unfoldase component and as such they are commonly referred to as AAA+ proteases [64]. A unifying feature of the AAA+ protein superfamily is the presence of an AAA+ domain spanning 200–250 amino acids. This domain is composed of two subdomains – the small and large subdomain. The nucleotide is bound, in a cleft created by the large and small subdomains of a single subunit and the large subdomain of the adjacent subunit [65, 66]. Each domain contains several highly conserved sequence motifs required for the binding and hydrolysis of ATP (e.g. Walker-A and Walker-B), the binding and translocation of substrates (e.g. pore-1 and pore-2) and the binding of the peptidase, ClpP (e.g. IGF loop) [67–70]. For a detailed description of the different AAA+ proteases in *E. coli*, refer to [71].

In *E. coli*, a single protease (ClpXP) is responsible for the turnover of σ^S. Like other AAA+ proteases, ClpXP is composed of two components. The peptidase component (ClpP) is composed of two heptameric rings that stack back-to-back to form a barrel-shaped oligomer. The catalytic residues of ClpP are sequestered away from cytosolic proteins, within an aqueous chamber. Access to this chamber is limited by a narrow axial portal (~10 Å in diameter), which only allows entry of short peptides and unfolded proteins [72, 73]. As a consequence of this narrow entry portal, the degradation of folded proteins requires an additional component – the unfoldase (ClpX), which is responsible, not only for the recognition of the substrate but also for its unfolding and translocation into ClpP. Complexes of ClpXP, as illustrated by electron micrographs, can be either single- or double-headed [74–76]. Single-headed ClpXP complexes contain a single hexamer of ClpX stacked onto one end of the ClpP dodecamer, while double-headed complexes of ClpXP contain a ClpX hexamer bound to both ends of ClpP [76]. The interaction between ClpX and ClpP is mediated by two structural elements. The primary interaction occurs between the IGF loop (located on ClpX) and a hydrophobic pocket (composed of Tyr60 and Tyr62 from one subunit and Phe82 from the adjacent subunit) located on the apical surface of ClpP [68, 73]. Interestingly, these loops are disordered in the crystal structure of ClpX, hence they are likely to exhibit a high degree of flexibility required to facilitate the asymmetric connection between the ClpX hexamer and the heptameric ring of ClpP [65, 66, 68]. Consistent with the role of these loops in docking to ClpP, mutation of specific residues within the IGF motif, inhibits ClpP binding without affecting the ClpP-independent activity of ClpX [68, 77]. Indeed binding of the IGF loop to the hydrophobic pocket on ClpP is believed to open the axial channel of the protease [78]. Likewise, several recently identified antibiotics (acyldepsipeptides (ADEPs) and activators of self-compartmentalising proteases (ACPs)) that activate ClpP for unregulated degradation, have also been shown to open the axial pore of ClpP through binding to the hydrophobic pocket [79–83]. The second interaction site, between ClpX and ClpP, is mediated by the N-terminal β-hairpin loop (~20 residues) of ClpP [84, 85] and the pore-2 loop of ClpX. This interaction is dynamic and sensitive to the nucleotide-bound state of ClpX [78] and mutation of either region results in the destabilisation of the ClpXP complex [84, 86–88].

Protein degradation by ClpXP can be divided into four fundamental steps, (substrate recognition, unfolding and translocation), which are performed by the unfoldase ClpX, and hydrolysis of the protein into short peptide fragments which is performed by the associated peptidase. In general, the first step (substrate recognition) requires nucleotide binding by the unfoldase, but not its hydrolysis. In this state, ClpX is able to recognise a wide variety of different protein substrate, largely through short sequence motifs (commonly referred to as tags or degrons). These tags are often located at the N- or C-terminus of the substrate protein [89]. While, the majority of these motifs are intrinsic to the protein, some proteins require processing or modification (e.g. attachment of a tag such as the SsrA tag) for ClpX recognition to occur. For a more detailed description of the SsrA tagging system, refer to [71]. Following recognition, the substrate is unfolded by ClpX and translocated into ClpP, in an ATP-dependent fashion. Finally, the translocated polypeptide is degraded into small peptides, by ClpP [76, 90–92]. In the case of ClpX, many of the molecular details of substrate recognition and translocation have been defined. In general, substrates are recognised by a conserved aromatic-hydrophobic motif (GYVG) located on the pore-1 loop [87]. These loops protrude from each subunit into the central cavity of the hexamer [93–96]. Mutations in the highly conserved aromatic residue of the pore-1 loop have been shown to impair substrate binding and processing, with little to no effect on oligomerisation or ATPase activity of the AAA+protein [93, 94, 96, 97]. Cycles of ATP binding and hydrolysis, drive rigid-body movements in ClpX, which translate to a pulling force on the substrate, resulting in unfolding and translocation of the substrate [87, 88, 94]. For a detailed analysis of the mechanism of action of AAA+proteases refer to [71].

The Adaptor Protein–RssB

Although the vast majority of substrates are recognised directly by the unfoldase, some substrates require the assistance of an adaptor protein for their recognition and hence their degradation by the protease. In the case of ClpXP, three adaptor proteins have been identified – SspB, UmuD and RssB. SspB is the best characterised of these adaptor proteins and is required for the enhanced delivery and degradation of SsrA-tagged proteins as well as the delivery of a fragment of the anti-sigma factor (RseA). In contrast to SspB, both UmuD and RssB are essential for the delivery of their respective substrates [58, 98]. UmuD is essential for the *in trans* delivery of UmuD′ [98], while RssB (also referred to as SprE) is essential for the recognition and delivery of σ^S [54, 56, 57]. Interestingly, despite little overall homology between the three adaptor proteins, all appear to contain a short sequence motif in common. This motif is located at the C-terminus of SspB and RssB, and near the N-terminus of UmuD. In SspB, this motif was termed the ClpX binding region (XBR) as it was shown to be critical for binding to ClpX and hence delivery of its cargo to the protease [99–101]. Specifically, the XBR of SspB docks onto the N-terminal domain of ClpX, placing the adaptor protein SspB in an ideal position to deliver its bound substrate [99, 101]. The increased local concentration of the substrate (tethered to ClpX, by the adaptor protein) enhances its recognition by the pore residues of ClpX, where it is unfolded and translocated into ClpP. Although RssB was identified over 15 years ago, the mechanism by which it binds to and delivers its substrate (σ^S) to ClpXP for degradation still remains elusive. However, based on the sequence similarity of the XBR region of SspB and RssB, a model for the RssB-mediated delivery of σ^S has been proposed (Fig. 5.3).

Although RssB shares little-to-no sequence similarity with other adaptor proteins, it does share considerable homology with a family of proteins known as two-component response regulators (RRs). These proteins are generally composed of two domains, an N-terminal receiver domain and a C-terminal output (or effector) domain. In contrast to the majority of RRs (which contain a C-terminal DNA-binding domain and serve as transcriptional regulators) the C-terminal effector domain of RssB is a PP2C-type Ser/Thr phosphatase [102]. Interestingly, this region in RssB lacks the critical residues required for phosphatase activity and hence the precise role of this domain remains unclear [102]. The receiver domain on the other hand, is highly conserved amongst all RRs, both in sequence and structure. In the case of RssB, this domain is phosphorylated at a highly conserved aspartic acid residue (Asp58), which is proposed to trigger a conformational change in RssB resulting in an improved interaction with σ^S [55, 58, 59, 61]. Consistent with this idea, mutation of Asp58 prevents RssB phosphorylation and reduces the rate of σ^S turnover *in vivo* [55, 59, 103]. However, the role of RssB phosphorylation remains controversial, as σ^S is still degraded in an *E. coli* strain containing a non-phosphorylatable mutant of RssB [103]. Similarly, given that phosphorylation of Asp is transient, any structural changes that occur to RssB remain undefined. Nevertheless, the effect of phosphorylation has been examined at the molecular level for some RRs [104, 105]. Indeed in these cases, phosphorylation has been shown to trigger both local changes to the N-terminal receiver domain, as well as long-range changes to the RR [106–110]. From these data several models have been proposed. One possibility is that the receiver domain exists in an equilibrium, between two-states (an active and an inactive state) that is influenced by phosphorylation. In most cases, phosphorylation of the RR is linked to activation of the protein, while in a handful of cases phosphorylation appears to inhibit the activity of the RR.

Regardless of the role of RssB phosphorylation, the molecular details of substrate interaction and delivery to ClpXP are also poorly defined. Currently, two different models of substrate delivery have been proposed. The first model implies a direct interaction between the adaptor and ClpX, for delivery of the substrate, while the second model suggests that RssB is only required to "activate" the substrate for binding to ClpX and does not, itself, dock to ClpX. Both models nevertheless, converge to suggest that binding of RssB to σ^S, triggers a conformational change in σ^S that exposes a concealed "low affinity" ClpX binding site on the substrate. Exposure of this site then permits the downstream recognition of σ^S by ClpX, when presented by RssB (Fig. 5.3). Consistent with this idea, the N-terminus of σ^S does not contribute to RssB binding, but is predicted to contain a ClpX binding motif [89, 111]. In addition to the predicted ClpX binding site, located on the N-terminus of σ^S, a "turnover element" in σ^S, located downstream of the promoter recognising region 2.4, is also required for its degradation [60]. Mutations introduced into the "turnover element" of σ^S, in particular Lys173, have been shown to inhibit the turnover of σ^S in growing cells [60]. Consistently, mutation of Lys173 also inhibited the binding of σ^S *in vitro* [60]. Collectively these data suggest that the "turnover element" in σ^S is an important region for interaction with RssB. Hence, in the absence of RssB, the N-terminal region of σ^S is occluded, possibly by the C-terminal region of σ^S which upon binding of RssB becomes exposed for recognition by ClpX [111].

Anti-adaptors (Inhibitors of RssB Activity, Ira)

Interestingly, in the last 5 years another level of σ^S regulation was discovered (see below). In this case, a group of unrelated proteins were shown to inhibit the ClpXP-mediated turnover of σ^S. These proteins were termed anti-adaptors, and as the name suggests they inhibit or antagonise the activity of the adaptor protein, RssB. The first anti-adaptor to be characterised was identified as a regulator of competence development in *B. subtilis* [112–114]. In non-competent cells, the adaptor protein MecA, recognises the competence transcription factor ComK, and delivers it to the ClpCP protease for degradation [115, 116]. When competence development is initiated, by a quorum sensing mechanism, the levels of ComS increase [112, 117]. ComS then acts as a "suicide" anti-adaptor binding to MecA and thereby preventing the turnover of ComK [114, 116, 117]. More recently however, similar proteins were also identified in *E. coli*. Consistent with the regulatory role of ComS in the development of competence, these novel *E. coli* anti-adaptor proteins function to regulate the stationary-phase stress response. As such, the following section will describe recent insights into these small, yet interesting proteins that work to stabilise σ^S.

Using an *E. coli* genomic DNA library, Gottesman and colleagues identified three different genes of unknown function that specifically affected the activity of an *rpoS-lacZ* translational fusion [62, 63]. Through a series of elegant genetic and biochemical experiments, the proteins encoded by these genes were shown to act as specific inhibitors of RssB activity and hence were collectively termed anti-adaptors. Interestingly, deletion of each gene did not affect the stability of σ^S under all starvation conditions; rather stabilisation of σ^S was limited to a specific condition. The first identified anti-adaptor, YiaB renamed IraP (Inhibitor of RssB activity during phosphate starvation) is a small 86 amino acid protein, which is transcribed in response to phosphate starvation, and mediated by ppGpp [63, 118]. A multi-copy plasmid carrying the *iraP* gene demonstrated that expression of IraP, driven from the plasmid, resulted in an approximately three-fold increase in σ^S stability in the exponential-phase in comparison to a seven-fold increase in the stationary phase [63]. These results, in conjunction with *in vitro* 'pull-down' experiments confirmed IraP as a *bona fide* regulator of σ^S, which prevents σ^S turnover through direct interaction with RssB [63].

Following the identification of the first anti-adaptor (i.e. IraP), two additional genes (*yjiD* and *ycgW*) were also shown to stabilise σ^S. These gene products were renamed IraD and IraM respectively, because of their ability to stabilise σ^S in response to DNA damage (IraD) and during magnesium starvation (IraM) [62]. Consistently, both IraD and IraM were able to inhibit the RssB-mediated degradation of σ^S *in vitro* [62]. Although all three anti-adaptors seem to perform the same role, their interaction with RssB and/or σ^S seems to vary [62]. Current data suggests that IraP functions by binding directly to RssB (forming an RssB-IraP complex) which sequesters RssB from σ^S, thereby preventing its turnover [63]. *In vitro* 'pull-down' experiments using IraD and IraM suggest that IraD, like IraP, interacts directly with RssB, whilst the mode of action of IraM remains unclear [62] (Fig. 5.3). Interestingly, the transcriptional regulator AppY, was also able to stabilise σ^S in a mutant strain lacking all three anti-adaptors, which suggests a putative role for AppY in activating the transcription of a yet to be identified anti-adaptor [62]. This has physiological importance when considering the action of multiple stresses on the cell. Based on our current understanding of IraP, the presence of multiple stresses may induce the expression of multiple anti-adaptors. Given that each anti-adaptor may exhibit a different mechanistic approach to inhibit σ^S degradation, this may cause an avidity effect, which could culminate in rapid stabilisation of σ^S. As such, this provides an efficient way of coupling external stress stimuli to a rapid survival response.

The Heat-Shock Response

In contrast to the general stress response, the heat shock response (HSR) is a specific cellular response to a rapid, sub-lethal, increase in temperature. This response was first observed in the salivary glands of *Drosophila melanogaster*, where the synthesis of a small group of proteins (termed heat-shock proteins (HSPs)) increased in response to a temperature upshift [119] and was later shown to be a universal response. In *E. coli*, the HSR is controlled by a single transcription factor, σ^{32} (also known as σ^H) and results in the expression of HSPs (i.e. molecular chaperones and proteases). The molecular chaperones (e.g. DnaK/J, GroEL/S and the small

HSPs – IbpA and IbpB) and the ATP-dependent proteases (e.g. Lon and FtsH), help to maintain a productive protein-folding environment in the cell by refolding or removing the misfolded proteins, thereby returning the cell to its pre-stressed state. The response is controlled not only by σ^{32} translation, but also through its turnover, and can be divided into three distinct phases; induction, adaptation and a final steady state phase.

Regulated Turnover of σ^{32}

Similar to σ^{S}, the steady-state levels of σ^{32} are low under non-stressed conditions. In the absence of stress (i.e. at 30 °C), low levels of σ^{32} (~50 molecules/cell) are maintained by (a) inefficient initiation of translation due to base pairing within the *rpoH* mRNA which occludes the Shine-Dalgarno sequence and (b) rapid degradation of σ^{32} primarily by FtsH (half-life of~1 min.), which is mediated by both the DnaK and GroE chaperone systems [120, 121]. Upon temperature upshift, the inhibitory structure of the *rpoH* mRNA is opened and translation of σ^{32} increases. Simultaneously, the accumulation of unfolded proteins in the cell sequesters chaperones and proteases, albeit transiently (~5–10 min.) thereby stabilising σ^{32} [122] (see Fig. 5.4). As a result, there is a rapid increase in the levels of σ^{32} and hence HSPs during the induction phase of this response. During the next phase – adaptation – the synthesis of HSPs is blocked, as the activity of σ^{32} becomes inhibited through (a) the presence of high levels of chaperones and proteases and (b) the accelerated turnover of σ^{32} at higher temperatures, resulting in a new steady-state level of σ^{32} (and HSPs) during the final "steady-state" phase. As such, the cellular levels of both chaperones and proteases, not only control induction of the HSR, but also the shutdown of this response.

Although several cytoplasmic proteases including HslUV (also known as ClpYQ) contribute to the turnover of σ^{32}, the metabolic stability of σ^{32} *in vivo* is primarily controlled by the membrane bound protease FtsH [123–125]. Both proteases (HslUV and FtsH) belong to the AAA+ protein superfamily [69]. Each protease is composed of two components (an unfoldase and a peptidase). In the case of HslUV, the two components are located on separate polypeptides, while in the case of FtsH both components are located on a single polypeptide. While both machines exhibit a six fold symmetry, the HslUV complex is formed by one or two ring-shaped unfoldase components (composed of six subunits of HslU), that stack onto either or both ends of the peptidase component (i.e. HslV), which is composed of two hexameric rings stacked back-to-back [126]. By contrast, FtsH forms a homohexameric ring-shaped complex that is embedded in the periplasmic membrane with its active sites exposed to the cytoplasm. For a detailed description of FtsH structure and function refer to the accompanying review by Okuno and Ogura [127].

Interestingly, the FtsH-mediated degradation of σ^{32} is accelerated, not only by increased temperature [128], but also by the presence of molecular chaperones such as the GroEL-GroES (ELS) chaperone system and the DnaK-DnaJ-GrpE (KJE)

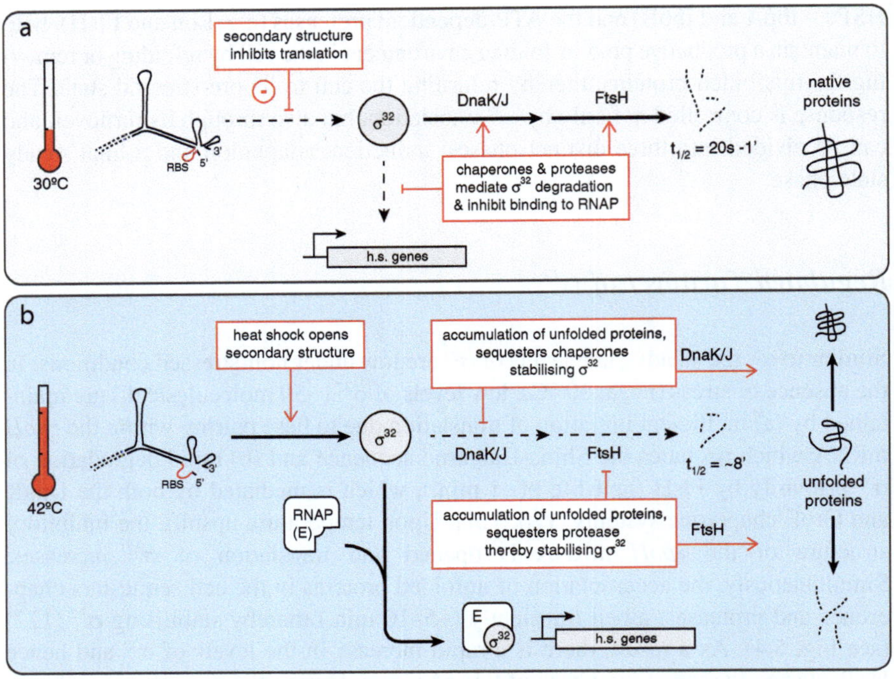

Fig. 5.4 The heat-shock response is controlled by the transcription factor, Sigma32 (σ^{32}). (**a**) Under non-stressed conditions the RBS of *rpoH* is occluded and hence translation of σ^{32} is low. In the absence of unfolded proteins chaperones and proteases are free to binding to σ^{32} mediating its rapid degradation and inhibiting its binding to RNAP. (**b**) Under heat shock conditions, the secondary structure of *rpoH* is melted and transcription increases. The accumulation of unfolded proteins, sequesters the chaperones and proteases from σ^{32}, which results in an increase in the half-life of σ^{32}. The increased cellular concentration of σ^{32} allows binding to RNAP and hence transcription of the h.s. genes

chaperone system [125]. Although the exact role of these chaperone systems in promoting the FtsH-mediated degradation of σ^{32} is yet to be determined, hydrogen-deuterium exchange (HDX) experiments have shown that both DnaK and DnaJ are able to promote unfolding of σ^{32} [129]. Binding of DnaJ to region 2.1 triggers a conformational change in σ^{32}, which facilitates DnaK binding to region 3.2 (Fig. 5.5) and further unfolding of σ^{32}, which is believed to mediate delivery to, and degradation by, FtsH. Interestingly, in contrast to most other bacterial proteases (e.g. ClpXP or ClpAP) in which substrate recognition involves a single N- or C-terminal motif (see [71] for further details), the FtsH-mediated degradation of σ^{32} appears to require two distinct regions; region 2.1 (Leu47, Ala50 and Ile54) and region C (Ala131 and Lys134) [130, 131], both of which are located internally (Fig. 5.5). Given that (a) these "turnover elements" are located within the middle of the polypeptide, and (b) that FtsH lacks a robust unfoldase activity [132], it is likely that molecular chaperones facilitate the FtsH-mediated degradation of σ^{32}, by either triggering a

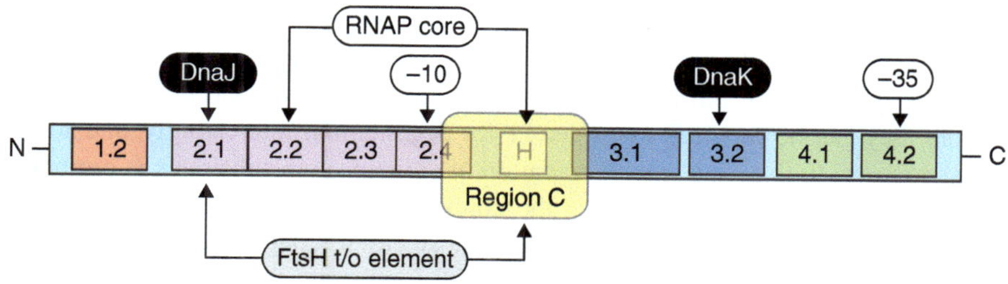

Fig. 5.5 Domain organisation of σ^{32}. Sigma factors are divided into five functional regions (region 1, 2, 3, 4 and H/region C). These regions can be further divided into subregions (e.g. 2.1, 2.2, 2.3 and 2.4). The RNAP core enzyme binds to regions 2.2 and H/region C. This partially overlaps with the FtsH turnover element, which has be mapped to region 2.1 and H/region C. DnaJ and DnaK bind to regions 2.1 and 3.2, respectively

local conformational change in the substrate or by "unfolding" it. Consistently, the binding sites for both DnaJ and DnaK are located on, or adjacent to, the FtsH turnover elements on the substrate. Interestingly, although still somewhat speculative, molecular modelling of σ^{32} using known sigma factor structures has revealed that the residues implicated in the two turnover elements may form a single discontinuous "motif" for recognition by FtsH [133, 134].

Removal of Misfolded and/or Aggregated Proteins – Degradation by AAA+Proteases as a Last Resort

As mentioned above, heat shock results in the accumulation of unfolded proteins, which may be detrimental to the viability of the cell. As such, the primary aim of the heat-shock response is to maintain cell viability by restoring the protein-folding environment of the cell. This is achieved, through the expression of chaperones and proteases, which either refold or remove the misfolded proteins. Interestingly, in contrast to oxidative stress, which results in the irreversible damage to proteins, heat stress largely results in a "reversible" damage to proteins. Moreover, given that it is generally more energetically favourable to refold a protein than to degrade and resynthesize it, the primary strategy of the heat-shock response is to refold the misfolded proteins. As such, it is important for the cell to discriminate between unfolded proteins that can be refolded by chaperones, and terminally damaged proteins that must be removed from the cell by proteases. One possibility is that the final fate of a misfolded (or aggregated) protein is controlled by the kinetics of chaperone and protease binding [135, 136]. Consistent with this view, chaperones such as DnaK and DnaJ recognize largely hydrophobic residues, which are commonly exposed in unfolded proteins, while the proteases such as Lon, bind primarily to sequences rich in aromatic residues, which are less common in unfolded proteins [137]. Importantly, chaperones such as DnaK and DnaJ are significantly more abundant than proteases, especially under heat-shock conditions [138–140] and hence refolding is generally favoured over degradation. Therefore not surprisingly, following heat-shock most unfolded proteins are refolded directly by "folder" chaperones (i.e. KJE or ELS) before they can be captured by proteases for degradation. Interestingly, and somewhat contrary to this view, the degradation of some protein substrates is promoted by chaperones, possibly by altering the confirmation of the substrate, exposing a protease binding site [129]. However in most cases, chaperones and proteases appear to compete for binding to the misfolded substrate to deliver their specific activities (refolding versus degradation, respectively).

Interestingly, even during conditions of prolonged stress when protein folding chaperones are sequestered, the refolding arm of protein quality control network is favoured. Under these conditions, the accumulation of misfolded proteins results in their aggregation. However these aggregated proteins can be refolded by a specialised bi-chaperone system, which combines the "disaggregation-power" of the AAA+unfoldase ClpB, with the refolding activity of the KJE chaperone system [140, 141]. Interestingly, despite the fact that some proteases can degrade aggregated proteins *in vitro* [138, 142, 143] and that several *B. subtilis* Clp components (ClpC, ClpE and ClpP) have been implicated in protein disaggregation [144, 145], there is currently little evidence in *E. coli* to suggest that aggregated proteins are degraded *in vivo*. This is, partly due to binding of "folder" chaperones (i.e. DnaK) to protein aggregates, which restricts the binding of proteases [138, 146], and partly due to the action of "holder" chaperones (i.e. inclusion body proteins A and B, IbpA and IbpB, respectively), which trap the substrates in a "folding" competent state. Indeed, the "holder" chaperones that bind to misfolded and aggregated proteins, appear to facilitate their subsequent reactivation and refolding by the ClpB/KJE bi-chaperone system [147, 148]. It is likely that the competitive edge, of chaperones over proteases, in the recognition of most misfolded and/or aggregated proteins is energetically advantageous to bacteria.

Interestingly, both "holder" chaperones (IbpA and IbpB) are degraded by Lon, [149], and despite the high overall sequence similarity of both proteins, IbpB is degraded significantly faster by Lon, than IbpA is [149]. Surprisingly however, the rate of IbpA degradation, by Lon, was substantially increased under heat-shock conditions suggesting an intriguing link between protein aggregation and degradation. Therefore from these data, Baker and colleagues have proposed several interesting models whereby (a) free inclusion bodies are degraded (b) both the aggregated proteins and the Ibp's are degraded or alternatively (c) only the Ibp's are degraded by Lon (Fig. 5.6).

Conclusion

It has been long known, that proteases play a crucial role in the removal of unwanted proteins from the cell under a variety of different cellular conditions. However, recent findings have highlighted that these cellular machines also play an important role in controlling several different stress response pathways. These studies have

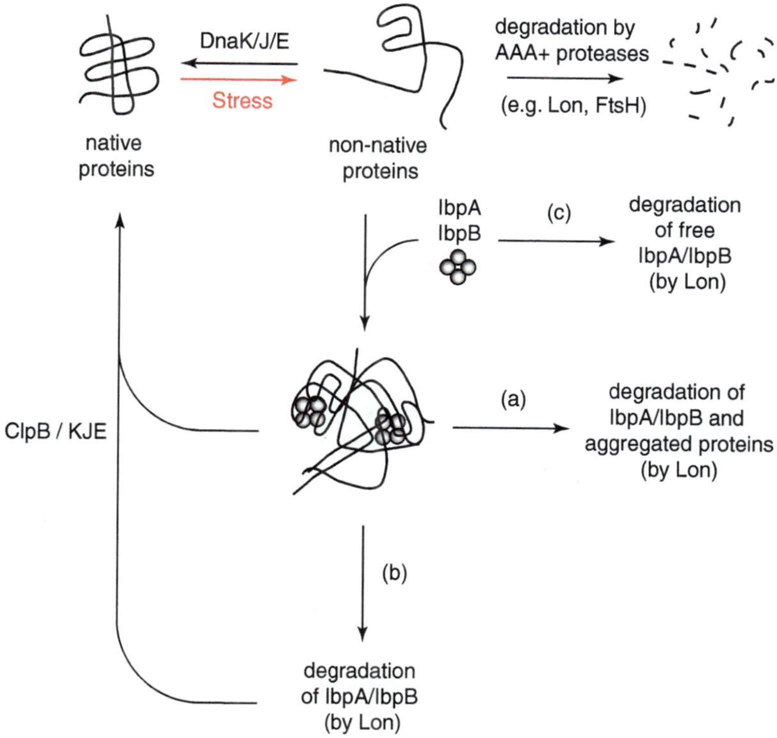

Fig. 5.6 Misfolded and aggregated proteins are refolded by chaperones and/or removed by proteases. Following thermal stress, native proteins unfold. Unfolded proteins are preferentially refolded by chaperones, however in order to maintain a productive folding environment in the cell, some unfolded proteins may be degraded by AAA+proteases. If stress is prolonged, misfolded proteins tend to aggregate, either in the absence or presence of small heat-shock proteins (i.e. IbpA and IbpB). These aggregated proteins can be refolded by a specialised bi-chaperone system; ClpB together with the DnaK chaperone system (ClpB/KJE). ClpB/KJE-mediated reactivation of the aggregate can be accelerated by the presence of IbpA and IbpB. Recently, it was proposed that these Ibp's are degraded by the AAA+protease Lon, either (**a**) together with the aggregated protein, (**b**) to release the aggregated protein and further accelerate the ClpB/KJE-reactivation or (**c**) to free proteins

illustrated that the proteases responsible for the degradation of these transcription factors, are not only highly regulated, but also exhibit an exquisite specificity. Despite these advances many questions remain unanswered, currently little is known regarding the regulation or removal of the adaptor or anti-adaptor proteins that regulate the turnover of these transcription factors. Similarly at a structural level, our current understanding of how each component interacts with one another, remains limited. As such many important challenges, for current and future researchers, still remain in this field.

Acknowledgments Work in the DAD laboratory is funded by the Australian Research Council.

References

1. Barchinger SE, Ades SE (2013) Regulated proteolysis: control of the Escherichia coli σE-dependent cell envelope stress response. In: Dougan DA (ed) Regulated proteolysis in microorganisms. Springer, Subcell Biochem 66:129–160
2. Sledjeski DD, Gupta A, Gottesman S (1996) The small RNA, DsrA, is essential for the low temperature expression of RpoS during exponential growth in Escherichia coli. EMBO J 15(15):3993–4000
3. Muffler A, Traulsen DD, Lange R, Hengge-Aronis R (1996) Posttranscriptional osmotic regulation of the sigma(s) subunit of RNA polymerase in Escherichia coli. J Bacteriol 178(6):1607–1613
4. Lange R, Hengge-Aronis R (1991) Growth phase-regulated expression of bolA and morphology of stationary-phase Escherichia coli cells are controlled by the novel sigma factor sigma S. J Bacteriol 173(14):4474–4481
5. Hengge-Aronis R (1993) Survival of hunger and stress: the role of rpoS in early stationary phase gene regulation in E. coli. Cell 72(2):165–168
6. McCann MP, Kidwell JP, Matin A (1991) The putative sigma factor KatF has a central role in development of starvation-mediated general resistance in Escherichia coli. J Bacteriol 173(13):4188–4194
7. Farewell A, Kvint K, Nystrom T (1998) Negative regulation by RpoS: a case of sigma factor competition. Mol Microbiol 29(4):1039–1051
8. Lange R, Hengge-Aronis R (1994) The cellular concentration of the sigma S subunit of RNA polymerase in Escherichia coli is controlled at the levels of transcription, translation, and protein stability. Genes Dev 8(13):1600–1612
9. Atlung T, Brondsted L (1994) Role of the transcriptional activator AppY in regulation of the cyx appA operon of Escherichia coli by anaerobiosis, phosphate starvation, and growth phase. J Bacteriol 176(17):5414–5422
10. Altuvia S, Almiron M, Huisman G, Kolter R et al (1994) The dps promoter is activated by OxyR during growth and by IHF and sigma S in stationary phase. Mol Microbiol 13(2):265–272
11. Dong T, Schellhorn HE (2009) Global effect of RpoS on gene expression in pathogenic Escherichia coli O157:H7 strain EDL933. BMC Genomics 10:349
12. Weber H, Polen T, Heuveling J, Wendisch VF et al (2005) Genome-wide analysis of the general stress response network in Escherichia coli: sigmaS-dependent genes, promoters, and sigma factor selectivity. J Bacteriol 187(5):1591–1603
13. Lacour S, Landini P (2004) SigmaS-dependent gene expression at the onset of stationary phase in Escherichia coli: function of sigmaS-dependent genes and identification of their promoter sequences. J Bacteriol 186(21):7186–7195
14. Patten CL, Kirchhof MG, Schertzberg MR, Morton RA et al (2004) Microarray analysis of RpoS-mediated gene expression in Escherichia coli K-12. Mol Genet Genomics 272(5):580–591
15. Bishop RE, Leskiw BK, Hodges RS, Kay CM et al (1998) The entericidin locus of Escherichia coli and its implications for programmed bacterial cell death. J Mol Biol 280(4):583–596
16. Dong T, Schellhorn HE (2010) Role of RpoS in virulence of pathogens. Infect Immun 78(3):887–897
17. Lange R, Fischer D, Hengge-Aronis R (1995) Identification of transcriptional start sites and the role of ppGpp in the expression of rpoS, the structural gene for the sigma S subunit of RNA polymerase in Escherichia coli. J Bacteriol 177(16):4676–4680
18. Lange R, Hengge-Aronis R (1994) The nlpD gene is located in an operon with rpoS on the Escherichia coli chromosome and encodes a novel lipoprotein with a potential function in cell wall formation. Mol Microbiol 13(4):733–743
19. Hengge-Aronis R (2002) Signal transduction and regulatory mechanisms involved in control of the sigma(S) (RpoS) subunit of RNA polymerase. Microbiol Mol Biol Rev 66(3):373–395
20. Muffler A, Fischer D, Hengge-Aronis R (1996) The RNA-binding protein HF-I, known as a host factor for phage Qbeta RNA replication, is essential for rpoS translation in Escherichia coli. Genes Dev 10(9):1143–1151

21. Argaman L, Hershberg R, Vogel J, Bejerano G et al (2001) Novel small RNA-encoding genes in the intergenic regions of Escherichia coli. Curr Biol 11(12):941–950
22. Eddy SR (2002) Computational genomics of noncoding RNA genes. Cell 109(2):137–140
23. Storz G (2002) An expanding universe of noncoding RNAs. Science 296(5571):1260–1263
24. Wassarman KM, Repoila F, Rosenow C, Storz G et al (2001) Identification of novel small RNAs using comparative genomics and microarrays. Genes Dev 15(13):1637–1651
25. Storz G, Opdyke JA, Zhang A (2004) Controlling mRNA stability and translation with small, noncoding RNAs. Curr Opin Microbiol 7(2):140–144
26. Updegrove T, Wilf N, Sun X, Wartell RM (2008) Effect of Hfq on RprA-rpoS mRNA pairing: Hfq-RNA binding and the influence of the 5′ rpoS mRNA leader region. Biochemistry 47(43):11184–11195
27. Waters LS, Storz G (2009) Regulatory RNAs in bacteria. Cell 136(4):615–628
28. Zhang A, Wassarman KM, Rosenow C, Tjaden BC et al (2003) Global analysis of small RNA and mRNA targets of Hfq. Mol Microbiol 50(4):1111–1124
29. Brown L, Elliott T (1996) Efficient translation of the RpoS sigma factor in Salmonella typhimurium requires host factor I, an RNA-binding protein encoded by the hfq gene. J Bacteriol 178(13):3763–3770
30. Moller T, Franch T, Hojrup P, Keene DR et al (2002) Hfq: a bacterial Sm-like protein that mediates RNA-RNA interaction. Mol Cell 9(1):23–30
31. Muffler A, Traulsen DD, Fischer D, Lange R et al (1997) The RNA-binding protein HF-I plays a global regulatory role which is largely, but not exclusively, due to its role in expression of the sigmaS subunit of RNA polymerase in Escherichia coli. J Bacteriol 179(1):297–300
32. Zhang A, Wassarman KM, Ortega J, Steven AC et al (2002) The Sm-like Hfq protein increases OxyS RNA interaction with target mRNAs. Mol Cell 9(1):11–22
33. Majdalani N, Chen S, Murrow J, St John K et al (2001) Regulation of RpoS by a novel small RNA: the characterization of RprA. Mol Microbiol 39(5):1382–1394
34. Sledjeski DD, Whitman C, Zhang A (2001) Hfq is necessary for regulation by the untranslated RNA DsrA. J Bacteriol 183(6):1997–2005
35. Ueguchi C, Misonou N, Mizuno T (2001) Negative control of rpoS expression by phosphoenolpyruvate: carbohydrate phosphotransferase system in Escherichia coli. J Bacteriol 183(2):520–527
36. Zhang A, Altuvia S, Tiwari A, Argaman L et al (1998) The OxyS regulatory RNA represses rpoS translation and binds the Hfq (HF-I) protein. EMBO J 17(20):6061–6068
37. Kawamoto H, Koide Y, Morita T, Aiba H (2006) Base-pairing requirement for RNA silencing by a bacterial small RNA and acceleration of duplex formation by Hfq. Mol Microbiol 61(4):1013–1022
38. Lease RA, Cusick ME, Belfort M (1998) Riboregulation in Escherichia coli: DsrA RNA acts by RNA:RNA interactions at multiple loci. Proc Natl Acad Sci U S A 95(21):12456–12461
39. Majdalani N, Cunning C, Sledjeski D, Elliott T et al (1998) DsrA RNA regulates translation of RpoS message by an anti-antisense mechanism, independent of its action as an antisilencer of transcription. Proc Natl Acad Sci U S A 95(21):12462–12467
40. Majdalani N, Hernandez D, Gottesman S (2002) Regulation and mode of action of the second small RNA activator of RpoS translation, RprA. Mol Microbiol 46(3):813–826
41. Altuvia S, Weinstein-Fischer D, Zhang A, Postow L et al (1997) A small, stable RNA induced by oxidative stress: role as a pleiotropic regulator and antimutator. Cell 90(1):43–53
42. Repoila F, Gottesman S (2001) Signal transduction cascade for regulation of RpoS: temperature regulation of DsrA. J Bacteriol 183(13):4012–4023
43. Sledjeski D, Gottesman S (1995) A small RNA acts as an antisilencer of the H-NS-silenced rcsA gene of Escherichia coli. Proc Natl Acad Sci U S A 92(6):2003–2007
44. Repoila F, Gottesman S (2003) Temperature sensing by the dsrA promoter. J Bacteriol 185(22):6609–6614
45. Lease RA, Woodson SA (2004) Cycling of the Sm-like protein Hfq on the DsrA small regulatory RNA. J Mol Biol 344(5):1211–1223
46. Cunning C, Brown L, Elliott T (1998) Promoter substitution and deletion analysis of upstream region required for rpoS translational regulation. J Bacteriol 180(17):4564–4570

47. Soper T, Mandin P, Majdalani N, Gottesman S et al (2010) Positive regulation by small RNAs and the role of Hfq. Proc Natl Acad Sci U S A 107(21):9602–9607
48. Soper TJ, Woodson SA (2008) The rpoS mRNA leader recruits Hfq to facilitate annealing with DsrA sRNA. RNA 14(9):1907–1917
49. Hwang W, Arluison V, Hohng S (2011) Dynamic competition of DsrA and rpoS fragments for the proximal binding site of Hfq as a means for efficient annealing. Nucleic Acids Res 39(12):5131–5139
50. Arluison V, Mutyam SK, Mura C, Marco S et al (2007) Sm-like protein Hfq: location of the ATP-binding site and the effect of ATP on Hfq– RNA complexes. Protein Sci 16(9):1830–1841
51. McCullen CA, Benhammou JN, Majdalani N, Gottesman S (2010) Mechanism of positive regulation by DsrA and RprA small noncoding RNAs: pairing increases translation and protects rpoS mRNA from degradation. J Bacteriol 192(21):5559–5571
52. Papenfort K, Said N, Welsink T, Lucchini S et al (2009) Specific and pleiotropic patterns of mRNA regulation by ArcZ, a conserved, Hfq-dependent small RNA. Mol Microbiol 74(1):139–158
53. Mandin P, Gottesman S (2010) Integrating anaerobic/aerobic sensing and the general stress response through the ArcZ small RNA. EMBO J 29(18):3094–3107
54. Schweder T, Lee KH, Lomovskaya O, Matin A (1996) Regulation of Escherichia coli starvation sigma factor (sigma s) by ClpXP protease. J Bacteriol 178(2):470–476
55. Klauck E, Lingnau M, Hengge-Aronis R (2001) Role of the response regulator RssB in sigma recognition and initiation of sigma proteolysis in Escherichia coli. Mol Microbiol 40(6):1381–1390
56. Muffler A, Fischer D, Altuvia S, Storz G et al (1996) The response regulator RssB controls stability of the sigma(S) subunit of RNA polymerase in Escherichia coli. EMBO J 15(6):1333–1339
57. Pratt LA, Silhavy TJ (1996) The response regulator SprE controls the stability of RpoS. Proc Natl Acad Sci U S A 93(6):2488–2492
58. Zhou Y, Gottesman S, Hoskins JR, Maurizi MR et al (2001) The RssB response regulator directly targets sigma(S) for degradation by ClpXP. Genes Dev 15(5):627–637
59. Becker G, Klauck E, Hengge-Aronis R (2000) The response regulator RssB, a recognition factor for sigmaS proteolysis in Escherichia coli, can act like an anti-sigmaS factor. Mol Microbiol 35(3):657–666
60. Becker G, Klauck E, Hengge-Aronis R (1999) Regulation of RpoS proteolysis in Escherichia coli: the response regulator RssB is a recognition factor that interacts with the turnover element in RpoS. Proc Natl Acad Sci U S A 96(11):6439–6444
61. Bouche S, Klauck E, Fischer D, Lucassen M et al (1998) Regulation of RssB-dependent proteolysis in Escherichia coli: a role for acetyl phosphate in a response regulator-controlled process. Mol Microbiol 27(4):787–795
62. Bougdour A, Cunning C, Baptiste PJ, Elliott T et al (2008) Multiple pathways for regulation of sigmaS (RpoS) stability in Escherichia coli via the action of multiple anti-adaptors. Mol Microbiol 68(2):298–313
63. Bougdour A, Wickner S, Gottesman S (2006) Modulating RssB activity: IraP, a novel regulator of sigma(S) stability in Escherichia coli. Genes Dev 20(7):884–897
64. Sauer RT, Baker TA (2011) AAA+ proteases: ATP-fueled machines of protein destruction. Annu Rev Biochem 80:587–612
65. Glynn SE, Martin A, Nager AR, Baker TA et al (2009) Structures of asymmetric ClpX hexamers reveal nucleotide-dependent motions in a AAA+ protein-unfolding machine. Cell 139(4):744–756
66. Kim DY, Kim KK (2003) Crystal structure of ClpX molecular chaperone from Helicobacter pylori. J Biol Chem 278(50):50664–50670
67. Iyer LM, Leipe DD, Koonin EV, Aravind L (2004) Evolutionary history and higher order classification of AAA+ ATPases. J Struct Biol 146(1–2):11–31
68. Kim YI, Levchenko I, Fraczkowska K, Woodruff RV et al (2001) Molecular determinants of complex formation between Clp/Hsp100 ATPases and the ClpP peptidase. Nat Struct Biol 8(3):230–233

69. Neuwald AF, Aravind L, Spouge JL, Koonin EV (1999) AAA+: a class of chaperone-like ATPases associated with the assembly, operation, and disassembly of protein complexes. Genome Res 9(1):27–43
70. Walker JE, Saraste M, Gay NJ (1982) E. coli F1-ATPase interacts with a membrane protein component of a proton channel. Nature 298(5877):867–869
71. Gur E, Ottofuelling R, Dougan DA (2013) Machines of destruction – AAA+proteases and the adaptors that control them. In: Dougan DA (ed) Regulated proteolysis in microorganisms. Springer, Subcell Biochem 66:3–33
72. Jennings LD, Lun DS, Medard M, Licht S (2008) ClpP hydrolyzes a protein substrate processively in the absence of the ClpA ATPase: mechanistic studies of ATP-independent proteolysis. Biochemistry 47(44):11536–11546
73. Wang J, Hartling JA, Flanagan JM (1997) The structure of ClpP at 2.3 A resolution suggests a model for ATP-dependent proteolysis. Cell 91(4):447–456
74. Grimaud R, Kessel M, Beuron F, Steven AC et al (1998) Enzymatic and structural similarities between the Escherichia coli ATP-dependent proteases, ClpXP and ClpAP. J Biol Chem 273(20):12476–12481
75. Maglica Z, Kolygo K, Weber-Ban E (2009) Optimal efficiency of ClpAP and ClpXP chaperone-proteases is achieved by architectural symmetry. Structure 17(4):508–516
76. Ortega J, Lee HS, Maurizi MR, Steven AC (2002) Alternating translocation of protein substrates from both ends of ClpXP protease. EMBO J 21(18):4938–4949
77. Joshi SA, Hersch GL, Baker TA, Sauer RT (2004) Communication between ClpX and ClpP during substrate processing and degradation. Nat Struct Mol Biol 11(5):404–411
78. Martin A, Baker TA, Sauer RT (2007) Distinct static and dynamic interactions control ATPase-peptidase communication in a AAA+protease. Mol Cell 27(1):41–52
79. Dougan DA (2011) Chemical activators of ClpP: turning Jekyll into Hyde. Chem Biol 18(9):1072–1074
80. Kirstein J, Hoffmann A, Lilie H, Schmidt R et al (2009) The antibiotic ADEP reprogrammes ClpP, switching it from a regulated to an uncontrolled protease. EMBO Mol Med 1(1):37–49
81. Lee BG, Park EY, Lee KE, Jeon H et al (2010) Structures of ClpP in complex with acyldepsipeptide antibiotics reveal its activation mechanism. Nat Struct Mol Biol 17(4):471–478
82. Leung E, Datti A, Cossette M, Goodreid J et al (2011) Activators of cylindrical proteases as antimicrobials: identification and development of small molecule activators of ClpP protease. Chem Biol 18(9):1167–1178
83. Li DH, Chung YS, Gloyd M, Joseph E et al (2010) Acyldepsipeptide antibiotics induce the formation of a structured axial channel in ClpP: A model for the ClpX/ClpA-bound state of ClpP. Chem Biol 17(9):959–969
84. Gribun A, Kimber MS, Ching R, Sprangers R et al (2005) The ClpP double ring tetradecameric protease exhibits plastic ring-ring interactions, and the N termini of its subunits form flexible loops that are essential for ClpXP and ClpAP complex formation. J Biol Chem 280(16):16185–16196
85. Kang SG, Maurizi MR, Thompson M, Mueser T et al (2004) Crystallography and mutagenesis point to an essential role for the N-terminus of human mitochondrial ClpP. J Struct Biol 148(3):338–352
86. Jennings LD, Bohon J, Chance MR, Licht S (2008) The ClpP N-terminus coordinates substrate access with protease active site reactivity. Biochemistry 47(42):11031–11040
87. Martin A, Baker TA, Sauer RT (2008) Pore loops of the AAA+ClpX machine grip substrates to drive translocation and unfolding. Nat Struct Mol Biol 15(11):1147–1151
88. Martin A, Baker TA, Sauer RT (2008) Diverse pore loops of the AAA+ClpX machine mediate unassisted and adaptor-dependent recognition of ssrA-tagged substrates. Mol Cell 29(4):441–450
89. Flynn JM, Neher SB, Kim YI, Sauer RT et al (2003) Proteomic discovery of cellular substrates of the ClpXP protease reveals five classes of ClpX-recognition signals. Mol Cell 11(3):671–683
90. Kim YI, Burton RE, Burton BM, Sauer RT et al (2000) Dynamics of substrate denaturation and translocation by the ClpXP degradation machine. Mol Cell 5(4):639–648

91. Ortega J, Singh SK, Ishikawa T, Maurizi MR et al (2000) Visualization of substrate binding and translocation by the ATP-dependent protease, ClpXP. Mol Cell 6(6):1515–1521
92. Thompson MW, Singh SK, Maurizi MR (1994) Processive degradation of proteins by the ATP-dependent Clp protease from Escherichia coli. Requirement for the multiple array of active sites in ClpP but not ATP hydrolysis. J Biol Chem 269(27):18209–18215
93. Schlieker C, Weibezahn J, Patzelt H, Tessarz P et al (2004) Substrate recognition by the AAA+chaperone ClpB. Nat Struct Mol Biol 11(7):607–615
94. Siddiqui SM, Sauer RT, Baker TA (2004) Role of the processing pore of the ClpX AAA+ATPase in the recognition and engagement of specific protein substrates. Genes Dev 18(4):369–374
95. Wang J, Song JJ, Franklin MC, Kamtekar S et al (2001) Crystal structures of the HslVU peptidase-ATPase complex reveal an ATP-dependent proteolysis mechanism. Structure 9(2):177–184
96. Yamada-Inagawa T, Okuno T, Karata K, Yamanaka K et al (2003) Conserved pore residues in the AAA protease FtsH are important for proteolysis and its coupling to ATP hydrolysis. J Biol Chem 278(50):50182–50187
97. Lum R, Tkach JM, Vierling E, Glover JR (2004) Evidence for an unfolding/threading mechanism for protein disaggregation by Saccharomyces cerevisiae Hsp104. J Biol Chem 279(28): 29139–29146
98. Gonzalez M, Rasulova F, Maurizi MR, Woodgate R (2000) Subunit-specific degradation of the UmuD/D′ heterodimer by the ClpXP protease: the role of trans recognition in UmuD′ stability. EMBO J 19(19):5251–5258
99. Dougan DA, Weber-Ban E, Bukau B (2003) Targeted delivery of an ssrA-tagged substrate by the adaptor protein SspB to its cognate AAA+protein ClpX. Mol Cell 12(2):373–380
100. Neher SB, Sauer RT, Baker TA (2003) Distinct peptide signals in the UmuD and UmuD′ subunits of UmuD/D′ mediate tethering and substrate processing by the ClpXP protease. Proc Natl Acad Sci U S A 100(23):13219–13224
101. Wah DA, Levchenko I, Rieckhof GE, Bolon DN et al (2003) Flexible linkers leash the substrate binding domain of SspB to a peptide module that stabilizes delivery complexes with the AAA+ClpXP protease. Mol Cell 12(2):355–363
102. Galperin MY (2006) Structural classification of bacterial response regulators: diversity of output domains and domain combinations. J Bacteriol 188(12):4169–4182
103. Peterson CN, Ruiz N, Silhavy TJ (2004) RpoS proteolysis is regulated by a mechanism that does not require the SprE (RssB) response regulator phosphorylation site. J Bacteriol 186(21):7403–7410
104. Bren A, Welch M, Blat Y, Eisenbach M (1996) Signal termination in bacterial chemotaxis: CheZ mediates dephosphorylation of free rather than switch-bound CheY. Proc Natl Acad Sci U S A 93(19):10090–10093
105. Hess JF, Bourret RB, Simon MI (1988) Histidine phosphorylation and phosphoryl group transfer in bacterial chemotaxis. Nature 336(6195):139–143
106. Bachhawat P, Stock AM (2007) Crystal structures of the receiver domain of the response regulator PhoP from Escherichia coli in the absence and presence of the phosphoryl analog beryllofluoride. J Bacteriol 189(16):5987–5995
107. Bachhawat P, Swapna GV, Montelione GT, Stock AM (2005) Mechanism of activation for transcription factor PhoB suggested by different modes of dimerization in the inactive and active states. Structure 13(9):1353–1363
108. Lee SY, Cho HS, Pelton JG, Yan D et al (2001) Crystal structure of activated CheY. Comparison with other activated receiver domains. J Biol Chem 276(19):16425–16431
109. Toro-Roman A, Mack TR, Stock AM (2005) Structural analysis and solution studies of the activated regulatory domain of the response regulator ArcA: a symmetric dimer mediated by the alpha4-beta5-alpha5 face. J Mol Biol 349(1):11–26
110. Toro-Roman A, Wu T, Stock AM (2005) A common dimerization interface in bacterial response regulators KdpE and TorR. Protein Sci 14(12):3077–3088

111. Studemann A, Noirclerc-Savoye M, Klauck E, Becker G et al (2003) Sequential recognition of two distinct sites in sigma(S) by the proteolytic targeting factor RssB and ClpX. EMBO J 22(16):4111–4120
112. D'Souza C, Nakano MM, Zuber P (1994) Identification of comS, a gene of the srfA operon that regulates the establishment of genetic competence in Bacillus subtilis. Proc Natl Acad Sci U S A 91(20):9397–9401
113. Hamoen LW, Eshuis H, Jongbloed J, Venema G et al (1995) A small gene, designated comS, located within the coding region of the fourth amino acid-activation domain of srfA, is required for competence development in Bacillus subtilis. Mol Microbiol 15(1):55–63
114. Turgay K, Hamoen LW, Venema G, Dubnau D (1997) Biochemical characterization of a molecular switch involving the heat shock protein ClpC, which controls the activity of ComK, the competence transcription factor of Bacillus subtilis. Genes Dev 11(1):119–128
115. Persuh M, Turgay K, Mandic-Mulec I, Dubnau D (1999) The N- and C-terminal domains of MecA recognize different partners in the competence molecular switch. Mol Microbiol 33(4):886–894
116. Turgay K, Hahn J, Burghoorn J, Dubnau D (1998) Competence in Bacillus subtilis is controlled by regulated proteolysis of a transcription factor. EMBO J 17(22):6730–6738
117. Ogura M, Liu L, Lacelle M, Nakano MM et al (1999) Mutational analysis of ComS: evidence for the interaction of ComS and MecA in the regulation of competence development in Bacillus subtilis. Mol Microbiol 32(4):799–812
118. Bougdour A, Gottesman S (2007) ppGpp regulation of RpoS degradation via anti-adaptor protein IraP. Proc Natl Acad Sci U S A 104(31):12896–12901
119. Tissieres A, Mitchell HK, Tracy UM (1974) Protein synthesis in salivary glands of Drosophila melanogaster: relation to chromosome puffs. J Mol Biol 84(3):389–398
120. Gamer J, Bujard H, Bukau B (1992) Physical interaction between heat shock proteins DnaK, DnaJ, and GrpE and the bacterial heat shock transcription factor sigma 32. Cell 69(5):833–842
121. Guisbert E, Herman C, Lu CZ, Gross CA (2004) A chaperone network controls the heat shock response in E. coli. Genes Dev 18(22):2812–2821
122. Straus DB, Walter WA, Gross CA (1987) The heat shock response of E. coli is regulated by changes in the concentration of sigma 32. Nature 329(6137):348–351
123. Herman C, Thevenet D, D'Ari R, Bouloc P (1995) Degradation of sigma 32, the heat shock regulator in Escherichia coli, is governed by HflB. Proc Natl Acad Sci U S A 92(8):3516–3520
124. Kanemori M, Nishihara K, Yanagi H, Yura T (1997) Synergistic roles of HslVU and other ATP-dependent proteases in controlling in vivo turnover of sigma32 and abnormal proteins in Escherichia coli. J Bacteriol 179(23):7219–7225
125. Tomoyasu T, Gamer J, Bukau B, Kanemori M et al (1995) Escherichia coli FtsH is a membrane-bound, ATP-dependent protease which degrades the heat-shock transcription factor sigma 32. EMBO J 14(11):2551–2560
126. Sousa MC, Trame CB, Tsuruta H, Wilbanks SM et al (2000) Crystal and solution structures of an HslUV protease-chaperone complex. Cell 103(4):633–643
127. Okuno T, Ogura T (2013) FtsH protease-mediated regulation of various cellular functions. In: Dougan DA (ed) Regulated proteolysis in microorganisms. Springer, Subcell Biochem 66:53–69
128. Kanemori M, Yanagi H, Yura T (1999) Marked instability of the sigma(32) heat shock transcription factor at high temperature. Implications for heat shock regulation. J Biol Chem 274(31):22002–22007
129. Rodriguez F, Arsene-Ploetze F, Rist W, Rudiger S et al (2008) Molecular basis for regulation of the heat shock transcription factor sigma32 by the DnaK and DnaJ chaperones. Mol Cell 32(3):347–358
130. Horikoshi M, Yura T, Tsuchimoto S, Fukumori Y et al (2004) Conserved region 2.1 of Escherichia coli heat shock transcription factor sigma32 is required for modulating both metabolic stability and transcriptional activity. J Bacteriol 186(22):7474–7480

131. Obrist M, Milek S, Klauck E, Hengge R et al (2007) Region 2.1 of the Escherichia coli heat-shock sigma factor RpoH (sigma32) is necessary but not sufficient for degradation by the FtsH protease. Microbiology 153(Pt 8):2560–2571
132. Herman C, Prakash S, Lu CZ, Matouschek A et al (2003) Lack of a robust unfoldase activity confers a unique level of substrate specificity to the universal AAA protease FtsH. Mol Cell 11(3):659–669
133. Langklotz S, Baumann U, Narberhaus F (2012) Structure and function of the bacterial AAA protease FtsH. Biochim Biophys Acta 1823(1):40–48
134. Obrist M, Langklotz S, Milek S, Fuhrer F et al (2009) Region C of the Escherichia coli heat shock sigma factor RpoH (sigma 32) contains a turnover element for proteolysis by the FtsH protease. FEMS Microbiol Lett 290(2):199–208
135. Tomoyasu T, Arsene F, Ogura T, Bukau B (2001) The C terminus of sigma(32) is not essential for degradation by FtsH. J Bacteriol 183(20):5911–5917
136. Wickner S, Maurizi MR, Gottesman S (1999) Posttranslational quality control: folding, refolding, and degrading proteins. Science 286(5446):1888–1893
137. Gur E, Sauer RT (2008) Recognition of misfolded proteins by Lon, a AAA(+) protease. Genes Dev 22(16):2267–2277
138. Dougan DA, Reid BG, Horwich AL, Bukau B (2002) ClpS, a substrate modulator of the ClpAP machine. Mol Cell 9(3):673–683
139. Farrell CM, Grossman AD, Sauer RT (2005) Cytoplasmic degradation of ssrA-tagged proteins. Mol Microbiol 57(6):1750–1761
140. Mogk A, Tomoyasu T, Goloubinoff P, Rudiger S et al (1999) Identification of thermolabile Escherichia coli proteins: prevention and reversion of aggregation by DnaK and ClpB. EMBO J 18(24):6934–6949
141. Goloubinoff P, Mogk A, Zvi AP, Tomoyasu T et al (1999) Sequential mechanism of solubilization and refolding of stable protein aggregates by a bichaperone network. Proc Natl Acad Sci U S A 96(24):13732–13737
142. Schlothauer T, Mogk A, Dougan DA, Bukau B et al (2003) MecA, an adaptor protein necessary for ClpC chaperone activity. Proc Natl Acad Sci U S A 100(5):2306–2311
143. Wawrzynow A, Wojtkowiak D, Marszalek J, Banecki B et al (1995) The ClpX heat-shock protein of Escherichia coli, the ATP-dependent substrate specificity component of the ClpP-ClpX protease, is a novel molecular chaperone. EMBO J 14(9):1867–1877
144. Kock H, Gerth U, Hecker M (2004) The ClpP peptidase is the major determinant of bulk protein turnover in Bacillus subtilis. J Bacteriol 186(17):5856–5864
145. Miethke M, Hecker M, Gerth U (2006) Involvement of Bacillus subtilis ClpE in CtsR degradation and protein quality control. J Bacteriol 188(13):4610–4619
146. Dougan DA, Mogk A, Bukau B (2002) Protein folding and degradation in bacteria: to degrade or not to degrade? That is the question. Cell Mol Life Sci 59(10):1607–1616
147. Matuszewska M, Kuczynska-Wisnik D, Laskowska E, Liberek K (2005) The small heat shock protein IbpA of Escherichia coli cooperates with IbpB in stabilization of thermally aggregated proteins in a disaggregation competent state. J Biol Chem 280(13):12292–12298
148. Mogk A, Deuerling E, Vorderwulbecke S, Vierling E et al (2003) Small heat shock proteins, ClpB and the DnaK system form a functional triade in reversing protein aggregation. Mol Microbiol 50(2):585–595
149. Bissonnette SA, Rivera-Rivera I, Sauer RT, Baker TA (2010) The IbpA and IbpB small heat-shock proteins are substrates of the AAA+Lon protease. Mol Microbiol 75(6):1539–1549

Chapter 6
Regulated Proteolysis: Control of the *Escherichia coli* σ^E-Dependent Cell Envelope Stress Response

Sarah E. Barchinger and Sarah E. Ades

Abstract Over the past decade, regulatory proteolysis has emerged as a paradigm for transmembrane signal transduction in all organisms, from bacteria to humans. These conserved proteolytic pathways share a common design that involves the sequential proteolysis of a membrane-bound regulatory protein by two proteases. Proteolysis releases the regulator, which is inactive in its membrane-bound form, into the cytoplasm where it performs its cellular function. One of the best-characterized examples of signal transduction via regulatory proteolysis is the pathway governing the σ^E-dependent cell envelope stress response in *Escherichia coli*. In unstressed cells, σ^E is sequestered at the membrane by the transmembrane anti-sigma factor, RseA. Stresses that compromise the cell envelope and interfere with the proper folding of outer membrane proteins (OMPs) activate the proteolytic pathway. The C-terminal residues of unfolded OMPs bind to the inner membrane protease, DegS, to initiate the proteolytic cascade. DegS removes the periplasmic domain of RseA creating a substrate for the next protease in the pathway, RseP. RseP cleaves RseA in the periplasmic region in a process called regulated intramembrane proteolysis (RIP). The remaining fragment of RseA is released into the cytoplasm and fully degraded by the ATP-dependent protease, ClpXP, with the assistance of the adaptor protein, SspB, thereby freeing σ^E to reprogram gene expression. A growing body of evidence indicates that the overall proteolytic framework that governs the σ^E response is used to regulate similar anti-sigma factor/sigma factor pairs throughout

S.E. Barchinger
Graduate Program in BMMB, Department of Biochemistry and Molecular Biology, The Pennsylvania State University, University Park, PA 16802, USA

S.E. Ades (✉)
Department of Biochemistry and Molecular Biology,
The Pennsylvania State University, University Park, PA 16802, USA
e-mail: ades@psu.edu

D.A. Dougan (ed.), *Regulated Proteolysis in Microorganisms*, Subcellular Biochemistry 66,
DOI 10.1007/978-94-007-5940-4_6, © Springer Science+Business Media Dordrecht 2013

the bacterial world and has been adapted to recognize a wide variety of signals and control systems as diverse as envelope stress responses, sporulation, virulence, and iron-siderophore uptake. In this chapter, we review the extensive physiological, biochemical, and structural studies on the σ^E system that provide remarkable insights into the mechanistic underpinnings of this regulated proteolytic signal transduction pathway. These studies reveal design principles that are applicable to related proteases and regulatory proteolytic pathways in all domains of life.

Introduction

In biological systems, proteolysis serves both as a cellular housekeeper and an orchestrator of regulatory pathways. As housekeepers, proteases rid the cell of damaged proteins in a relatively non-specific manner, helping to preserve the efficiency of cellular physiology. In contrast, when acting as regulators, proteases degrade or process only select substrates in response to distinct signals, yielding a defined change in the activity of their targets. Unlike regulatory systems in which protein activity is reversibly modulated by ligand binding, proteolysis provides a rapid and irreversible change in the activity of target proteins. Regulated proteolysis rivals post-translational modification in the diversity of cellular pathways it controls, ranging from transcriptional regulation to modulation of enzyme activity.

Regulated proteolysis is the key component of the solution to one of the most common biological problems shared by all cells: how a signal generated on one side of a membrane can be communicated across the membrane to elicit the required response. This regulatory paradigm governs transmembrane signaling responses ranging from the cell envelope stress response of *Escherichia coli*, the subject of this chapter, to the SREBP (sterol regulatory element binding proteins) pathway of humans. In its simplest form, the overall design of the proteolytic system consists of two proteases that sequentially cleave a regulatory, transmembrane (TM) protein on either side of the membrane, resulting in release of the regulatory domain with a distinct biological function (Fig. 6.1) [1]. These systems are most commonly found in compartmental membranes in eukaryotes and the cytoplasmic membrane of prokaryotes. The signal is generally sensed on the lumenal or extracytoplasmic side of the membrane and the regulatory protein is released into the cytoplasm [1–3]. The first protease cleaves the target protein in its lumenal domain to remove most of this domain, but only after receiving the inducing signal. The second protease then cleaves the remaining fragment of the protein in the membrane-spanning region in a process termed regulated intramembrane proteolysis, or RIP [1, 2]. Intramembrane cleavage releases the regulatory domain from the membrane so that it can fulfill its biological mission [2]. Not only is the overall framework of this proteolytic pathway conserved, but the proteases are also evolutionarily related. The initiating proteases are often serine or aspartate proteases [1, 2]. Several conserved families of proteases have been identified that perform intramembrane cleavage: the site-2 protease (S2P),

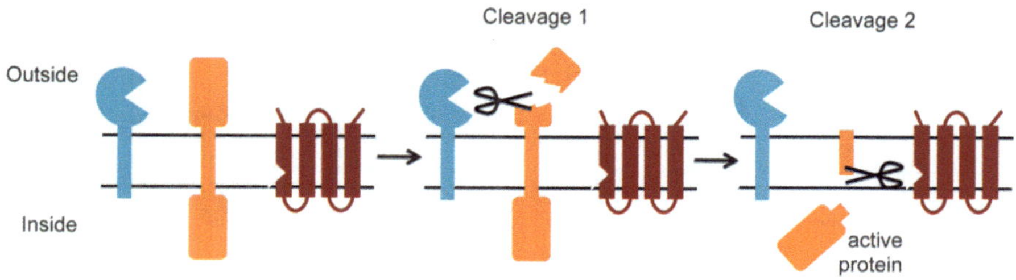

Fig. 6.1 Overview of transmembrane signal transduction by regulated intramembrane proteolysis. The first cleavage is signal-dependent and removes the lumenal domain of the substrate (*orange*). The RIP protease (*red*) then cuts the substrate in the membrane-spanning region only after the first cleavage event and releases the biologically active domain of the substrate into the cytoplasm

rhomboid serine proteases, and pre-senilin aspartate proteases [2]. Only S2P and rhomboid proteases have been found thus far in prokaryotes.

Among the first examples of regulatory proteolysis involving RIP was the SREBP pathway that controls lipid metabolism in animals and a step in the regulatory cascade that controls sporulation in *Bacillus subtilis* [1]. SREBP is a transcription factor that activates the expression of genes required for lipid synthesis and uptake. It is synthesized as a membrane-bound precursor that is localized to the endoplasmic reticulum when sterols are present [4]. The two proteases, site-1 protease (S1P) and S2P, are also integral membrane proteins localized in the Golgi apparatus, and therefore unable to access SREBP. When sterols are depleted, SREBP translocates to the Golgi, where it is cleaved first by S1P, a subtilisin-like serine protease [5]. This cleavage creates a substrate for S2P, a zinc metalloprotease, that cleaves SREBP in the first transmembrane segment, releasing the amino terminal transcription factor domain into the cytoplasm where it rapidly translocates into the nucleus [5].

In *B. subtilis*, an analogous proteolytic system regulates the activity of the transcription factor σ^K, which controls genes required for forespore development. σ^K is synthesized in an inactive, membrane-anchored form, similar to SREBP [6]. At the appropriate time during sporulation, pro-σ^K is cleaved within the membrane-spanning region by SpoIVFB, an ortholog of mammalian S2P, releasing σ^K to direct transcription [6]. Just as S2P can only cleave SREBP after S1P acts, cleavage of pro-σ^K is reliant on an upstream proteolytic event. Two proteins, SpoIVFA and BofA, bind to SpoIVFB and prevent it from cleaving pro-σ^K [7, 8]. Signals from the developing forespore lead to production of the serine proteases, SpoVB and CtpB, which cleave SpoIVFA, alleviating inhibition of SpoIVFB [9, 10] (for further description of this system, see also [11]). The overall similarities of the design of these two proteolytic systems, the polytopic membrane-bound metalloprotease S2P and their signal-dependent cleavage of a membrane-bound substrate to release a biologically active regulatory protein, led to the realization that this framework is a conserved solution for transmembrane signaling. Diverse proteolytic regulatory systems with this overall design have since been found in all domains of life.

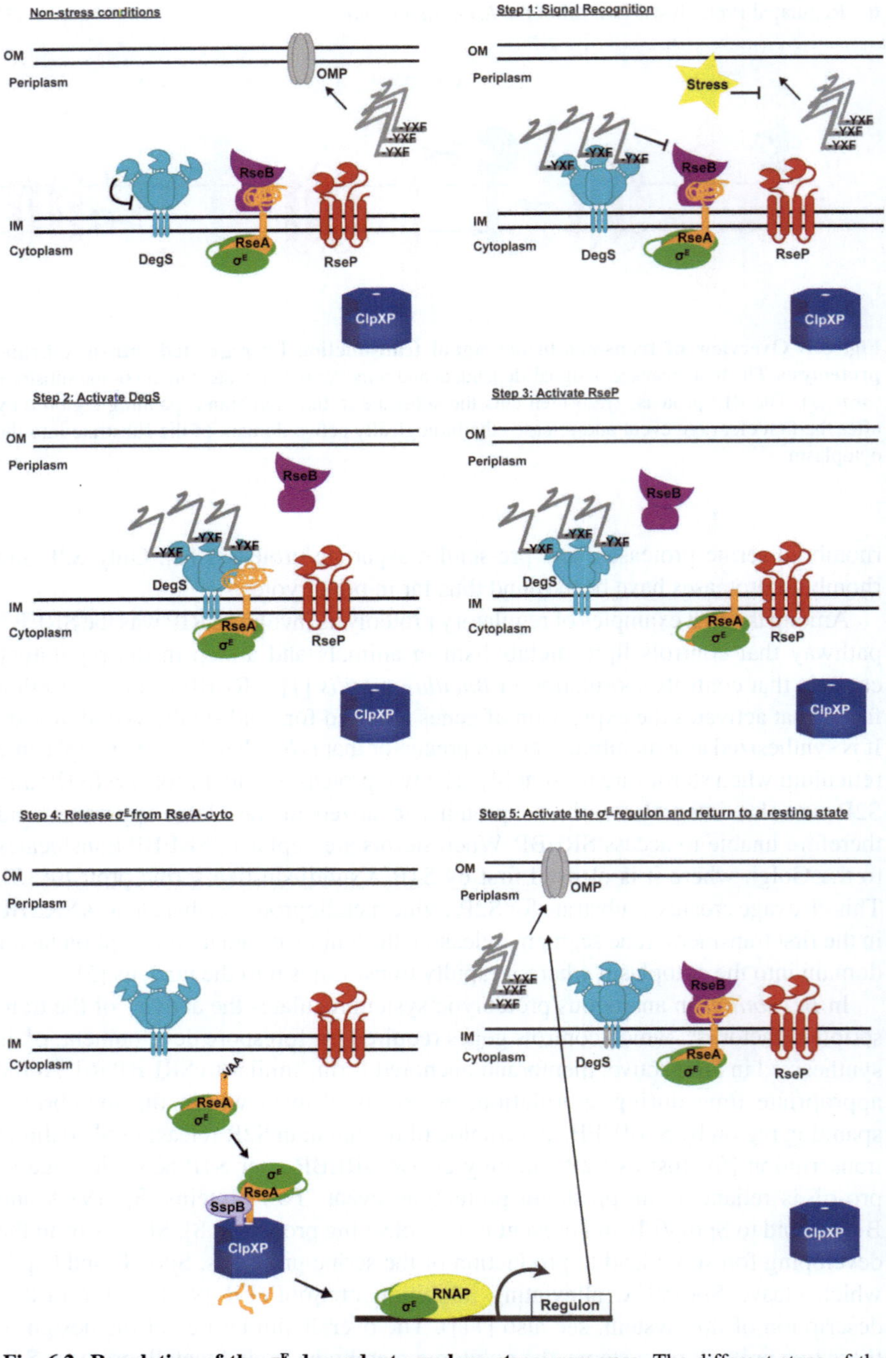

Fig. 6.2 Regulation of the σ^E-dependent envelope stress response. The different steps of the proteolytic cascade regulating σ^E are outlined. *Non-stress conditions.* In the absence of envelope stress, OMPs fold efficiently, and the low steady-state level of proteolysis of RseA provides the cell with sufficient free σ^E to support viability. *Step 1: Signal recognition.* Stress conditions that disrupt OMP folding lead to the accumulation of unfolded OMPs with exposed C-termini that bind to the PDZ domain of DegS. *Step 2: Activate DegS.* Peptide binding to the DegS PDZ domains activates DegS and an unknown signal alleviates inhibition of proteolysis by RseB. DegS cleaves RseA

In this chapter, we describe one of the best-studied systems that utilizes RIP, the proteolytic cascade that governs the σ^E-dependent cell envelope stress response in *E. coli* (Fig. 6.2). The key players in the σ^E system are the transcription factor, σ^E, its regulators, RseA and RseB, and the proteases, DegS, RseP, and ClpXP. The response is activated by stresses that affect the integrity of the outer compartment of the bacterium, the cell envelope [12, 13]. The proteolytic cascade serves to communicate this information across the inner membrane to σ^E in the cytoplasm. The inner membrane proteases, DegS and RseP, correspond to S1P and S2P, respectively [14, 15]. Their target is another inner membrane protein, RseA (Regulator of SigmaE), which is an anti-sigma factor that binds tightly to σ^E sequestering it at the membrane [13, 16]. Signals that induce the pathway activate DegS to initiate the proteolytic cascade. The sequential action of DegS followed by RseP releases the cytoplasmic domain of RseA, still bound to σ^E, from the membrane [14, 15]. The adaptor protein, SspB, binds near the new C-terminus of RseA and delivers it to the cytoplasmic protease, ClpXP. ClpXP then degrades the remaining portion of RseA, freeing σ^E to bind to the RNA polymerase (RNAP) core enzyme [17]. Extensive biochemical, structural, and physiological studies on this system have provided a remarkably detailed understanding of the proteolytic pathway, revealing fundamental principles of regulatory proteolysis that are applicable to all such systems.

Cell Envelope Stress and σ^E

The cell envelope is the hallmark of Gram-negative bacteria. It is composed of the inner membrane, outer membrane, and periplasmic space between the membranes that contains a thin layer of peptidoglycan. This compartment not only is critical for the structural integrity of the cell, but also serves as the interface by which the bacterium interacts with its surroundings. The cell envelope is not a static structure, but is actively remodeled in response to changes in the environment [18]. Cell envelope stress responses have evolved to maintain the integrity of this important compartment, to assist in the elaboration of complex cell envelope structures such as pili and fimbriae, and to protect the envelope from damaging stresses [19, 20]. The σ^E-dependent envelope stress response is one of the key pathways that monitors the state of the cell envelope [21, 22].

removing the periplasmic domain. *Step 3: Activate RseP.* Cleavage of the periplasmic domain of RseA generates a substrate for RseP, and inhibitory interactions that keep RseP from cutting RseA are removed. RseP cleaves RseA in the TM domain to release the cytoplasmic domain of RseA bound to σ^E into the cytoplasm. *Step 4: Release σ^E from RseA-cyto.* Cleavage of RseA by RseP generates a fragment of RseA with recognition sequences for the adaptor protein SspB near the C-terminus and for ClpXP at the extreme C-terminus. ClpXP degrades RseA-cyto releasing σ^E to bind to core RNAP. *Step 5: Activate the σ^E regulon and return to a resting state.* σ^E transcribes the genes in its regulon that restore the proper folding environment in the periplasm. The concentration of unfolded OMPs with exposed C-termini decreases, RseA stability increases, and the system resets

In bacteria, promoter recognition is conferred by the sigma subunit of RNAP. While σ^{70} is the primary sigma subunit in the cell, σ^E is one of an array of alternative sigma factors that displace σ^{70} and direct RNAP to the appropriate promoters of genes required for response to that stress [23, 24]. As such, activation of σ^E rapidly reprograms transcription to focus on genes that allow the cell to cope with cell envelope stress. σ^E was first identified not for its role in extracytoplasmic stress, but for its role in transcribing the gene encoding another alternative sigma factor at high temperatures, the cytoplasmic heat shock factor σ^{32} [25]. For further details of the heat shock response refer to [26]. A connection between σ^E and the cell envelope was found in genetic screens demonstrating that σ^E activity increased following overproduction of outer membrane porins (OMPs), inactivation of genes encoding periplasmic chaperones and proteases, and deletion of genes involved in lipopolysaccharide biosynthesis [27–30]. These findings, along with work on the CpxAR two-component system, were among the first pieces of evidence indicating that Gram-negative bacteria use distinct stress response pathways to combat damage in the cell envelope and the cytoplasm [31–33]. Compartmentalization of stress responses in *E. coli* was strikingly demonstrated for the σ^E system by the observation that accumulation of the porin OmpX in the cytoplasm due to disruption of its signal sequence resulted in activation of σ^{32}, but not σ^E, while overproduction of OmpX with an intact signal sequence activated σ^E, but not σ^{32} (in this case the post-transcriptional regulation of σ^{32} prevents its activation, despite increased transcription by σ^E) [34].

σ^E systems have now been identified in at least 112 sequenced bacterial genomes and, where investigated, share a number of basic properties [35]. The gene encoding σ^E, *rpoE*, is essential in *E. coli* and *Yersinia* spp., and *Vibrio cholerae rpoE* mutants rapidly accumulate suppressor mutations suggesting that *rpoE* is also essential in *V. cholerae* [21, 36, 37]. In other bacteria, such as *Salmonella enterica* serovar typhimurium and *Bordetella bronchiseptica*, the σ^E system is important for interactions with the host immune system during infection [38, 39, 40]. Despite differences in the responses to specific stress conditions across species, in all bacteria where the σ^E system has been studied in any detail, it has been found to be involved in cell envelope-associated processes [20, 24, 35, 41].

The major group of conserved genes in the σ^E regulon encode a series of proteins, including chaperones and proteases, that are central to the synthesis, assembly, and maintenance of OMPs and LPS [42]. In addition to the proteins that serve to fold or degrade misfolded OMPs, σ^E regulates the expression of several sRNAs that target mRNAs encoding OMPs for degradation [43–45]. Unfolded OMPs, which can be toxic when they accumulate in the cell, serve as a barometer for the overall state of the cell envelope [13]. Their proper folding and assembly depends on lipoprotein and LPS biosynthesis, as well as chaperones that prevent aggregation and escort unfolded OMPs across the periplasm and the Bam complex that assembles OMPs in the outer membrane [18, 46, 47]. Therefore, disruption of any component of the cell envelope that hinders OMP folding activates the σ^E response, which, in turn, increases the levels of proteins important for both OMP and LPS synthesis, while decreasing the load on the envelope by reducing *de novo* synthesis of OMPs via the sRNA regulators. In addition to the aforementioned regulon members, σ^E

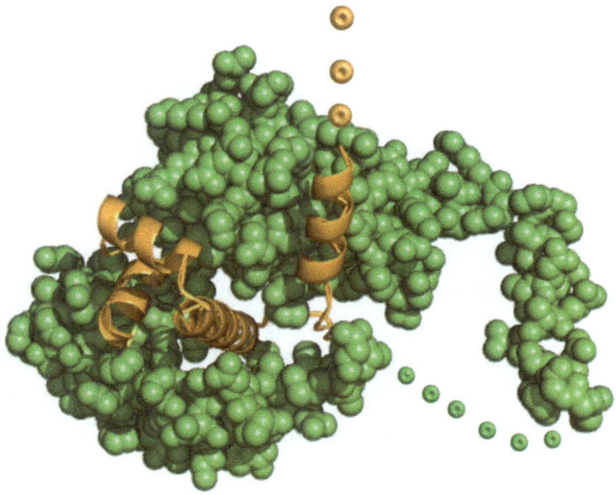

Fig. 6.3 σ^E bound to the cytoplasmic domain of RseA. Ribbon representation of the cytoplasmic domain of RseA (*orange*) embedded between conserved regions 2 and 4 of σ^E (*green*), shown in space-filling mode (1OR7 [54], generated using PyMol [55]). The surfaces used by σ^E to contact core RNAP are buried in the interface of the complex. *Green dots* represent the linker between regions 2 and 4 of σ^E, which was not ordered in the crystal structure. *Orange dots* represent the residues of RseA that lead to the transmembrane domain in the cytoplasmic membrane

transcribes its own gene in most bacteria in which it is found [29, 37, 48–51]. Activation of σ^E, therefore, results in an autoregulatory loop ensuring that σ^E continues to be made as long as the inducing stress remains.

Regulation of σ^E Activity

σ^E activity is controlled by two proteins, RseA and RseB, which are encoded in an operon together with the gene encoding σ^E [52, 53]. This operon structure is widely conserved amongst σ^E orthologs [35], suggesting that the regulatory pathway is also conserved. RseA is the central player in the regulatory system that controls σ^E activity. It is a single-pass TM protein located in the inner membrane. The cytoplasmic domain is a σ^E-specific anti-sigma factor [53, 54]. Early work on the σ^E system found that deletion of *rseA* resulted in constitutively elevated σ^E activity and rendered σ^E insensitive to signals in the cell envelope [52]. These data provided strong evidence that RseA forms the critical link between events in the cell envelope and σ^E in the cytoplasm. In addition, overexpression of the cytoplasmic domain of RseA in a Δ*rseA* mutant greatly reduced σ^E activity and did not restore the response to envelope stress, demonstrating that the inducing signal is generated in the periplasm [53]. The structure of the cytoplasmic domain of RseA in complex with σ^E revealed the molecular basis of RseA's anti-sigma factor activity (Fig. 6.3) [54]. RseA forms a compact helical structure that is sandwiched between the two conserved domains of σ^E that are responsible for promoter recognition and binding to RNAP core enzyme [54, 56]. Biochemical studies complement the structural data, and show that RseA is a strong competitive inhibitor of core RNAP for binding to σ^E. In fact, the K_d of the RseA:σ^E complex is estimated to be <10 pM, compared to ~1 nM for σ^E binding to core RNAP [57, 58].

RseB is the second key regulator of σ^E. It is a soluble periplasmic protein and binds to the periplasmic domain of RseA [52, 53]. RseB does not regulate σ^E in the absence of RseA [52, 53]. Deletion of *rseB* results in a modest two- to three-fold increase in σ^E activity, and σ^E activity is still induced in response to envelope stress in cells lacking *rseB*, suggesting that RseB fine-tunes the response [52, 59]. In contrast, *in vitro* experiments suggested that RseB plays a greater role in regulating the pathway, because RseB protects RseA from DegS-dependent cleavage in a reconstituted purified system, even in the presence of inducing signals [60]. Recent work indicates that inactivation of RseB and activation of DegS are both required before the proteolytic cascade can begin [61]. Therefore, as described below, it is now thought that RseB plays a major role in maintaining the uninduced state of the system.

The Proteolytic Cascade

The discovery that the signal transduction pathway is controlled by proteolysis was uncovered through a series of experiments establishing that RseA is an unstable protein whose half-life in the cell is correlated with σ^E activity [62, 63]. The half-life of RseA decreased under conditions of envelope stress, elevated temperature and over-expression of outer membrane proteins, when σ^E activity was high [62, 63]. The half-life increased in strains lacking the regulator OmpR, when σ^E activity was low ([64] and S.E. Ades unpublished observations). RseA was degraded, presumably to completion, since no fragments were observed by Western blotting with antibodies raised against the periplasmic or cytoplasmic domains [62]. Once the stress was removed, the system was reset, *i.e.* the stability of RseA returned to that in the absence of stress [63]. A survey of strains lacking periplasmic and inner membrane proteases revealed that RseA was stable, and σ^E was no longer inducible in a strain in which the gene encoding the inner membrane serine protease, DegS, was inactivated [62]. Genetic studies demonstrated that DegS is encoded by an essential gene whose function is to degrade RseA so that sufficient σ^E will be available to support viability [64]. Strains lacking DegS accumulated suppressor mutations that could also suppress the requirement of σ^E for viability, providing further evidence of the close connection between DegS and σ^E [64].

Participation of a second protease in the pathway was found in studies on the protein YaeL, later termed RseP (Regulator of SigmaE, Protease), which was first identified as an S2P zinc metalloprotease ortholog and shown to be an integral inner membrane protein [65]. Like DegS, RseP is encoded by an essential gene [14].

Genetic studies identified *rpoE* as a multicopy suppressor that allowed *E. coli* to grow in the absence of *rseP*, establishing a genetic connection between RseP and the σ^E pathway [15]. Both the Ito and Gross groups demonstrated that DegS and RseP worked in tandem to cleave RseA in a scheme analogous to the SREBP system [14, 15]. DegS cuts first and releases most of the periplasmic domain of RseA (Fig. 6.2) [14]. The remaining fragment of RseA (residues 1-148) stays in the inner membrane and retains anti-sigma factor activity [60]. This fragment is a substrate for RseP, which cuts RseA in the transmembrane region, releasing the cytoplasmic domain of RseA (residues 1-108), still bound to σ^E, into the cytoplasm (Fig. 6.2) [14, 15, 17, 65]. The protease(s) responsible for degradation of the periplasmic and inner membrane fragments of RseA have yet to be identified.

The final step in the proteolytic pathway that completes the degradation of RseA (Fig. 6.2) was found not from studies of the σ^E pathway, but from a proteomic analysis of substrates of the cytoplasmic protease, ClpXP [17, 66]. The cytoplasmic domain of RseA was identified as one several proteins trapped in the cavity of a catalytically inactive ClpP variant. RseP cleavage exposes recognition signals for ClpXP at the C-terminus of the soluble RseA fragment ($RseA^{1-108}$), released from the membrane [66, 67]. *In vitro* and *in vivo* experiments verified that ClpXP degraded $RseA^{1-108}$, but not σ^E, thereby freeing σ^E to bind core RNAP and transcribe the genes in its regulon to combat cell envelope stress [17].

Thus, three proteases are required for the complete degradation of RseA and initiation of the envelope stress response. DegS is the sole protease to sense the inducing signal and each cleavage event generates a substrate for the next protease in the proteolytic cascade [57]. As a result, σ^E is both released from RseA and prevented from re-binding to RseA, so that it is free to bind core RNAP and transcribe the genes in its regulon. DegS cleavage of RseA is the rate-limiting step in the complete degradation of RseA, which specifically tunes the system to the folding state of OMPs [57]. In the sections below, we outline in detail the structural, biochemical, and physiological details of the proteolytic pathway that controls σ^E activity.

Activating σ^E via Regulated Proteolysis

Step 1: Signal Recognition

The proteolytic pathway that activates σ^E has many built-in checkpoints to ensure that σ^E is properly regulated, only activated when necessary, and only to the extent required. The system is held in the "off" state with minimal signal-independent proteolysis by multiple inhibitory interactions that control the activity of DegS and RseP [59, 61, 68–70]. These inhibitory interactions are alleviated as the proteolytic cascade progresses, resulting in complete degradation of RseA.

Two independent signal recognition events are currently known to be required to initiate the proteolytic cascade: (1) DegS, which exists primarily in an inactive conformation, must be activated, and (2) RseB, which protects RseA from proteolysis, must be inactivated (Fig. 6.2, top row). We are now just beginning to understand how RseB activity is modulated. In contrast, structural and biochemical studies have yielded a wealth of information about how DegS activity is regulated.

DegS is a member of the HtrA family of serine proteases in which the protease domain is followed by one or more peptide-binding PDZ domains [71]. DegS is anchored in the inner membrane by a single transmembrane helix and the majority of the protein, including the active site and its single PDZ domain, project from the membrane into the periplasm [72, 73]. DegS lacking the transmembrane domain can be expressed as a soluble protein that retains the same properties as the intact protein in a purified *in vitro* system [70]. However, DegS is inactive *in vivo* without the transmembrane region, suggesting that it must be localized to the inner membrane near RseA to function in the cell [64].

A survey of peptides that bind to the DegS PDZ domain revealed that DegS preferentially bound to peptides with the C-terminal sequence, YxF (where x = any amino acid). *In vitro* experiments demonstrated that binding of these peptides converted DegS from a proteolytically inactive state into an active state [70]. These findings provided the critical piece of information to explain how envelope stress is sensed by the σ^E pathway. The YxF tripeptide motif is found at the extreme C-terminus of many of the major OMPs in *E. coli* [70]. These C-terminal residues are buried between beta strands in the correctly folded beta-barrel structure of OMPs and are inaccessible in the properly folded protein [74]. However, when OMP folding is disrupted, the residues are exposed.

OMP folding and insertion into the outer membrane is a complex process [47, 75]. OMPs are translocated as unfolded polypeptides from the cytoplasm into the periplasm via the Sec machinery. Following secretion, periplasmic chaperones bind to the unfolded OMPs to prevent their aggregation in the periplasm. The chaperones then deliver the OMPs to the Bam complex in the outer membrane, which assists in folding and assembly of the properly folded OMPs in the membrane [46, 47]. If any of these steps are disrupted or if the system is overwhelmed, folding intermediates containing exposed C-terminal residues accumulate. The majority of the known inducers of the σ^E pathway have the potential to disrupt the proper folding of OMPs and include conditions that stress both the cytoplasm and cell envelope, such as heat and addition of ethanol, as well as those that specifically stress the cell envelope, such as deletion of periplasmic folding catalysts and chaperones, deletion of genes required for proper elaboration of LPS, and overexpression of outer membrane proteins [27, 28, 50, 53]. Most of the folding catalysts, chaperones, and members of the Bam complex are encoded by genes in the σ^E regulon [42, 76]. As such, problems associated with OMP maturation, due to cell envelope conditions that lead to overload or failure of the folding and assembly pathway, trigger increased expression via σ^E of the very proteins needed to restore the flux of OMPs to the outer membrane.

In the second event required to induce the σ^E response, RseB must be inactivated. It was originally proposed that unfolded proteins in the periplasm competed with RseA for binding to RseB and titrated RseB away from RseA when they accumulated [77]. However, this model is not well supported either *in vivo* or *in vitro* [59]. Recent work suggests that, in addition to activating DegS, unassembled OMPs

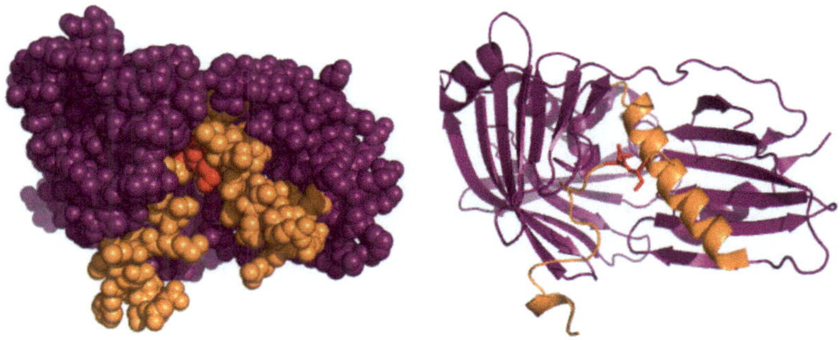

Fig. 6.4 RseB bound to the periplasmic domain of RseA. Space filling (*left*) and ribbon (*right*) representations of RseB (*purple*) bound to the periplasmic domain of RseA (*orange*) are shown (3M4W [79], generated using PyMOL [55]). Two regions of RseA were ordered in the crystal structure. $RseA^{169\text{-}190}$ forms a helix and binds to the smaller domain of RseB. $RseA^{132\text{-}159}$ includes the cleavage site and binds in the cleft between the two domains of RseB. The peptide bond cleaved by DegS, Val148-Ser149 (side chains shown in *red*), is almost completely occluded

antagonize inhibition by RseB through different parts of their C-termini [61]. Residues approximately 10–20 amino acids upstream of the C-terminal YxF motif were found to be required for maximal activation of the response *in vivo* in the presence of RseB, but not in its absence. In addition, sequences in this region from different OMPs activated the response to slightly different extents, again only in the presence of RseB. These data led to the model that residues upstream of the YxF motif are specifically required to antagonize RseB, not to activate DegS [61]. However, the peptides did not alleviate inhibition *in vitro*, indicating that another component is required to antagonize RseB [61]. Therefore, currently, the precise mechanism for release of RseB inhibition remains unclear. Based on structural homology between RseB and the lipid binding domains of LolA, LolB, and LppX [60, 61, 78], it has been proposed that the second signal required for activation of the response is either a lipid, free lipoprotein, or LPS that has not been correctly delivered to the outer membrane [61]. This model, however, has yet to be tested experimentally. If the inducing signal for RseB proves to be a lipophilic molecule, then DegS and RseB integrate distinct signals from the cell envelope to control the stress response.

The crystal structure of RseB bound to the periplasmic domain of RseA ($RseA_{peri}$) revealed how RseB protects RseA from proteolysis by DegS (Fig. 6.4) [79]. RseB consists of two domains, a smaller C-terminal domain and a larger N-terminal domain, which has homology to lipoprotein-binding proteins [78–80]. $RseA_{peri}$ is largely unstructured, with the exception of two regions encompassing residues 132-151 and 169-190 [79]. The structural observations are consistent with previous experiments indicating that $RseA_{peri}$ assumed a molten globule-like conformation [70]. In biochemical studies, $RseA^{169\text{-}190}$ was found to be necessary and sufficient for binding to RseB [60]. In the $RseA_{peri}$:RseB complex, these residues form a helical structure that binds to the smaller domain of RseB (Fig. 6.4). The other structured region of RseA in the complex, $RseA^{132\text{-}151}$, includes the site where RseA is cleaved

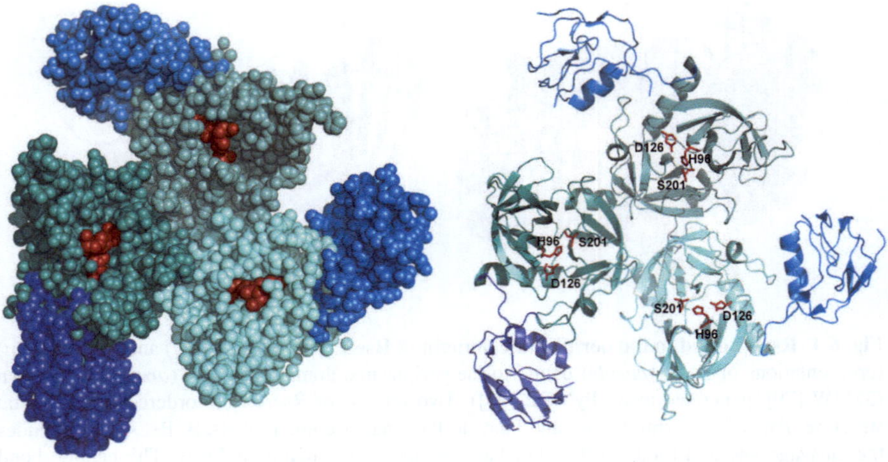

Fig. 6.5 DegS trimer. Space-filling (*left*) and ribbon (*right*) representations of the DegS trimer in the inactive state are shown viewed from the *top* (1SOT [72], generated using PyMOL [55]). The cytoplasmic membrane is *below* the molecule. The active site residues, His96, Asp126, and Ser201 (side chains shown in *red*) are accessible from the periplasm. The PDZ domains that are arranged around the perimeter of the trimer are shown in *blue*. The protease domains of each trimer are in different shades of *green*

by DegS [79]. These residues bind in the cleft between the two domains of RseB, largely burying the cleavage site, suggesting that RseB prevents cleavage of RseA by blocking access to the sessile bond (Fig. 6.4). RseB also protects RseA from proteolysis by RseP in a process that is less well understood and involves the PDZ domains of RseP [59].

Step 2: Activate DegS

DegS is the gate-keeper of the σ^E response (Fig. 6.2, middle left). Once it cleaves RseA, all the subsequent cleavage events occur in a signal-independent manner and with kinetics that are faster than the initial cleavage by DegS [57]. Therefore, the level of DegS activity, which is set by the amount and identity of the unfolded OMPs, determines how much σ^E is released from RseA and the extent of the response. Structural and biochemical studies have provided amazingly detailed views of DegS and how it functions.

Given the destructive nature of proteases, including DegS, their activity must be controlled to prevent rampant degradation of cellular proteins. Like other serine proteases, DegS is held in an inactive state until a specific activating event occurs [3]. In structures of DegS, the protease domains form a funnel-shaped trimer with the PDZ domains decorating the edges (Fig. 6.5). Although the active sites are fully

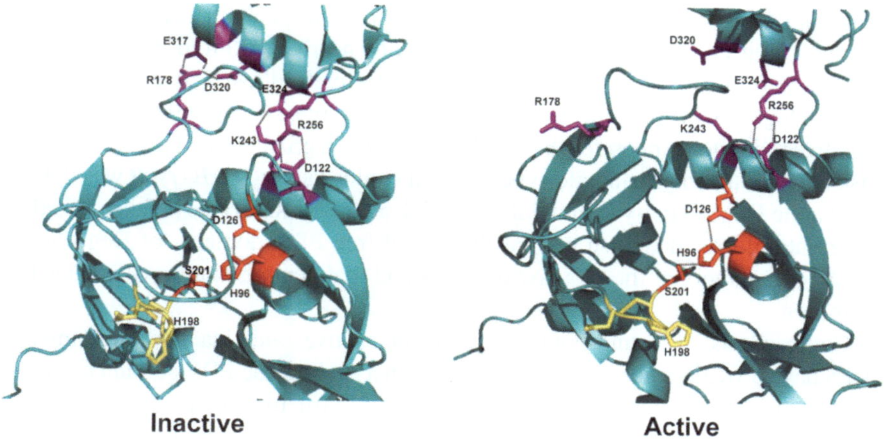

Fig. 6.6 Free and peptide-bound DegS monomer. *Left*, Ribbon representation of the inactive, peptide-free monomer of DegS (1TE0 [73], generated using PyMOL [55]). *Right*, Ribbon representation of the active, peptide-bound monomer of DegS (1SOZ [72], generated using PyMOL [55]). The protease domain and portions of the PDZ domain that form key interdomain salt bridges are shown. Side chains of the amino acids forming salt bridges (Arg178-Glu317/Asp320, Lys243-Glu324, and Asp122-Arg256) between the protease and PDZ domains are in *purple*, and salt bridges are indicated by *dashed lines*. Residues participating in the formation of the oxyanion hole are in *yellow* (His198, Gly199, Asn200, and Ser201). Side chains of the active site catalytic triad (His96, Asp126, Ser201) are in red. Hydrogen bonds between side chains in the active site are indicated by *dashed lines*

exposed on the inner surfaces of the trimer facing the periplasm, DegS has extremely low activity in the absence of inducing peptide [70, 72, 73]. The crystal structures of ligand-free DegS provide a ready explanation for this apparent dichotomy. The active site Ser-His-Asp triad is not appropriately aligned for catalysis, and the oxyanion hole is not in the proper conformation to form the requisite hydrogen bonds needed to stabilize the tetrahedral intermediate of the peptide cleavage reaction (Fig. 6.6) [72, 73]. Therefore, although the active sites are exposed, they can do no damage in this inactive conformation. In contrast, structures of the active form of the enzyme (in complex with inducing peptides), illustrate that the catalytic triad moves to the appropriate position for catalysis (Fig. 6.6) [72, 81]. The oxyanion hole is also properly formed due to rotation of His198 (Fig. 6.6), which repositions the backbone amide so that it can form a hydrogen bond with the carbonyl group of the sessile peptide bond [72, 73, 81, 82].

Role of the PDZ Domains

As with many other members of the HtrA family of proteases, the PDZ domains of DegS regulate its proteolytic activity [71]. In a purified system, DegS cleaved RseA very slowly in the absence of peptide [81]. Peptides that bound to the PDZ domain dramatically increased the rate of cleavage, with the YYF tripeptide being the strongest inducer, increasing the rate of cleavage by nearly 1,000-fold [82]. In contrast, DegS lacking the PDZ domain was only four to fivefold less active than peptide-bound DegS (~200-fold more active than ligand-free DegS) and was no longer sensitive to inducing peptides [81, 83, 84]. *In vivo*, DegSΔPDZ had a 12-fold higher basal level of activity than wild-type DegS in strains lacking *rseB* (*rseB* was deleted to separate DegS activity from inhibition of cleavage by RseB) [61]. Taken together, these data clearly demonstrate that the ligand-free PDZ domains of DegS act as negative regulators of the protease domains and that peptide binding alleviates this inhibition.

Comparisons of ligand-bound, ligand-free, active, and inactive structures of DegS have provided much detailed information as to how the PDZ domain regulates DegS activity and how peptide binding leads to structural changes required for proteolysis. The interface between the PDZ and protease domains appears to be very flexible, particularly in the peptide-bound forms of DegS [83]. Therefore, detailed biochemical experiments using specific variants of DegS have been critical in determining which residues, predicted to form key contacts based on structural data, are important for regulated proteolysis.

In the ligand-free state, the PDZ domain stabilizes the inactive conformation of DegS [72, 81, 83, 85]. The PDZ domain is anchored to the protease domain through a series of interactions that include three salt bridge pairs across the two domains: Asp122-Arg256, Arg178-Glu317/Asp320, and Lys243-Glu324 (Fig. 6.6) [72, 81]. Two of these three salt bridges are absent in the peptide-bound DegS structures (Fig. 6.6), and the PDZ domains appear to be more mobile (relative to the protease domain) in the ligand-bound enzyme, suggesting that the PDZ domains are no longer tightly associated with the protease domains in the active enzyme [72, 81, 85]. Consistent with a role for the PDZ domain in stabilizing the inactive form of the enzyme, disruption of the salt bridges increased DegS activity [61, 81]. In addition to holding the enzyme in an inactive conformation, two of the amino acids (Arg178 and Asp122) that form salt bridges with the PDZ domain also participate in interactions that stabilize the active site when peptide is bound [72, 81, 84]. In particular, Arg178 appears to be a key residue in the transition. In the inactive enzyme, Arg178 forms a salt bridge with a pair of amino acids in the PDZ domain (Fig. 6.6) [72, 73]. In the ligand-bound form of the enzyme, it rotates 90° and makes a different set of hydrogen bonds that link it to the functional conformation of the oxyanion hole [72, 81, 82, 84]. Asp122 is part of a hydrogen-bonding network that includes the peptide backbone adjacent to the catalytic site and stabilizes the active conformation in the ligand-bound enzyme [72, 81]. Other amino acids also change conformation in the ligand-bound structure and help to promote the active conformation. These residues include, among others, His198, mentioned above, and Tyr162, which also moves allowing the formation of hydrogen bonds that stabilize the active conformation of the peptide backbone around His198 and the oxyanion hole [72].

Mechanism of Activation

Two models have been proposed to explain how peptide binding leads to enzyme activation. The first model, called the scaffolding or peptide-activation model, proposes that the penultimate residue of the activating peptide contacts the protease domain and directly participates in a network of interactions that serve to remodel the active site [72, 81]. This model is based on crystallographic evidence showing that this amino acid interacts with the L3 loop in the protease domain, reorienting the stem of the loop through a series of interactions that ultimately stabilize the active conformation of the catalytic site [72, 81]. Since DegS can be activated by peptides with several different amino acids of chemically diverse nature at the penultimate position, proponents of this model propose that each peptide forms a slightly different set of interactions with the L3 loop to accommodate the different amino acids [85]. Each of these interactions ultimately alters the conformation of the L3 loop in a way that leads to stabilization of the active enzyme. As such, the L3 loop acts as a sensor of peptide binding via contacts with the inducing peptide bound to the PDZ domain. The PDZ domains in the scaffolding model not only stabilize the inactive state, but are also required for the transition to the active state because they position the peptide to interact with the L3 loop [71, 72, 81].

Several observations oppose the scaffolding model. The finding that the active form of the enzyme could be obtained by deleting the PDZ domain altogether suggests that the PDZ domain is not needed to form or stabilize the active conformation [81, 84]. Additionally, peptides that vary only at the penultimate residue activated DegS to nearly the same extent under saturating concentrations [81, 82]. More variation would be expected because different amino acids at the penultimate position must form different contacts with the L3 loop that have different energies [81, 82]. Substitutions at other positions in inducing peptides actually caused larger changes in activation than those at the penultimate positions [81–83]. Finally, in the different structures of DegS bound to various different peptides, the orientation of the PDZ domains varied substantially even within the same trimer, and no contacts were seen between the peptide and the protease domain in many of the structures [83].

The second model, called the relief of inhibition model, proposes that the PDZ domain holds the protease domain in the inactive conformation [81]. Peptide binding leads to an allosteric rearrangement that relieves the inhibitory interactions and stabilizes the active conformation of the protease domain. However, the PDZ domain itself is not required to stabilize the active enzyme. This model is supported by detailed biochemical analyses of peptide activation and the kinetics of proteolysis by DegS using an optimized *in vitro* degradation assay with purified components and is coupled with crystallographic studies of many DegS variants [81–84]. At this point, there are over 20 different structures of wild-type DegS and DegS variants that provide many views of the protein. The main caveat of this model is that it is based on structural and biochemical studies with mutant enzymes, and these mutations may introduce changes that are not representative of the wild-type enzyme [71].

Positive Cooperativity

A wide range of peptides ending in the YxF motif can activate DegS, which is likely to ensure that the system can monitor the folding of different OMPs in the cell [70]. However, not all peptides are equivalent activators of DegS. Peptides differ in the maximal extent of activation (V_{max}) and the concentration required for half-maximal activation (K_{act}) [81, 82]. In addition, the V_{max} and concentration dependency of activation are not correlated, so that a peptide that activates DegS to a lesser extent than another peptide may do so at a lower concentration [81, 82]. Despite differences in their kinetic parameters, all peptides exhibit positively cooperative activation of DegS [81, 82]. In other words, peptide binding to a DegS protomer facilitates binding of peptides to the remaining subunits of the trimer resulting in a sensitive and rapid switch from the inactive to the active form of the enzyme. In addition to exhibiting positive cooperativity with respect to peptide binding, DegS also exhibits positive cooperativity in substrate degradation [81, 84]. When the concentration of RseA was varied in the presence of saturating peptide, the Michaelis-Menton plot was sigmoidal with a Hill constant >1 [81, 84]. Therefore, RseA facilitates its own degradation. Positive cooperativity in RseA degradation was also seen with the DegSΔPDZ variant, indicating that allosteric regulation of DegS is not confined solely to the PDZ domain, but is also an inherent property of the protease domain [84].

Allosteric Regulation- The MWC Model

The presence of interactions that stabilize the inactive form of an enzyme, combined with positively cooperative ligand binding, are hallmarks of the Monod Wyman Changot (MWC) model of allostery (Fig. 6.7) [86]. This model has been used to explain allosteric regulation of diverse proteins ranging from enzymes such as aspartate decarbamylase to hemoglobin to G-protein coupled receptors [87]. The central tenet of this model is that the protein exists in two conformations, tense (inactive) and relaxed (active), which are in a dynamic equilibrium [86, 87]. The tense state is more stable, and predominates in the absence of ligand. Ligands bind preferentially to the relaxed form of the protein, thereby shifting the equilibrium toward the relaxed state (Fig. 6.7). Data from peptide activation experiments could be fit to the MWC model, and the model explained the variations in activating potential of different peptides [81–83]. Because peptides can bind to both tense and relaxed DegS, the difference in the affinity of a given peptide for each state determines the overall amount of DegS in the active form (Fig. 6.7). The stronger the preference for the relaxed state compared to the tense state, the stronger the cooperativity and the greater the extent of activation.

For the MWC model to be an accurate description of DegS, mutations that lower the energy barrier between the tense and relaxed states, shifting the equilibrium toward the relaxed state, should increase activity of the ligand-free enzyme and reduce cooperativity. These mutations could either destabilize the tense state or stabilize

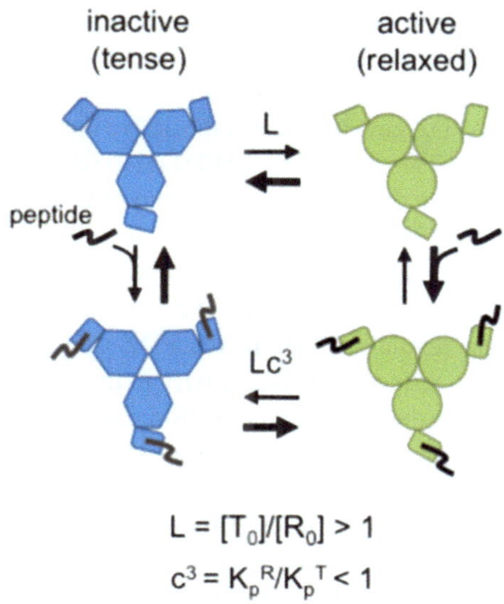

Fig. 6.7 MWC model for allosteric regulation of DegS by activating peptides. The cartoon depicts regulation of DegS by YxF peptides according to the MWC model of allostery [86]. The tense form (*blue* hexagons) of DegS is not proteolytically active and binds to inducing peptides with a lower affinity (K_p^T) than the proteolytically active relaxed form (*green circles*) of the enzyme (K_p^R). Without bound peptide, the ratio of the tense to relaxed forms of the enzyme (L) is greater than one, and the tense form of the enzyme predominates. The ratio of the two forms of DegS with bound peptide is given by Lc^3, where c is the ratio of the affinities of tense and relaxed DegS for peptide. Because c is less than one, Lc^3 is less than one and the relaxed active form predominates. The direction that is favored in the equilibria among the different states is shown with thick arrows. Only the fully peptide-bound forms of DegS are shown for simplicity

the relaxed state. For DegS, key candidates for interactions that stabilize the tense form are the salt bridges formed between the PDZ domain and protease domain in ligand-free DegS. Disruption of each of these salt bridges by mutation, significantly increased the basal rate of proteolysis, although not to the level of the fully ligand-bound enzyme indicating that other interactions remain that stabilize the tense state [81–83]. These mutations also reduced the cooperativity of both peptide activation and RseA degradation, providing strong evidence in support of the MWC model [81–83].

Although reorientation of the PDZ domains is critical for the allosteric activation of DegS, the protease domains themselves exhibit allostery. Cleavage of RseA by DegSΔPDZ is cooperative, suggesting that the tense state is still significantly populated and not all of the enzyme is in the relaxed form [84]. Further evidence that the inactive state is populated even in the absence of the PDZ domain comes from experiments with the H198P variant of DegS. This mutation eliminated nearly all of the cooperativity of RseA degradation by DegSΔPDZ and increased the activity of DegSΔPDZ, as it does in the full-length enzyme [84]. Crystal structures of both DegSΔPDZ(H198P) and full-length DegS(H198P) revealed that the proline makes packing interactions that should stabilize the active conformation of the oxyanion hole [81, 84]. These stabilizing interactions would shift the equilibrium to favor the active, relaxed form. Indeed, a fit of the data to the MWC model predicted that 1% of the wild-type DegSΔPDZ enzymes are in the active conformation in the absence of substrate, compared 87% of the DegSΔPDZ(H198P) variants [84]. One confounding piece of data is that the catalytic site is properly formed in crystal structures of wild-type DegSΔPDZ [81, 84]. However, it is believed that the active form is trapped by crystal packing conditions, given the biochemical data [84]. The cooperative substrate activation inherent in the protease domain may reflect an evolutionarily early form of allosteric regulation, before the PDZ domains were acquired, that provided a mechanism to reduce protease activity in the absence of substrate.

Taken together, what do these data and models mean? What benefit can be gained by having an allosteric system control the envelope stress response? Several answers to these questions have been proposed [81, 82, 84]. DegS is essential in *E. coli* because it must initiate degradation of RseA to release sufficient σ^E to maintain cell viability. Because DegS is in equilibrium between an active and an inactive form, a small number of DegS enzymes will assume the active conformation, even in the absence of inducing peptides. This small population of active enzyme, combined with cooperative binding by RseA itself, may ensure sufficient basal level cleavage of RseA to maintain viability [81, 82, 84]. The basal level of uninduced degradation may also be supported by a low level of peptide-induced degradation due to stochastic fluctuations in the OMP folding pathways that expose a small number of inducing peptides at any given time. In terms of activating the response, positive cooperativity ensures that during stress, the pathway can be rapidly activated over a narrow concentration range of inducers. Activation is also reversible, so that once the unfolded OMPs have been cleared, DegS will quickly return to the inactive state. Since different peptides activate the enzyme to different extents, it is tantalizing to speculate that the response is tuned to monitor the folding of different OMPs.

Step 3: Activate RseP

Release of the periplasmic domain of RseA by DegS generates a substrate for RseP, the next enzyme in the proteolytic cascade (Fig. 6.2, middle right). RseP is a zinc metalloprotease and belongs to the S2P group of RIP proteases that are found in a wide range of organisms [2, 88]. RseP is an inner membrane protein with four TM segments [65]. The active site is formed by the HExxH motif (where x = any amino acid) in TM1 and the LDG motif in TM3, which are conserved amongst S2P family members [65]. TM3 is also thought to be critical for substrate binding by RseP [89]. Two circularly permuted PDZ domains are located in the periplasmic domain between TM2 and TM3 [90, 91]. Because they are polytopic membrane proteins, the S2P family of proteases has been far more difficult to characterize biochemically and structurally than proteases such as DegS, which can be readily expressed as soluble active enzymes.

A major question that is relevant to all families of intramembrane proteases is how peptide bond cleavage takes place in the lipid environment of the membrane. Proteolysis is thought to occur through nucleophilic attack by a water molecule that is bound to the active site zinc and activated for peptide bond hydrolysis by the conserved glutamate of the HExxH motif [88]. Therefore, water must be able to access the active site of the enzyme. The structure of the catalytic core of a S2P family member from *Methanocaldococcus jannaschii* has been solved in the presence of detergents and provides a high-resolution view of how intramembrane peptide bond cleavage is thought to occur [92]. This S2P has six transmembrane segments, and the active site is positioned so that it lies within the plane of the membrane. The zinc ion is coordinated by the two histidine residues of the HExxH motif in TM2 and the aspartate of the LDG motif in TM4, as predicted from biochemical and genetic data [14, 64, 91]. The glutamate residue is also properly aligned for catalysis. A narrow channel lined with hydrophilic amino acids connects the active site to the cytosolic side of the membrane providing a way for water molecules to access the active site.

Although a structure of the catalytic domain of RseP is not yet available, biochemical experiments provide some insights. The environment of the active site of RseP was analyzed by determining the accessibility of a membrane-impermeable alkylating reagent to cysteine residues engineered in the active site of the enzyme [93]. In the native state (in membrane vesicles), the cysteine residues of the engineered RseP remained unmodified. However, when protein denaturant (guanidine HCl) was added, accessibility increased. Full modification of the cysteines only occurred when the membrane vesicles were completely solubilized using detergent and guanidine HCl. The increased accessibility of cysteines in the presence of guanidine HCl indicates that the active site sits within a proteinaceous structure in or closely associated with the membrane that can be accessed by the denaturant and at least partially unfolded [93]. These results, together with cleavage site studies using RseA and model substrates, suggest that the active site of RseP is found within a folded protein structure sequestered from the extramembrane environment and probably from the membrane lipids as well [93, 94].

Substrate Selectivity

In comparison to DegS (whose only known substrate is RseA), RseP appears to have a relatively broad substrate specificity. For example, RseP can cleave TM segments TM1 and TM5 from LacY and the signal sequence from beta-lactamase in model substrates that do not contain RseA-related sequences [94]. Experiments varying the sequence of target TM segments, including the TM of RseA, indicated that the major requirement for binding to, and efficient cleavage by, RseP was the presence of helix-destabilizing residues within the transmembrane region of substrates, as opposed to a sequence-specific recognition motif [89, 93]. Destabilizing residues in the TM helix may make the peptide backbone more accessible to RseP for hydrolysis compared to the peptide backbone of a stable helix that is fully engaged in hydrogen bonds. The ability of RseP to cleave substrates other than RseA and the relatively low sequence specificity suggest that RseP plays a role in the cell beyond the envelope stress response. Indeed, recent work indicates that RseP is also responsible for the degradation of signal peptides of secreted proteins, once they have been cleaved by signal protein peptidase [95]. A connection between the envelope stress response and degradation of signal peptides has not been established and this new activity may be an independent function of RseP.

Regulation of RseP

Similar to DegS, RseP cleaves full-length RseA very slowly [14, 15, 65, 90]. In contrast, the N-terminal fragment of RseA generated by DegS cleavage is rapidly cleaved by RseP [14, 15, 69]. This second cleavage event by RseP happens around three-fold faster than the initial cleavage by DegS, such that the signal-sensitive step is the rate-limiting step in the signaling pathway [57]. To further insulate the system from RseP-mediated uninduced degradation of RseA, RseP activity is blocked by a series of inhibitory interactions involving the PDZ domains of RseP, a Gln-rich region in the periplasmic domain of RseA, RseB, and DegS [59, 68, 69]. The mechanism by which these different factors act to restrain RseP is not fully understood, especially compared to the wealth of information about DegS. RseB and the Gln-rich regions of RseA, but not DegS, protect RseA from cleavage by RseP only when the PDZ domains of RseP are intact [59]. These data suggest that either RseB and/or RseA interact with the RseP PDZ domains. Structural studies indicate that the second PDZ domain of RseP can bind to the C-terminal amino acid of RseA$^{1\text{-}148}$, although binding is too weak to be detected with biochemical assays [91]. While the mechanism of inhibition remains unclear, the framework for how the inhibitory interactions are relieved is evident. The RseB binding sites and the Gln-rich regions of RseA are on the C-terminal side of the DegS cleavage site, so they will be removed when DegS acts [70]. How inhibition by DegS and the PDZ domains is alleviated is not as easily explained and is not yet known. DegS may sequester RseA from RseP or directly interact with RseP in an inhibitory manner.

Role of the PDZ Domains

The role of the PDZ domains in RseP remains an intriguing puzzle. Originally, RseP was predicted to have a single PDZ domain, but it was later shown through sequence alignments and crystallography to have two circularly permuted PDZ domains [68, 69, 88, 90, 91]. The RsePΔPDZ deletion used in the early studies was thought to have excised the single PDZ domain [68, 69], although in fact this deletion removed part of each of the two circularly permuted PDZ domains. Nevertheless, the deletion disrupted the native PDZ domains and resulted in DegS-independent cleavage of RseA that was no longer inhibited by RseB [59, 68, 69, 90]. These results support a model in which the PDZ domains block RseP function. Additional evidence supporting a regulatory function for the PDZ domains came from genetic studies isolating mutations in RseP that increase the basal level of σ^E activity *in vivo* [90]. Most of these mutations fell in the predicted peptide-binding regions of both PDZ domains, with the strongest mutations in the N-terminal PDZ domain, suggesting that it plays a critical role in regulating RseP. The mutations, which are predicted to disrupt ligand binding, did not increase the intrinsic proteolytic activity of RseP, but instead increased the basal level of σ^E activity by alleviating the requirement for upstream cleavage by DegS. Interestingly, variants of RseP lacking either PDZ domain did not degrade full-length RseA, although they could still cleave $RseA^{1\text{-}140}$, which lacks most of its periplasmic domain. Therefore, it appears that either PDZ domain can regulate RseP.

How do the RseP PDZ domains block protease activity? The most straightforward hypothesis is; that one (or both) of the PDZ domains bind to RseA to keep RseP in the "off" state. Although PDZ domains often bind to the C-termini of proteins, it is unlikely that the C-terminus of intact RseA is the ligand because RseP does not degrade RseA variants that contain different C-termini [69]. If a specific binding interaction between RseA and the PDZ domains were required for inhibition, then changes to the C-terminus should have abrogated the interaction and led to DegS-independent cleavage by RseP. Other experiments suggested that the C-terminal valine of $RseA^{1\text{-}148}$ was required to activate cleavage [91] because replacement of Val148 in $RseA^{1\text{-}148}$ with dissimilar amino acids reduced cleavage by RseP [91]. This model is attractive, since similar results have been found for the intramembrane protease of animals, γ-secretase, suggesting that this mechanism is evolutionarily conserved [96]. However, RseP will cleave a variety of model substrates and RseA fragments with different C-terminal amino acids, indicating that peptide binding to the PDZ domain is not a prerequisite for activity, or that the specificity of binding is quite broad [69, 94]. It is possible that RseA is a unique substrate for RseP and interacts with the protein somewhat differently than other substrates. Future experiments will surely clarify the role of the PDZ domains and illuminate the mechanistic underpinnings of the proteolytic activity of RseP and its regulation.

Step 4: Releasing σ^E from $RseA_{cyto}$

After RseP cleaves RseA, the remaining RseA fragment ($RseA^{1\text{-}108}$) is released from the membrane with σ^E still tightly bound [91]. The final step in the proteolytic cascade completes the degradation of RseA, releasing σ^E to bind to core RNAP (Fig. 6.2, bottom left). Because the interaction between σ^E and the cytoplasmic domain of RseA is extremely stable and the dissociation rate is extremely slow, proteolysis is the predominant mechanism to free σ^E [57]. The fragment of RseA remaining after RseP cleavage, $RseA^{1\text{-}108}$, contains the σ^E binding domain (residues 1-66), followed by residues that are not required for σ^E binding, but target the protein for degradation. $RseA^{1\text{-}108}$ terminates with the amino acid sequence VAA, which is a recognition sequence for ClpX, the ATP-dependent unfoldase of the ClpXP protease [17, 66]. Upstream of the ClpX binding site in RseA$^{1\text{-}108}$ is a binding site for the adaptor protein, SspB, which facilitates proteolysis by ClpXP [66]. Both *in vitro* and *in vivo* experiments demonstrated that RseA$^{1\text{-}108}$ is rapidly degraded by ClpXP and that its degradation is enhanced by the adaptor protein SspB [17] (see also [97]). Although ClpXP is the major protease for degradation of RseA$^{1\text{-}108}$, other cytoplasmic proteases also contribute to the degradation and hence to the release of σ^E [17, 57]. Thus the final step in the degradation of RseA is relatively non-specific, a marked contrast to the initial cleavage events that are wholly dependent on DegS and RseP. Presumably the redundancy in the final cytoplasmic degradation step is important to ensure that once σ^E is released, it will be free to direct transcription rather than rebind to the cytoplasmic domain of RseA.

Step 5: Activation of the σ^E Regulon and Return to a Resting State

When σ^E is released from RseA, it binds to core RNAP and transcribes the genes in its regulon. Included among these genes are chaperones and proteases that help refold or degrade aberrantly folded OMPs, and sRNAs that target OMP mRNAs for degradation [42–45]. Together, these regulon members serve to simultaneously restore the OMP folding pathway and prevent continued load on the system from newly synthesized proteins (Fig. 6.2, bottom right). As a result, the overall concentration of unfolded OMPs with exposed C-termini decreases and DegS returns to the inactive state, effectively shutting off the proteolytic pathway and the rapid degradation of RseA. Because the σ^E:RseA complex is extremely stable, σ^E will be bound by RseA and the response will quickly return to basal levels.

A Common System for Regulation of Membrane Localized Sigma/Anti-sigma Modules

σ^E belongs to a large group of sigma factors, the group 4 or extracytoplasmic function (ECF) sigma factors, that is widely distributed throughout the bacterial world [35, 56]. Many of these sigma factors are regulated by membrane-bound anti-sigma factors [35], and the regulatory proteolytic scheme used in the regulation of σ^E is emerging as a paradigm for the signal transduction pathways governing these systems. The proteolytic pathway controlling the σ^E-dependent stress response of *E. coli* is the best characterized of these signaling systems, especially at the structural and biochemical levels. Ongoing studies in other systems are also shedding light on the themes and variations associated with the regulatory proteolytic pathways.

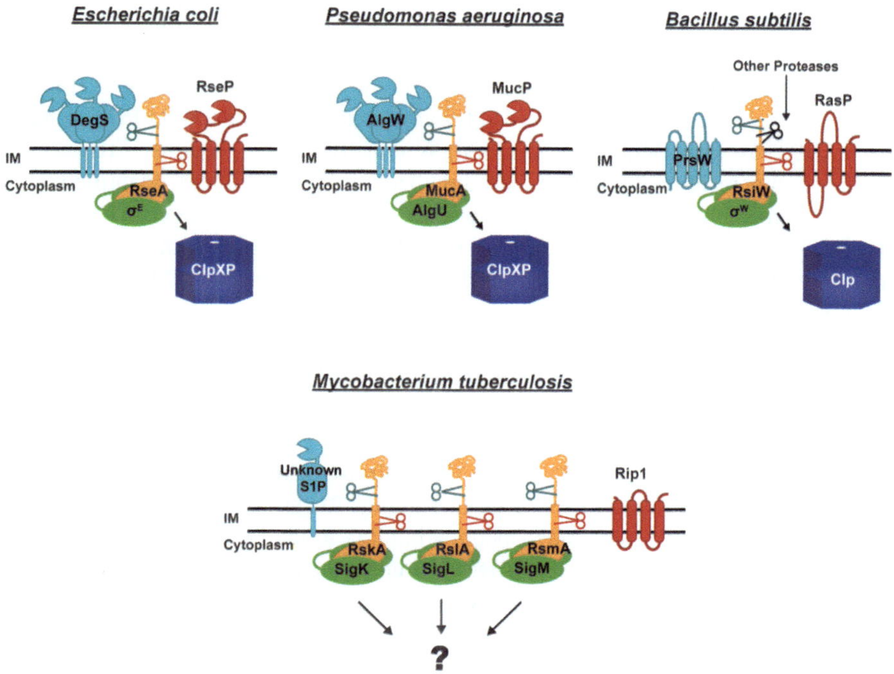

Fig. 6.8 Sequential proteolysis of transmembrane anti-sigma factors as a regulatory paradigm. The overall design of the regulated proteolytic pathways controlling σ^E in *E. coli*, AlgU in *P. aeruginosa*, σ^W in *B. subtilis*, and SigK, L, and M in *M. tuberculosis* is shown. The proteases on the *left* perform the first cleavage (*blue* scissors). The S2P proteases are on the *right* and perform the intramembrane cleavage step (*red* scissors)

Regulation of Pseudomonas aeruginosa AlgU

In *P. aeruginosa*, the proteolytic pathway governing AlgU is analogous to the σ^E pathway in *E. coli*. Work in this system provides instructive comparisons with the *E. coli* system [98]. AlgU, the σ^E ortholog in *P. aeruginosa*, mediates a cell envelope stress response and transcribes genes that control the expression of the exopolysaccharide, alginate [99]. The regulatory pathway controlling AlgU activity is very similar to that in *E. coli* (Fig. 6.8). AlgU activity is inhibited by the RseA-like anti-sigma factor MucA [100]. MucA is degraded by a proteolytic cascade that is initiated by AlgW, a DegS homologue [98, 101, 102]. Following AlgW-dependent cleavage, the S2P, MucP, cleaves MucA in its transmembrane domain [101]. The cytoplasmic domain of MucA bound to AlgU is released into the cytoplasm following MucP cleavage, where it is degraded by ClpP, releasing AlgU [103]. MucB, a homologue of RseB, binds to the periplasmic domain of MucA and blocks cleavage by AlgW [98].

Cleavage by AlgW is the most extensively investigated part of the pathway, and there are interesting similarities and differences when compared to *E. coli* DegS. The pathway is initiated in a manner similar to the *E. coli* system. AlgW is inactive until a protein with the appropriate C-terminal sequence binds to its PDZ domain [98, 101]. As with DegS of *E. coli*, AlgW is activated by proteins with specific hydrophobic residues at their C-termini. The strongest inducer, identified in a genetic screen, was the MucE protein whose C-terminal WVF sequence was critical for activation of AlgU. *In vitro*, a WVF peptide activated AlgW with positive cooperativity [98]. In contrast to *E. coli*, in which a large percentage of OMPs contain the C-terminal YxF motif, the only *P. aeruginosa* protein known to contain a C-terminal WVF motif is MucE, a periplasmic protein. Currently, however, very little is known about MucE, and why it would be an inducer of AlgU. Interestingly, the C-termini of two of the most abundant porins in *P. aeruginosa* are not inducers of AlgW. In contrast, inducing sequences were identified at the C-termini of two phosphate/pyrophosphate specific OMPs [101]. These findings suggest that the *P. aeruginosa* system is not tuned to sense OMP folding. Regardless of the source of the activating peptide, the outcome of peptide binding to the PDZ domain of AlgW is the same as for DegS, activation of AlgW to initiate the proteolytic cascade.

The PDZ domain of AlgW negatively regulates its activity, similar to that of DegS, but it also appears to be a positive regulator [60, 101]. Deletion of the PDZ domain increased AlgW activity. However, the increase was far less than that seen for DegSΔPDZ, suggesting that the PDZ domain is required for full activity of AlgW [98]. In addition to regulation via the PDZ domain, AlgW uses a second inhibitory mechanism, which is not used by DegS. In sequence alignments of AlgW homologues, AlgW contains an extended active site LA loop in the protease domain. In contrast, DegS has a much shorter LA loop. In the crystal structure of *Thermotoga maritima* DegP, the extended LA loop blocks access to the active site and is thought to act as a regulator of protease activity [104]. The LA loop of AlgW also appears to have regulatory activities, because truncation of this loop increases AlgW activity [98]. Therefore, additional regulators may be needed to alter the conformation of the LA loop for maximal activation of AlgW, adding an additional level of regulation to AlgW that is not found in DegS.

Regulation of B. subtilis *σ^W*

In *B. subtilis*, the ECF sigma factor σ^W is activated by antimicrobial peptides, cell-wall active antibiotics, and alkaline shock [105, 106]. It is regulated by the membrane-bound anti-sigma factor RsiW [107]. As with the σ^E system, release of σ^W is controlled by a proteolytic cascade that starts with the inner membrane protease, PrsW, followed by the S2P, RasP, and concludes with ClpP in the cytoplasm (Fig. 6.8) [107–109]. Although the overall layout of this pathway is similar to the proteolytic cascade that leads to release of σ^E, the first protease in this pathway is unrelated to DegS [107]. Instead it is thought that PrsW, a multi-pass inner membrane protein, performs the first cleavage [108, 109]. Although orthologs of PrsW are wide-spread, PrsW does not contain a PDZ domain, nor is it a member of any of the canonical families of proteases [108, 109]. PrsW-dependent cleavage of RsiW occurs following one of the inducing stresses, although currently the nature of the signal is unknown. It has been proposed that degradation of RsiW involves two proteolytic modules [110]. Following cleavage of the RsiW periplasmic domain, the C-terminus of RsiW is trimmed by another periplasmic protease until it becomes a substrate for RasP. RsiW is then cleaved within the TM region, by RasP, releasing the anti-sigma factor domain containing the C-terminal AAA motif. Similar to the σ^E and AlgU systems, this sequence motif then targets the anti-sigma factor for degradation by the Clp protease system [111] (see also [11]).

Regulation of Virulence in Mycobacterium tuberculosis

In *M. tuberculosis*, the S2P protease, Rip1, has been shown to be a major virulence factor [112]. Its role in virulence is proposed, in part, to stem from its regulation of three anti-sigma/sigma factor systems (Fig. 6.8) [113]. Deletion of *rip1* in *M. tuberculosis*, leads to the accumulation of C-terminally truncated anti-sigma factors, similar to that seen for RseA in *E. coli* lacking RseP, MucA in *P. aeruginosa* lacking MucP, and RsiW in *B. subtilis* lacking RasP [101, 102, 113]. Although S2P proteases are known to process other substrates, in addition to anti-sigma factors, Rip1 is the first S2P shown to cleave multiple anti-sigma factors [113]. The initiating protease(s) and inducing signal(s) have not been identified yet for these proteolytic pathways. It will be of interest to learn whether the three systems all use the same initiating protease, such that a single regulatory proteolytic module has been adapted to control three different anti-sigma/sigma factor systems, or if the initiating proteases are unique and integrate distinct inducing signals.

Perspectives

The overall design of this proteolytic system has been adapted in different bacteria to regulate alternative sigma factors with a variety of cellular roles via proteolysis of their membrane-bound anti-sigma factors. In keeping with its role as the signal sensor, the first protease in the pathway is the most variable component of the system, while the other proteases are more highly conserved. Therefore, the overall proteolytic module appears to have evolved to sense different signals by varying the initiating protease. Even when the initiating protease is conserved, for example DegS and AlgW, subtle differences in nature of peptide binding, specificity, and allosteric activation serve to tailor the protease to the specific needs of the bacterium [70, 98].

While the initiating proteases appear to be variable, the downstream proteases are more highly conserved. The second cleavage is performed by a RIP protease, most often a member of the S2P family, and a Clp protease degrades the released anti-sigma factor domain. In all cases examined thus far, it appears that only the first cleavage is signal-dependent, and each cleavage generates a substrate for the next protease in the pathway [14, 15, 66, 101, 102, 106, 109, 112]. As more information is elucidated from related regulatory proteolytic cascades, it will be of interest to determine whether the somewhat intricate inhibitory interactions seen in the σ^E system are a hallmark of the overall regulatory design or if they are adaptations specific to the *E. coli* system. Many of these membrane-bound anti-sigma factor/ECF sigma factor systems also have an ortholog of RseB [35], suggesting modulation of the proteolytic cascade by RseB-like proteins is a conserved and important part of the regulatory pathways.

Why is such a complex hierarchical proteolytic cascade needed to control the activity of individual alternative sigma factor, like σ^E? From a design perspective, the proteolytic cascade provides a fast response to an inducing signal [57]. No step in the pathway is dependent on the synthesis of the next component, a much slower process, or on additional outside inputs. Once the signaling pathway is triggered, other signals are not required, ensuring that σ^E is rapidly released to promote cell survival. The numerous inhibitory interactions throughout the system prevent proteolysis in the absence of an inducing signal, yet allow the system to be poised to proceed as soon as a signal occurs. As a result, regulatory proteolysis provides a solution to the intercompartmental signaling problem that not only generates an on-off switch, but that can also finely tune biological activity to the strength of the inducing signal.

Notes

The molecular mechanisms that control RseP activity remain one of the outstanding questions about the proteolytic pathway that governs σ^E activity in *E. coli*. Conflicting results have been published about whether RseA must bind to the PDZ domains of RseP to activate the protease [68, 89, 90, 97]. However, recent work while this chapter was in production demonstrated that neither the identity of the C-terminal amino acid of the RseA$^{1-148}$ fragment nor the RseP PDZ domains were important for degradation of RseA by RseP in the cell and activation of the σ^E pathway in response to overproduction of OmpC [114]. The apparent requirement of the C-terminal valine of RseA$^{1-148}$ for cleavage by RseP appears to occur only *in vitro*. The authors suggest that the PDZ domains serve as a size filter, preventing RseP from degrading full-length RseA [114 and personal communication]. These data lead to the intriguing hypothesis that an alternative stress signal could bind to the RseP PDZ domains and induce DegS-independent cleavage of intact RseA by RseP.

References

Literature Cited

1. Brown MS, Ye J, Rawson RB, Goldstein JL (2000) Regulated intramembrane proteolysis: a control mechanism conserved from bacteria to humans. Cell 100(4):391–398
2. Chen G, Zhang X (2010) New insights into S2P signaling cascades: regulation, variation, and conservation. Protein Sci 19(11):2015–2030
3. Ehrmann M, Clausen T (2004) Proteolysis as a regulatory mechanism. Annu Rev Genet 38:709–724
4. Brown MS, Goldstein JL (1997) The SREBP pathway: regulation of cholesterol metabolism by proteolysis of a membrane-bound transcription factor. Cell 89(3):331–340
5. Brown MS, Goldstein JL (1999) A proteolytic pathway that controls the cholesterol content of membranes, cells, and blood. Proc Natl Acad Sci U S A 96(20):11041–11048
6. Rudner DZ, Fawcett P, Losick R (1999) A family of membrane-embedded metalloproteases involved in regulated proteolysis of membrane-associated transcription factors. Proc Natl Acad Sci U S A 96(26):14765–14770
7. Cutting S, Roels S, Losick R (1991) Sporulation operon spoIVF and the characterization of mutations that uncouple mother-cell from forespore gene expression in Bacillus subtilis. J Mol Biol 221(4):1237–1256
8. Ricca E, Cutting S, Losick R (1992) Characterization of bofA, a gene involved in intercompartmental regulation of pro-sigma K processing during sporulation in Bacillus subtilis. J Bacteriol 174(10):3177–3184
9. Campo N, Rudner DZ (2007) SpoIVB and CtpB are both forespore signals in the activation of the sporulation transcription factor sigmaK in Bacillus subtilis. J Bacteriol 189(16):6021–6027
10. Zhou R, Kroos L (2005) Serine proteases from two cell types target different components of a complex that governs regulated intramembrane proteolysis of pro-sigmaK during Bacillus subtilis development. Mol Microbiol 58(3):835–846
11. Molière N, Turgay K (2013) General and regulatory proteolysis in Bacillus subtilis. In: Dougan DA (ed) Regulated proteolysis: from bacteria to yeast. Springer, Subcell Biochem 66:73–103
12. Ades SE (2006) AAA+molecular machines: firing on all cylinders. Curr Biol 16(2): R46–R48
13. Alba BM, Gross CA (2004) Regulation of the Escherichia coli sigma-dependent envelope stress response. Mol Microbiol 52(3):613–619
14. Alba B, Leeds J, Onufryk C, Lu C et al (2002) DegS and YaeL participate sequentially in the cleavage of RseA to activate the sigma'E-dependent extracytoplasmic stress response. Genes Dev 16:2156–2168
15. Kanehara K, Ito K, Akiyama Y (2002) YaeL (EcfE) activates the sigmaE pathway of stress response through a site-2 cleavage of anti-sigma'E, RseA. Genes Dev 16:2147–2155
16. Ades SE (2004) Control of the alternative sigma factor sigmaE in Escherichia coli. Curr Opin Microbiol 7(2):157–162
17. Flynn JM, Levchenko I, Sauer RT, Baker TA (2004) Modulating substrate choice: the SspB adaptor delivers a regulator of the extracytoplasmic-stress response to the AAA+protease ClpXP for degradation. Genes Dev 18(18):2292–2301
18. Silhavy TJ, Kahne D, Walker S (2010) The bacterial cell envelope. Cold Spring Harb Perspect Biol 2(5):a000414
19. MacRitchie DM, Buelow DR, Price NL, Raivio TL (2008) Two-component signaling and gram negative envelope stress response systems. Adv Exp Med Biol 631:80–110

20. Rowley G, Spector M, Kormanec J, Roberts M (2006) Pushing the envelope: extracytoplasmic stress responses in bacterial pathogens. Nat Rev Microbiol 4(5):383–394
21. De Las Penas A, Connolly L, Gross CA (1997) SigmaE is an essential sigma factor in Escherichia coli. J Bacteriol 179(21):6862–6864
22. Hayden JD, Ades SE (2008) The extracytoplasmic stress factor, sigma, is required to maintain cell envelope integrity in Escherichia coli. PLoS One 3(2):e1573
23. Gruber TM, Gross CA (2003) Multiple sigma subunits and the partitioning of bacterial transcription space. Annu Rev Microbiol 57:441–466
24. Helmann JD (2002) The extracytoplasmic function (ECF) sigma factors. Adv Microb Physiol 46:47–110
25. Erickson JW, Gross CA (1989) Identification of the sigma E subunit of Escherichia coli RNA polymerase: a second alternate sigma factor involved in high-temperature gene expression. Genes Dev 3(9):1462–1471
26. Micevski D, Dougan DA (2013) Proteolytic regulation of stress response pathways in Escherichia coli. In: Dougan DA (ed) Regulated proteolysis: from bacteria to yeast. Springer, Subcell Biochem 66:105–128
27. Mecsas J, Rouviere PE, Erickson JW, Donohue TJ et al (1993) The activity of sigma E, an Escherichia coli heat-inducible sigma-factor, is modulated by expression of outer membrane proteins. Genes Dev 7(12B):2618–2628
28. Missiakas D, Betton JM, Raina S (1996) New components of protein folding in extracytoplasmic compartments of Escherichia coli SurA, FkpA and Skp/OmpH. Mol Microbiol 21(4): 871–884
29. Raina S, Missiakas D, Georgopoulos C (1995) The rpoE gene encoding the sigma E (sigma 24) heat shock sigma factor of Escherichia coli. EMBO J 14(5):1043–1055
30. Rouviere PE, Gross CA (1996) SurA, a periplasmic protein with peptidyl-prolyl isomerase activity, participates in the assembly of outer membrane porins. Genes Dev 10(24): 3170–3182
31. Connolly L, De Las PA, Alba BM, Gross CA (1997) The response to extracytoplasmic stress in Escherichia coli is controlled by partially overlapping pathways. Genes Dev 11(15):2012–2021
32. Raivio TL, Silhavy TJ (1999) The sigmaE and Cpx regulatory pathways: overlapping but distinct envelope stress responses. Curr Opin Microbiol 2(2):159–165
33. Ruiz N, Silhavy TJ (2005) Sensing external stress: watchdogs of the Escherichia coli cell envelope. Curr Opin Microbiol 8(2):122–126
34. Mecsas J, Welch R, Erickson JW, Gross CA (1995) Identification and characterization of an outer membrane protein, OmpX, in Escherichia coli that is homologous to a family of outer membrane proteins including Ail of Yersinia enterocolitica. J Bacteriol 177(3):799–804
35. Staron A, Sofia HJ, Dietrich S, Ulrich LE et al (2009) The third pillar of bacterial signal transduction: classification of the extracytoplasmic function (ECF) sigma factor protein family. Mol Microbiol 74(3):557–581
36. Davis BM, Waldor MK (2009) High-throughput sequencing reveals suppressors of Vibrio cholerae rpoE mutations: one fewer porin is enough. Nucleic Acids Res 37(17):5757–5767
37. Heusipp G, Schmidt MA, Miller VL (2003) Identification of rpoE and nadB as host responsive elements of Yersinia enterocolitica. FEMS Microbiol Lett 226(2):291–298
38. Humphreys S, Stevenson A, Bacon A, Weinhardt AB et al (1999) The alternative sigma factor, sigmaE, is critically important for the virulence of Salmonella typhimurium. Infect Immun 67(4):1560–1568
39. Testerman TL, Vazquez-Torres A, Xu Y, Jones-Carson J et al (2002) The alternative sigma factor sigmaE controls antioxidant defences required for Salmonella virulence and stationary-phase survival. Mol Microbiol 43(3):771–782
40. Barchinger SE, Zhang X, Hester SE, Rodriguez ME, Harvill ET, Ades SE (2012) sigE facilitates the adaptation of Bordetella bronchiseptica to stress conditions and lethal infection in immunocompromised mice. BMC Microbiol 12:179

41. Raivio TL (2005) Envelope stress responses and Gram-negative bacterial pathogenesis. Mol Microbiol 56(5):1119–1128
42. Rhodius VA, Suh WC, Nonaka G, West J et al (2006) Conserved and variable functions of the sigmaE stress response in related genomes. PLoS Biol 4(1):e2
43. Johansen J, Rasmussen AA, Overgaard M, Valentin-Hansen P (2006) Conserved small non-coding RNAs that belong to the sigmaE regulon: role in down-regulation of outer membrane proteins. J Mol Biol 364(1):1–8
44. Thompson KM, Rhodius VA, Gottesman S (2007) SigmaE regulates and is regulated by a small RNA in Escherichia coli. J Bacteriol 189(11):4243–4256
45. Vogel J, Papenfort K (2006) Small non-coding RNAs and the bacterial outer membrane. Curr Opin Microbiol 9(6):605–611
46. Rigel NW, Silhavy TJ (2012) Making a beta-barrel: assembly of outer membrane proteins in Gram-negative bacteria. Curr Opin Microbiol 15(2):189–193
47. Ruiz N, Kahne D, Silhavy TJ (2006) Advances in understanding bacterial outer-membrane biogenesis. Nat Rev Microbiol 4(1):57–66
48. Kovacikova G, Skorupski K (2002) The alternative sigma factor sigma(E) plays an important role in intestinal survival and virulence in Vibrio cholerae. Infect Immun 70(10):5355–5362
49. Miticka H, Rowley G, Rezuchova B, Homerova D et al (2003) Transcriptional analysis of the rpoE gene encoding extracytoplasmic stress response sigma factor sigmaE in Salmonella enterica serovar Typhimurium. FEMS Microbiol Lett 226(2):307–314
50. Rouviere PE, De Las Penas A, Mecsas J, Lu CZ et al (1995) rpoE, the gene encoding the second heat-shock sigma factor, sigma E, in Escherichia coli. EMBO J 14(5):1032–1042
51. Schurr MJ, Yu H, Boucher JC, Hibler NS et al (1995) Multiple promoters and induction by heat shock of the gene encoding the alternative sigma factor AlgU (sigma E) which controls mucoidy in cystic fibrosis isolates of Pseudomonas aeruginosa. J Bacteriol 177(19):5670–5679
52. De Las PA, Connolly L, Gross CA (1997) The sigmaE-mediated response to extracytoplasmic stress in Escherichia coli is transduced by RseA and RseB, two negative regulators of sigmaE. Mol Microbiol 24(2):373–385
53. Missiakas D, Mayer MP, Lemaire M, Georgopoulos C et al (1997) Modulation of the Escherichia coli sigmaE (RpoE) heat-shock transcription-factor activity by the RseA, RseB and RseC proteins. Mol Microbiol 24(2):355–371
54. Campbell EA, Tupy JL, Gruber TM, Wang S et al (2003) Crystal structure of Escherichia coli sigmaE with the cytoplasmic domain of its anti-sigma RseA. Mol Cell 11(4):1067–1078
55. Schrodinger LLC (2010) The PyMOL molecular graphics system, Version 1.3r1
56. Paget MS, Helmann JD (2003) The sigma70 family of sigma factors. Genome Biol 4(1):203
57. Chaba R, Grigorova IL, Flynn JM, Baker TA et al (2007) Design principles of the proteolytic cascade governing the sigmaE-mediated envelope stress response in Escherichia coli: keys to graded, buffered, and rapid signal transduction. Genes Dev 21(1):124–136
58. Maeda H, Fujita N, Ishihama A (2000) Competition among seven Escherichia coli sigma subunits: relative binding affinities to the core RNA polymerase. Nucleic Acids Res 28(18):3497–3503
59. Grigorova IL, Chaba R, Zhong HJ, Alba BM et al (2004) Fine-tuning of the Escherichia coli sigmaE envelope stress response relies on multiple mechanisms to inhibit signal-independent proteolysis of the transmembrane anti-sigma factor, RseA. Genes Dev 18(21):2686–2697
60. Cezairliyan BO, Sauer RT (2007) Inhibition of regulated proteolysis by RseB. Proc Natl Acad Sci U S A 104(10):3771–3776
61. Chaba R, Alba BM, Guo MS, Sohn J et al (2011) Signal integration by DegS and RseB governs the σ E-mediated envelope stress response in Escherichia coli. Proc Natl Acad Sci U S A 108(5):2106–2111
62. Ades SE, Connolly LE, Alba BM, Gross CA (1999) The Escherichia coli sigma(E)-dependent extracytoplasmic stress response is controlled by the regulated proteolysis of an anti-sigma factor. Genes Dev 13(18):2449–2461

63. Ades SE, Grigorova IL, Gross CA (2003) Regulation of the alternative sigma factor sigma(E) during initiation, adaptation, and shutoff of the extracytoplasmic heat shock response in Escherichia coli. J Bacteriol 185(8):2512–2519
64. Alba BM, Zhong HJ, Pelayo JC, Gross CA (2001) degS (hhoB) is an essential Escherichia coli gene whose indispensable function is to provide sigma (E) activity. Mol Microbiol 40(6):1323–1333
65. Kanehara K, Akiyama Y, Ito K (2001) Characterization of the yaeL gene product and its S2P-protease motifs in Escherichia coli. Gene 281(1–2):71–79
66. Flynn JM, Neher SB, Kim YI, Sauer RT et al (2003) Proteomic discovery of cellular substrates of the ClpXP protease reveals five classes of ClpX-recognition signals. Mol Cell 11(3):671–683
67. Levchenko I, Grant RA, Flynn JM, Sauer RT et al (2005) Versatile modes of peptide recognition by the AAA+adaptor protein SspB. Nat Struct Mol Biol 12(6):520–525
68. Bohn C, Collier J, Bouloc P (2004) Dispensable PDZ domain of Escherichia coli YaeL essential protease. Mol Microbiol 52(2):427–435
69. Kanehara K, Ito K, Akiyama Y (2003) YaeL proteolysis of RseA is controlled by the PDZ domain of YaeL and a Gln-rich region of RseA. EMBO J 22(23):6389–6398
70. Walsh N, Alba B, Bose B, Gross C et al (2003) OMP peptide signals initiate the envelope-stress response by activating DegS protease via relief of inhibition mediated by its PDZ domain. Cell 113:61–71
71. Clausen T, Kaiser M, Huber R, Ehrmann M (2011) HTRA proteases: regulated proteolysis in protein quality control. Nat Rev Mol Cell Biol 12(3):152–162
72. Wilken C, Kitzing K, Kurzbauer R, Ehrmann M et al (2004) Crystal structure of the DegS stress sensor: how a PDZ domain recognizes misfolded protein and activates a protease. Cell 117(4):483–494
73. Zeth K (2004) Structural analysis of DegS, a stress sensor of the bacterial periplasm. FEBS Lett 569(1–3):351–358
74. Cowan SW, Garavito RM, Jansonius JN, Jenkins JA et al (1995) The structure of OmpF porin in a tetragonal crystal form. Structure 3(10):1041–1050
75. Hagan CL, Silhavy TJ, Kahne D (2011) beta-Barrel membrane protein assembly by the Bam complex. Annu Rev Biochem 80:189–210
76. Rezuchova B, Miticka H, Homerova D, Roberts M et al (2003) New members of the Escherichia coli sigmaE regulon identified by a two-plasmid system. FEMS Microbiol Lett 225(1):1–7
77. Collinet B, Yuzawa H, Chen T, Herrera C et al (2000) RseB binding to the periplasmic domain of RseA modulates the RseA:sigmaE interaction in the cytoplasm and the availability of sigmaE.RNA polymerase. J Biol Chem 275(43):33898–33904
78. Wollmann P, Zeth K (2007) The structure of RseB: a sensor in periplasmic stress response of E. coli. J Mol Biol 372(4):927–941
79. Kim DY, Kwon E, Choi J, Hwang H-Y et al (2010) Structural basis for the negative regulation of bacterial stress response by RseB. Protein Sci 19(6):1258–1263
80. Kim DY, Jin KS, Kwon E, Ree M et al (2007) Crystal structure of RseB and a model of its binding mode to RseA. Proc Natl Acad Sci U S A 104(21):8779–8784
81. Sohn J, Grant RA, Sauer RT (2007) Allosteric activation of DegS, a stress sensor PDZ protease. Cell 131(3):572–583
82. Sohn J, Sauer RT (2009) OMP peptides modulate the activity of DegS protease by differential binding to active and inactive conformations. Mol Cell 33(1):64–74
83. Sohn J, Grant RA, Sauer RT (2009) OMP peptides activate the DegS stress-sensor protease by a relief of inhibition mechanism. Structure 17(10):1411–1421
84. Sohn J, Grant RA, Sauer RT (2010) Allostery is an intrinsic property of the protease domain of DegS: implications for enzyme function and evolution. J Biol Chem 285(44): 34039–34047
85. Hasselblatt H, Kurzbauer R, Wilken C, Krojer T et al (2007) Regulation of the E stress response by DegS: how the PDZ domain keeps the protease inactive in the resting state and

allows integration of different OMP-derived stress signals upon folding stress. Genes Dev 21(20):2659–2670
86. Monod J, Wyman J, Changeux JP (1965) On the nature of allosteric transitions: a plausible model. J Mol Biol 12:88–118
87. Cui Q, Karplus M (2008) Allostery and cooperativity revisited. Protein Sci 17(8):1295–1307
88. Kinch LN, Ginalski K, Grishin NV (2006) Site-2 protease regulated intramembrane proteolysis: sequence homologs suggest an ancient signaling cascade. Protein Sci 15(1):84–93
89. Koide K, Ito K, Akiyama Y (2008) Substrate recognition and binding by RseP, an Escherichia coli intramembrane protease. J Biol Chem 283(15):9562–9570
90. Inaba K, Suzuki M, Maegawa K, Akiyama S et al (2008) A pair of circularly permutated PDZ domains control RseP, the S2P family intramembrane protease of Escherichia coli. J Biol Chem 283(50):35042–35052
91. Li X, Wang B, Feng L, Kang H et al (2009) Cleavage of RseA by RseP requires a carboxyl-terminal hydrophobic amino acid following DegS cleavage. Proc Natl Acad Sci U S A 106(35):14837–14842
92. Feng L, Yan H, Wu Z, Yan N et al (2007) Structure of a site-2 protease family intramembrane metalloprotease. Science 318(5856):1608–1612
93. Koide K, Maegawa S, Ito K, Akiyama Y (2007) Environment of the active site region of RseP, an Escherichia coli regulated intramembrane proteolysis protease, assessed by site-directed cysteine alkylation. J Biol Chem 282(7):4553–4560
94. Akiyama Y, Kanehara K, Ito K (2004) RseP (YaeL), an Escherichia coli RIP protease, cleaves transmembrane sequences. EMBO J 23(22):4434–4442
95. Saito A, Hizukuri Y, Matsuo E, Chiba S et al (2011) Post-liberation cleavage of signal peptides is catalyzed by the site-2 protease (S2P) in bacteria. Proc Natl Acad Sci U S A 108(33):13740–13745
96. Shah S, Lee SF, Tabuchi K, Hao YH et al (2005) Nicastrin functions as a gamma-secretase-substrate receptor. Cell 122(3):435–447
97. Gur E, Ottofuelling R, Dougan DA (2013) Machines of destruction – AAA+proteases and the adaptors that control them. In: Dougan DA (ed) Regulated proteolysis: from bacteria to yeast. Springer, Subcell Biochem 66:3–33
98. Cezairliyan BO, Sauer RT (2009) Control of Pseudomonas aeruginosa AlgW protease cleavage of MucA by peptide signals and MucB. Mol Microbiol 72(2):368–379
99. Martin DW, Holloway BW, Deretic V (1993) Characterization of a locus determining the mucoid status of Pseudomonas aeruginosa: AlgU shows sequence similarities with a Bacillus sigma factor. J Bacteriol 175(4):1153–1164
100. Schurr MJ, Yu H, Martinez-Salazar JM, Boucher JC et al (1996) Control of AlgU, a member of the sigma E-like family of stress sigma factors, by the negative regulators MucA and MucB and Pseudomonas aeruginosa conversion to mucoidy in cystic fibrosis. J Bacteriol 178(16):4997–5004
101. Qiu D, Eisinger VM, Rowen DW, Yu HD (2007) Regulated proteolysis controls mucoid conversion in Pseudomonas aeruginosa. Proc Natl Acad Sci U S A 104(19):8107–8112
102. Wood LF, Leech AJ, Ohman DE (2006) Cell wall-inhibitory antibiotics activate the alginate biosynthesis operon in Pseudomonas aeruginosa: roles of sigma (AlgT) and the AlgW and Prc proteases. Mol Microbiol 62(2):412–426
103. Qiu D, Eisinger VM, Head NE, Pier GB et al (2008) ClpXP proteases positively regulate alginate overexpression and mucoid conversion in Pseudomonas aeruginosa. Microbiology 154(Pt 7):2119–2130
104. Kim DY, Kim DR, Ha SC, Lokanath NK et al (2003) Crystal structure of the protease domain of a heat-shock protein HtrA from Thermotoga maritima. J Biol Chem 278(8):6543–6551
105. Cao M, Wang T, Ye R, Helmann JD (2002) Antibiotics that inhibit cell wall biosynthesis induce expression of the Bacillus subtilis sigmaW and sigmaM regulons. Mol Microbiol 45(5):1267–1276
106. Wiegert T, Homuth G, Versteeg S, Schumann W (2001) Alkaline shock induces the Bacillus subtilis sigma(W) regulon. Mol Microbiol 41(1):59–71

107. Schöbel S, Zellmeier S, Schumann W, Wiegert T (2004) The Bacillus subtilis sigmaW anti-sigma factor RsiW is degraded by intramembrane proteolysis through YluC. Mol Microbiol 52(4):1091–1105
108. Ellermeier CD, Losick R (2006) Evidence for a novel protease governing regulated intramembrane proteolysis and resistance to antimicrobial peptides in Bacillus subtilis. Genes Dev 20(14):1911–1922
109. Heinrich J, Wiegert T (2006) YpdC determines site-1 degradation in regulated intramembrane proteolysis of the RsiW anti-sigma factor of Bacillus subtilis. Mol Microbiol 62(2):566–579
110. Heinrich J, Hein K, Wiegert T (2009) Two proteolytic modules are involved in regulated intramembrane proteolysis of Bacillus subtilis RsiW. Mol Microbiol 74(6):1412–1426
111. Zellmeier S, Schumann W, Wiegert T (2006) Involvement of Clp protease activity in modulating the Bacillus subtilissigmaw stress response. Mol Microbiol 61(6):1569–1582
112. Makinoshima H, Glickman MS (2005) Regulation of Mycobacterium tuberculosis cell envelope composition and virulence by intramembrane proteolysis. Nature 436(7049):406–409
113. Sklar JG, Makinoshima H, Schneider JS, Glickman MS (2010) M. tuberculosis intramembrane protease Rip1 controls transcription through three anti-sigma factor substrates. Mol Microbiol 77(3):605–617
114. Hizukuri Y, Akiyama Y (2012) PDZ domains of RseP are not essential for sequential cleavage of RseA or stress-induced σ^E activation in vivo. Mol Microbiol 86(5):1232–1245

Chapter 7
Bacterial Proteases and Virulence

Dorte Frees, Lone Brøndsted, and Hanne Ingmer

Abstract Bacterial pathogens rely on proteolysis for variety of purposes during the infection process. In the cytosol, the main proteolytic players are the conserved Clp and Lon proteases that directly contribute to virulence through the timely degradation of virulence regulators and indirectly by providing tolerance to adverse conditions such as those experienced in the host. In the membrane, HtrA performs similar functions whereas the extracellular proteases, in close contact with host components, pave the way for spreading infections by degrading host matrix components or interfering with host cell signalling to short-circuit host cell processes. Common to both intra- and extracellular proteases is the tight control of their proteolytic activities. In general, substrate recognition by the intracellular proteases is highly selective which is, in part, attributed to the chaperone activity associated with the proteases either encoded within the same polypeptide or on separate subunits. In contrast, substrate recognition by extracellular proteases is less selective and therefore these enzymes are generally expressed as zymogens to prevent premature proteolytic activity that would be detrimental to the cell. These extracellular proteases are activated in complex cascades involving auto-processing and proteolytic maturation. Thus, proteolysis has been adopted by bacterial pathogens at multiple levels to ensure the success of the pathogen in contact with the human host.

D. Frees • L. Brøndsted • H. Ingmer (✉)
Department of Veterinary Disease Biology, Faculty of Life Sciences,
University of Copenhagen, Stigbøjlen 4, Frederiksberg C 1870, Denmark
e-mail: hi@life.ku.dk

D.A. Dougan (ed.), *Regulated Proteolysis in Microorganisms*, Subcellular Biochemistry 66, 161
DOI 10.1007/978-94-007-5940-4_7, © Springer Science+Business Media Dordrecht 2013

Clp Proteases

Distribution, Structure, and Characteristics

The ATP-dependent Clp proteases are highly conserved among bacteria and in chloroplasts and mitochondria of eukaryotic cells, but are absent from Archaea [1]. Clp proteases are compartmentalized proteases composed of two different components. The peptidase component (e.g. ClpP) contains the conserved (Ser-His-Asp) catalytic triad that is typical of serine-proteases [1]. Two heptameric rings of ClpP subunits stack back-to-back, to form a cylinder, where the proteolytic sites line the inner surface [2]. Access to the concealed proteolytic chamber, is restricted by narrow pores that allow passage of only small peptides of up to 7 amino acids [3]. In order to degrade larger peptides and protein substrates, ClpP must associate with the hexameric ring of one of several possible Clp ATPases. The Clp ATPases are directly responsible for substrate recognition however the binding of some substrates may be modulated by specific adaptor proteins [4]. Bound substrates are unfolded by the Clp ATPase in an ATP consuming reaction and the unfolded polypeptide is threaded through the entry pores of the ClpP into the proteolytic chamber, where the polypeptides are degraded to small peptides that exit the chamber by diffusion (Fig. 7.1). The Clp ATPases constitute a family of closely related proteins that are divided into distinct subfamilies, based on the presence of specific signature sequences and the number and spacing of the nucleotide binding domains [5]. Depending on subfamily, the Clp ATPases participate in a variety of different processes, including protein folding, activation, and disaggregation. Notably, some Clp ATPases can function as molecular chaperone independently of ClpP and only a subgroup of the Clp ATPases can interact with ClpP, a property that is associated with the presence of a ClpP recognition tripeptide [6]. In Gram-negative bacteria ClpP typically associates with ClpA and ClpX ATPase families, whereas in Gram-positive bacteria ClpP associate with ATPases of the ClpX, ClpC, or ClpE families [7, 8] see also [9].

Like other ATP-dependent proteases, Clp is required for both general and regulated proteolysis. General proteolysis removes damaged or excess proteins from the cell thereby ensuring protein quality and homeostasis. While several proteases perform this task in *Escherichia coli* and related bacteria, Clp proteases seem to be primarily responsible for degrading non-native damaged proteins in low GC Gram-positive bacteria [5]. Regulated proteolysis on the other hand is the specific and conditional degradation of regulatory proteins that allows the bacterium to control cellular adaptations and differentiations in response to extra- or intracellular signals.

To date, Clp proteases have been shown to be important for virulence in cell and animal models for a number of important Gram-positive and Gram-negative pathogens. Below we will summarize current knowledge on the molecular mechanism contributing to the importance of Clp proteases in virulence. We furthermore will cover the new and very promising field – Clp as a target of newly identified antibiotic compounds, and the use of Clp in vaccines against widespread pathogens like *Salmonella enterica* serovar Typhimurium and *Streptococcus pneumoniae.*

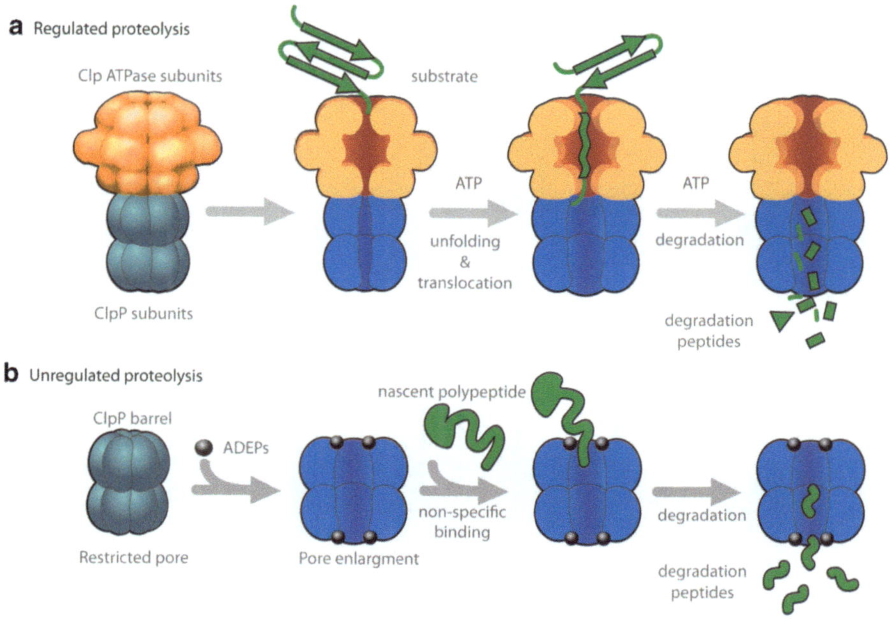

Fig. 7.1 Regulated and unregulated proteolysis by ClpP. (**a**) Regulated proteolysis. Substrates are specifically recognized and unfolded by the hexameric Clp ATPases (*orange*). The unfolded substrate is transferred into the gated chamber of the associated ClpP peptidase (*blue*), where proteolysis is carried out by active sites that line the inner surface of the barrel-like structure. (**b**) Unregulated proteolysis triggered by binding of acyldepsipeptide (ADEP) to ClpP. ADEP (*grey spheres*) binds between ClpP subunits (*blue*) thereby triggering an opening of the entrance pore of the ClpP barrel leading to unregulated degradation of nascent polypeptides and unfolded proteins

Clp in Gram-Positive Pathogens: Attenuated Virulence, Stress Sensitivity, and Control of Exoprotein Expression

Staphylococcus aureus

The opportunistic pathogen *Staphylococcus aureus* continues to be a leading cause of human infections ranging from minor skin infections to life-threatening endocarditis, pneumonia and septicemia. In *S. aureus*, inactivation of either *clpP* or *clpX* rendered the bacterium avirulent in a murine abscess model (Fig. 7.2) pointing to an important role for the ClpXP protease in virulence of *S. aureus* [10]. Inactivation of *clpP* also severely reduced the ability of *S. aureus* to cope with a wide range of stresses, and as the host represents a stressful environment, stress sensitivity may at least in part explain the attenuated virulence of the *clpP* mutant. However, the unexpected finding that the *clpX* mutant tolerates some stresses better than the wild type, and hence does not share the stress sensitivity of the *clpP* mutant, indicates that the ClpXP protease contributes to virulence not only by improving stress tolerance [5].

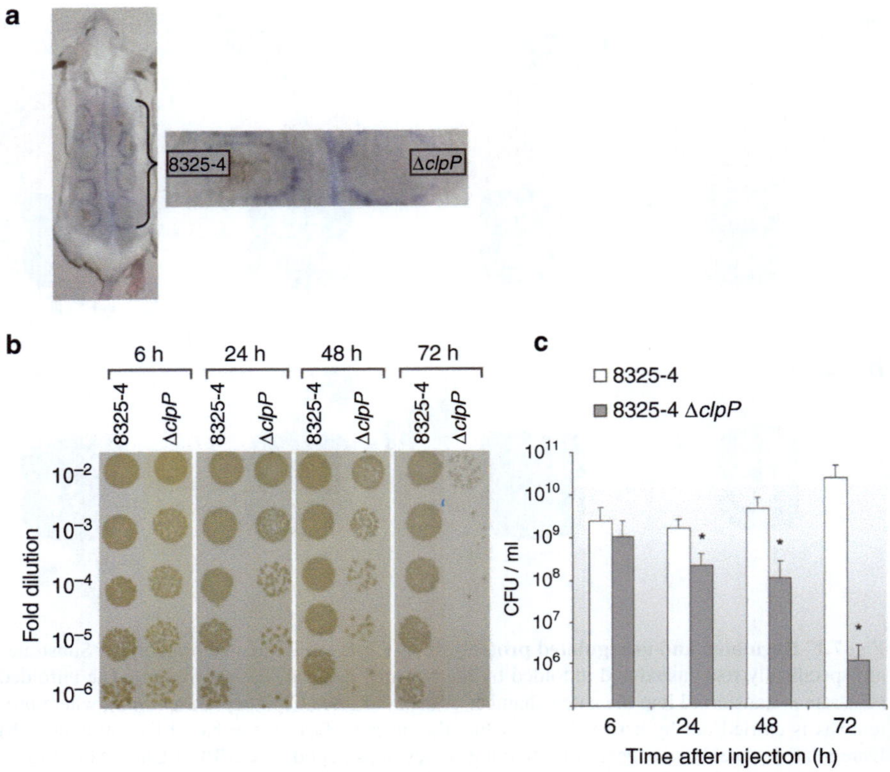

Fig. 7.2 ClpP is required for virulence of *S. aureus* in a mouse skin abscess model. (**a**) Ten Balb/cJ female mice were inoculated with 25 μl (1 × 10^9 CFU/ml) of bacterial suspension subcutaneously on eight spots along the back of the mice (four spots with the 8325-4 wild type strain and four spots with the 8325-4 *clpP* deletion mutant) [10]. (**b** and **c**) After 6, 24, 48, and 72 h, the mice were sacrificed and the lesions were aseptically removed, homogenized, diluted, and plated on TSA for CFU counting. Six hours after injection of bacteria, no obvious clinical signs of dermatitis were visible and equal amounts of CFU/ml were found in skin inoculated with wild-type and the *clpP* mutant strain. After 48 and 72 h, spots inoculated with the parental strain displayed severe dermatitis, while spots inoculated with the *clpP* mutant showed no signs of infection at any time point and after 72 h the CFU/ml of the *clpP* mutants strain at the site of inoculation was reduced 10^4-fold. In comparisons, the number of wild type bacteria did not decrease during the time frame of the experiment

The multifaceted pathogenicity of *S. aureus* relies on a wide array of surface-bound and secreted virulence factors that provide the bacterium with the ability of tissue binding, tissue destruction, and immune evasion. Interestingly, expression of hemolysins, extracellular proteases, and other virulence factors was reduced by as much as 100-fold in the absence *clpP* or *clpX*, as was expression of a number of global virulence regulators [8, 10, 11]. This finding indicates that ClpXP controls stability of one or more transcriptional virulence regulators. Alternatively, the link between ClpXP and virulence gene regulation is indirect in the sense that the cellular stress imposed by the lack of ClpP may create some general physiological or metabolic signals that are sensed by the regulatory network controlling virulence determinants. But undoubtly, the severely reduced expression of a range of important virulence factors is partly responsible for the avirulence of both the *clpP* and *clpX* mutants [10]. Interestingly, synthetic β-lactones were recently show to bind ClpP of *S. aureus* with high affinity [12]. Similar to the deletion of *clpP*, binding of β-lactones severely reduced expression of hemolysins and extracellular proteases and this effect was also seen in methicillin resistant *S. aureus* (MRSA) strains, suggesting that β-lactones can potentially be used as an anti-virulence drug against *S. aureus*.

Finally, the chaperone activity of ClpX, independent of ClpP, stimulates expression of the well known IgG binding protein, Protein A [8, 10, 13]. The effect of ClpX is dramatic as deletion of *clpX* virtually abolishes expression of Protein A both in a laboratory strains and in clinical isolates [13]. Interestingly, ClpX appears to perform dual roles in regulating Protein A expression [13]: First, ClpX stimulates transcription of *spa* (staphylococcal protein A) by enhancing translation of Rot, an activator of *spa* transcription. Secondly, ClpX is required for full translation of the *spa* mRNA. At present the mechanism whereby ClpX stimulates translation of both the *rot* and *spa* mRNAs is unknown.

Listeria monocytogenes

Listeria monocytogenes is a food borne pathogen that, when ingested by the host, adopts an intracellular lifestyle. Entry of *L. monocytogenes* into non-professional phagocytes is induced by binding of the bacterial surface proteins internalin A (InlA) and InlB to receptors on the host cells. *L. monocytogenes* then uses the pore-forming listeriolysin O (LLO) to escape the phagocytic vacuole that is formed after invasion. Then, with the help of ActA the bacterium can polymerize actin, which enables movement into the neighboring cell. In a mouse model, the *Listeria clpP* mutant was severely restricted in its ability to replicate in organs, and the bacteria were rapidly eliminated from the animal [14]. Confocal microscopy revealed that while the *clpP* mutant was taken up by macrophages at wild type frequencies, ClpP seems to be essential for the phagosomal escape and for the rapid intracytoplasmic replication of *L. monocytogenes* in macrophages [14]. The stress-sensitivity of the *Listeria clpP* mutant most likely contributes to the restricted growth of the mutant in the hostile environment of the host cells. Additionally, ClpP affects expression and activity of at least one essential virulence factor, namely listeriolysin (LLO). Berche and colleagues [14] observed that, while synthesis of LLO was only slightly reduced by the *clpP* mutation, activity of LLO was severely reduced [14, 15]. The loss of active LLO in the *clpP* mutant fits well with the inability of the mutant to escape the phagosome, and suggests that ClpP is required for export or activation of LLO. ClpP-mediated proteolysis may have a more general role in controlling expression of virulence factors in *Listeria*, as its partner Clp ATPase, ClpC, controls the level of transcription of *inlA*, *inlB*, and *actA* [16]. However, it is currently unclear if virulence gene expression is controlled by ClpC as part of a

ClpCP protease, or independently of ClpP. Finally, ClpCP is required for the turnover of the intracellular survival protein, Surface virulence-associated protein A (SvpA), in a process directed by the adaptor protein, MecA [17].

Streptococcus

The genus *Streptococcus* comprises a range of important pathogens. Worldwide, *S. pneumoniae* remains the most common cause of community-acquired pneumonia, bacterial meningitis, bacteremia, and otitis. In a murine lung and sepsis model, a *S. pneumoniae* strain D39 *clpP* mutant failed to colonize the lungs and in contrast to wild type cells did not cause mortality [18]. In the same experiment a *clpC* mutant was almost as virulent as the wild type strain, suggesting that ClpP acts in concert with an alternative Clp ATPase to control virulence of *S. pneumoniae* [18]. Invasion and dissemination of *S. pneumoniae* are usually accomplished from its natural niche, the nasopharynx, and importantly the D39 cells depleted of ClpP were unable to colonize the nasopharynx [19]. Interestingly, the effect of ClpP on virulence appears to be strain dependent as a *clpP* mutant in the TIGR4 background retained some virulence in a mouse pneumonia model, while the D39 *clpP* mutant was avirulent [20]. Presumably, many factors contribute to the strain dependent attenuation in virulence of the *clpP* mutants. Stress sensitivity may be part of the explanation but again an affect of ClpP on expression of major virulence factor seem to be involved [19]. Additional research is required to determine the proteolytic targets responsible for the observed phenotypes.

Clp in Gram-Negative Pathogens: Control of Motility and Type III Secretion

Salmonella enterica causes a wide spectrum of diseases. There are more than 2,000 different serovars of *S. enterica*, and serovar Typhimurium (*S. typhimurium*) and Enteritidis (*S. enteritidis*) are the most common cause of food-born gastroenteritis in humans, while other serovars such as *S. typhi* can cause fatal systemic diseases. In mice *S. typhimurium* causes a systemic typhoid-like disease, providing a useful model to study systemic salmonellosis. Interestingly in this mouse model, *S. typhimurium* mutants lacking *clpX* or *clpP* are avirulent demonstrating the importance of the ClpXP protease for the progression of systemic salmonellosis [21, 22]. Indeed, during infection *S. typhimurium* cells depleted of *clpP* or *clpX* are unable to invade and replicate inside professional phagocytic cells [22]. Several findings may explain the avirulence and the inability of intracellular replication of the *Salmonella clpP* mutant: Firstly, *S. typhimurium* cells lacking the activity of the Clp proteases have attenuated resistance towards stresses like high temperature, low pH, which presumably contributes to the inability of *S. typhimurium* to grow in the hostile environment of macrophages [23]. Secondly, the ClpXP protease is responsible for conditional degradation of the general stress sigma factor, σ^S that is essential for virulence [24] (see also [25]). Conceivably, the stabilization of σ^S in *clpP* and *clpX* mutants [21] interferes with the timely expression of virulence genes and thereby attenuates the virulence of these mutants. Finally, ClpXP forms part of the complex network that controls expression of the flagella and Type III secretion system (T3SS). The T3SS, conserved in many Gram-negative pathogens, forms a needle like structure that mediates direct transfer of bacterial virulence factors (called effector molecules) into the cytoplasm of the host cells. Once inside the host cell, these effector molecules specifically interfere with vital host functions such as actin polymerization, signal transduction and apoptosis. In *Salmonella*, the T3SS is encoded by multiple pathogenicity islands, which represent major virulence factors [26]. Flagella have, in some cases, been linked with pathogenesis either as adhesion factors or by promoting bacterial motility [27]. In *S. typhimurium*, inactivation of *clpP* caused a hyper-flagellated phenotype resulted from the overproduction of flagella proteins such as flagellin (the flagellum filament protein) encoded by *fliC* [28, 29]. The increased *fliC* expression results from the accumulation of FlhD/FlhC, the master regulator of flagella biosynthesis that is normally targeted to ClpXP for degradation [29]. In fact ClpXP degrades both proteins in the $FlhD_2FlhC_2$ complex, but not the isolated subunits FlhC or FlhD, suggesting that ClpX mediated recognition involves motifs distributed on both proteins [29].

In *Salmonella*, the type III secretion apparatus encoded by *Salmonella* pathogenicity island 1 (SPI1) is linked to flagellum biosynthesis, as one of the FlhD/FlhC regulated gene products, FliZ modulates the expression of HilA (hyper invasion locus A), the activator of T3SS genes [30]. The down-stream effects of ClpXP on T3SS was recently examined by Kage et al. [31] who showed that FliZ affects the expression of HilA through transcriptional and post-transcriptional effects on *hilC* and *hilD*. In ClpXP-depleted cells, the master regulator of the flagella regulon (FlhD/FlhC) accumulates resulting in the accumulation of the HilD by post-transcriptional control, which in turn leads to increased SPI1 expression. It has previously been reported that *Salmonella lon* mutant cells induce apoptosis within macrophages, due to increased expression of SPI1 gene products [32] and similarly, ClpXP could be required for suppression of SPI1 mediated apoptosis within macrophages [31].

The activity of Clp proteases have also been linked to expression of T3SS in other Gram-negative pathogens, however, currently the details of the underlying regulatory pathways are largely unknown. Enterohemorrhagic *Escherichia coli* (EHEC) strains are life-threatening human pathogens that cause hemorrhagic colitis, bloody diarrhea, and hemolytic uremic syndrome. The virulence factors of EHEC strains are encoded on the pathogenicity island termed the locus of enterocyte effacement (LEE). Central among these virulence factors are several Esp effector proteins that are secreted through the T3SS. Inactivation of either *clpP* or *clpX* dramatically reduced expression of LEE encoded Esp proteins [33]. The findings that (i) the level of GrlR, a negative regulator of *esp* transcription was enhanced in the *clpXP* mutant, and (ii) that *esp* transcription was restored in *grlR/clpXP* triple-mutant suggested that ClpXP controls *esp* transcription through GrlR [33].

Additionally, stabilization of RpoS in the *clp* mutant also appears to contribute to the lack of *esp* transcription [33]. Future analysis of regulated expression of T3SS in various pathogens will undoubtedly lead to new insights into the molecular mechanisms governing conditional proteolysis.

Antibiotics That Target Clp Proteases

Acyldepsipeptides

A very interesting finding in the field of Clp proteases was the recent discovery of a new class of antibiotics termed acyldepsipeptides (ADEPs). These antibiotics kill bacteria by interfering with the function of ClpP [34]. ADEPs are naturally produced by isolates of *Streptomyces hawaiiensis*, and are active against a number of important Gram-positive pathogens including *Streptococci*, *Staphylococci*, and *Listeria* sp. These compounds are also active against ClpP of Gram-negative bacteria, however, these bacteria are naturally protected from the function of ADEPs by efflux pumps and by the impermeability of the outer membrane [34]. The parent compound (ADEP1) has been pharmaceutically optimized, which has generated highly improved synthetic derivatives that exhibited impressive antibacterial activity both *in vitro* and in animal models [34, 35]. As ClpP represents an unprecedented target of antibiotics, ADEPs are effective against multi-drug resistant isolates of *S. aureus, S. pneumoniae*, and *Enterococcus*, which open the door to new possibilities for the treatment of infections generated by these important pathogens. But, as ADEP-resistant mutants are generated with moderately high frequencies (10^{-6}) in the laboratory, the use of ADEPs may be limited to combination therapy. Interestingly, ADEPs do not inhibit ClpP function [34] but rather the binding of ADEPs convert ClpP, from a highly regulated peptidase that can degrade substrates only with the help of an associated Clp ATPases, to an unregulated protease that indiscriminately degrades nascent polypeptides in the absence of its ATPase partner (Fig. 7.1) [36]. On the one hand, ADEP binding inhibits the ClpP-mediated degradation of the normal physiological substrates, while it stimulates the non-specific degradation of nascent or unfolded polypeptides such as those still associated with, or just released from, the ribosome [36]. The latter was recently shown to be responsible for the sensitivity of wild type cells to ADEPs, as it was demonstrated that uncontrolled proteolysis of the key cell division protein – FtsZ is responsible for the lethality of the compound [37]. Recently, two crystal structures of ADEP in complex with ClpP one from *B. subtilis* and the other from *E. coli* were solved, which has led to fascinating new insights into the mechanism by which ADEPs interfere with the function of ClpP [38, 39]. Both studies favor that ADEP binds to a hydrophobic pocket created by two adjacent subunits of ClpP. This binding of ADEP, then stimulates formation of the active ClpP tetradecamer, triggering a closed-to-open-gate transition in the pore of ClpP that allows ClpP to degrade substrates independently of it cognate ATPases [38, 39]. However, the two studies provide different explanation for how the "gate-opening" is accomplished. Intriguingly, the ADEP binding site on ClpP is identical to the proposed ATPase binding site on ClpP [38, 39]. Consequently, the ADEP-ClpP complex serves as a good model for studying the interactions between ClpP and its partner ATPases. However, the antibacterial activity of ADEPs may be challenged by the very recent finding that, in *Streptomyces lividans*, there are several mechanisms of ADEP resistance. Importantly, in one spontaneous mutant, the ClpP homolog (ClpP3) appears to be responsible for this resistance [40].

Cyclomarin A

Recently, another cyclic peptide produced by *Streptomyces* (called cyclomarin A) was reported to show bactericidal activity against *Mycobacterium tuberculosis*, the causative agent of tuberculosis. This bacterium grows very slowly, a feature that makes it naturally resistant to many antibiotics. In contrast to most know anti-tubercular drugs, cyclomarin A killed both growing and non-replicating mycobacteria, suggesting that cyclomarin A may function by a novel mechanism [41]. This assumption was strongly supported by the finding that cyclomarin A was effective against a panel of antibiotic resistant clinical isolates. Intriguingly, cyclomarin A was subsequently shown to bind with high affinity and specificity to the ClpC1 subunit of *M. tuberculosis* [41]. This is the first description of a Clp ATPase as a target of antibiotics. Importantly, *clpC1* appears to be essential in *M. tuberculosis*, and cyclomarin A resistant bacteria only arose with extremely low frequency ($<10^9$). Preliminary analysis revealed that cyclomarin A, similar to ADEPs, stimulates ClpP-mediated proteolysis, however, whether this increased proteolysis is responsible for the bactericidal effect of cyclomarin A on *M. tuberculosis* awaits further experimentation.

ClpP and Vaccines

To date, ClpP has been implicated in two different approaches to make vaccines, both of which have provided some very promising results. In the first case, purified ClpP was used to immunize mice against *S. pneumoniae*, while in the second case a *Salmonella clpP* mutant was used to immunize mice against *S. typhimurium.*

Streptococcus pneumoniae is a leading cause of potentially lethal diseases like pneumoniae, meningitis and septicaemia. High-risk groups include children below the age of 5. Current vaccines are based on immunity against capsular polysaccharide. Although this vaccine is effective in adults, it fails to protect immunocompromised people or children below the age of 2 years. Interestingly, a number of studies have identified ClpP as a promising vaccine candidate that can protect against a broad range of *S. pneumoniae* strains. ClpP was initially identified as the most promising candidate for vaccination from a proteomic screen, looking for surface exposed proteins [42]. This finding was rather surprising, since the ClpP protease is considered to be a cytoplasmic protease. However, consistent with this finding, Rhee and co-workers [19] found that ClpP is translocated into the cell wall after heat shock, suggesting that ClpP may in deed be exposed on the surface of *S. pneumoniae*. These authors also showed that immunization of mice with purified ClpP elicited a protective immune response against a fatal systemic challenge with *S. pneumoniae* D39. In a more comprehensive study, it was shown that immunization with ClpP and even passive immunization with ClpP antibodies could elicit serotype-independent protection against invasive *S. pneumoniae* in mice [43]. The potential of ClpP has also been examined in combination vaccine. Wu et al. [44] examined three protein candidates, ClpP, pneumolysin and a putative lipoate-protein ligase (Lpl), and found that maximum protection was achieved following intraperitoneal or intranasal challenge with a combination of all three proteins. Interestingly, a combination vaccine conferred complete protection against intranasal infections of three of the four most common pneumococcal strains [44].

In a different approach, Matsui et al. [45] found that oral immunization with a *S. typhimurium* strain carrying a disruption of the *clpPX* locus resulted in significant protection against a subsequent challenge using virulent *S. typhimurium* strains. Further analysis revealed that the challenge strain (and the strain used for immunization) was cleared from the spleen, mesenteric lymph nodes, and Peyer's patches in the immunized mice only 5 days after the challenge, suggesting that the immunization with the protease-defective mutant protected the mice from getting colonized by the challenge strain. Similar results were also obtained when mice were immunized with a strain deficient in the Lon protease. Although both of the protease mutant strains were considered to be avirulent, they did persist at low levels in several tissues for up to 12 weeks following administration. Before attenuated strains can be used for live-attenuated vaccines they must be shown to be safe not only in healthy individuals but also in very young children, elderly, and other immunocompromised groups. Surprisingly, the *clpP* mutant was shown to be virulent in orally infected 1 week old, germ-free pigs (a model for immunologically immature or immunocompromised individuals) [46] questioning the general use of the protease mutant strain as a live-vaccine.

Lon

Distribution, Structure, and Characteristics

The cytoplasmic protease, Lon, is highly conserved and widely distributed. It is found in all domains of life, from Bacteria and Archaea to Eucaryota where it is found in organelles. Interestingly, although Lon is absent from the genome of some important Gram-positive pathogens such as *M. tuberculosis*, *S. pneumoniae*, *L. monocytogenes*, and *S. aureus*, it appears to play an important role in the virulence of several Gram-negative pathogens such as *Salmonella* Typhimurium, *Yersinia pestis* and *Pseudomonas aeruginosa* (see below). Orthologues of Lon are divided into two subgroups, A-type Lon (here referred to as A-Lon, also known as LonA) and B-type Lon (here referred to as B-Lon, also known as LonB). A-Lon proteins contain a large multi-lobed N-terminal domain, while B-Lon proteins lack this N-terminal domain, but instead possess a membrane-anchoring region [47]. The crystal structure of B-Lon from *Thermococcus onnurineus* was recently solved and revealed that B-Lon protease shares the same general structure as FtsH protease [48]. Both proteins form compartmentalized proteases composed of six identical subunits in which each subunit contains both the ATPase and peptidase components, on a single polypeptide [48]. The structure has been described as a "bowl and lid" structure, where the proteolytic sites are lining the inner side of a relatively large bowl-like chamber that is covered with a "lid" made from the ATPase domains. The combined presence of ATPase and proteolytic sites in the chamber suggest that protein unfolding and degradation can occur simultaneously [48]. The chamber is accessible through a narrow axial pore, and in analogy with other ATP-dependent proteases a gate-opening mechanism may control the entrance of substrates through the channel.

In *E. coli*, Lon is primarily responsible for the degradation of non-native proteins [49]. It is therefore not surprising that in pathogenic relatives of *E. coli*, Lon also plays an important role in the stress response pathways and virulence (see below). However, in addition to these roles, Lon also plays an important role in controlling the stability of a number of regulatory proteins e.g. HilC and HilD, virulence regulators in *Salmonella* [50]. Hence, like other ATP-dependent proteases Lon must be capable of general recognition of damaged cellular proteins, while at the same time being capable of precise recognition of specific substrates. In a recent publication, Sauer and colleagues [51] demonstrated that unfolded proteins are not, *per se*, good Lon substrates. Rather, Lon dependent degradation relies on the recognition of specific sequences, rich in aromatic residues and poor in small polar amino acids that are normally exposed in unfolded proteins and buried in native proteins. It has been proposed that Lon synergistically recognizes multiple sequences in an unfolded polypeptide, which assures that irreversibly unfolded proteins are bound more strongly than proteins that transiently expose unfolded regions. Notably, folded proteins carrying an accessible recognition tag may also be degraded by Lon. Hence, this model elegantly explains how positioning of a recognition signal, within a native protein, allows Lon-mediated degradation, independent of the folding status of the substrate protein [51] and this substrate binding model may apply to other ATP-dependent proteases. For a more detailed description of this model please refer to [52].

Lon Controls Expression of Type Three Secretion Systems in Gram-Negative Pathogens

To proliferate in the vacuoles of macrophages, *S. typhimurium* colonize the small intestine and invades normally non-phagocytic epithelial cells. Surprising, mutants of *S. typimurium*, lacking Lon, exhibit enhanced invasion of epithelial cells, a massive induction of apoptosis in macrophages, and increased production of proteins expressed by SPI1 [53, 54]. Among the important virulence factors encoded by SPI1 is the T3SS that forms a needle like structure capable of injecting effector proteins directly into the host cell cytosol, forcing the uptake of the bacterium by rearranging actin cytoskeleton. Expression of T3SS proteins is tightly controlled by a complex cascade of transcriptional regulators. Central to this hierarchy is HilA, which activates transcription of SPI1 genes directly. Interestingly, HilA expression is controlled by the transcriptional activators (HilC and HilD), which are in turn regulated by Lon-mediated degradation [50, 55]. Consistently, inactivation of Lon stabilizes HilC and HilD, increases the cellular levels of HilA and consequently enhances the expression of SPI1 encoded genes.

From the enhanced expression of SPI1 genes, observed in the absence of Lon, one might predict that *lon* mutant cells are more virulent than wild type cells in an animal model. In contrast to this prediction, inactivation of *lon* in *S. typhimurium* not only dramatically increased the 50% lethal dose (LD50) in mice (i.e. greater that 1,000-fold) but also abolished proliferation and survival in murine macrophages [32]. Indeed the *lon* mutant turned out to be highly susceptible to acid and oxidative stress, suggesting that the most important contribution of Lon during infection, is to resist oxygen-dependent killing associated with the respiratory burst and the low pH of phagosomes [44].

In the related bacterium *Yersinia pestis* (the aetiological agent of plague), Lon also plays an important role in virulence, controlling expression of T3SS. In this case Lon is required for degradation of YmoA, a transcriptional repressor of a plasmid-encoded T3SS [56]. Interestingly, the Lon-mediated degradation of YmoA is temperature-dependent; at low temperatures YmoA is stabilized leading to repression of the T3SS, while at 37°C, the normal temperature of the host, YmoA is degraded and T3SS gene expression is de-repressed. Lon is also involved in temperature dependent regulation of a central transcriptional regulator, RovA, in *Yersinia pseudotuberculosis* that causes pseudotuberculosis in animals. RovA is a thermal-sensitive, DNA-binding protein, which at 37°C undergoes a conformational change releasing it from its DNA enhancing its susceptibility to Lon mediated degradation [57]. Since the temperature dependent conformational change of RovA is reversible, Lon is required to irreversible remove RovA from the cytoplasm and thereby prevent rebinding to DNA [57].

Lon in Quorum-Sensing

In some bacterial pathogens quorum-sensing (QS) systems are used to control the expression of virulence factors in response to bacterial population density. For example, the opportunistic Gram-negative pathogen, *Pseudomonas aeruginosa*, responsible for the infection of immunocompromised patients and known to colonize the lungs of cystic fibrosis patients, uses two different acyl-homoserine lactone (acyl-HSL) QS systems (LasR/LasI and RhlR/RhlI) to coordinate expression of virulence genes. Notably, the production of HSLs, is controlled by the Lon-mediated degradation of the HSL synthetases (LasI and RhlI) [58]. In the related organism, *Pseudomonas putida*, the transcriptional regulator, PpuR, of the PpuI/PpuR QS system is also modulated by Lon [59].

HtrA/DegP

Distribution, Structure, and Characteristics

High temperature requirement protein A (HtrA) – also referred to as DegP – is a highly conserved protein found in both prokaryotes and eukaryotes. In Gram-negative bacteria, HtrA is located in the periplasm, closely associated with the inner membrane, while in Gram-positive bacteria, HtrA is found at a single discrete micro domain on the cellular membrane [60, 61].

Similar to Lon protease, HtrA/DegP contains proteolytic and chaperone activity in the same polypeptide but in contrast to Lon, these activities are not ATP-dependent [62]. HtrA is a trypsin-like serine proteases, which contains a Ser-His-Asp catalytic triad. DegP also contains two carboxy-terminal PDZ (Postsynaptic density of 95 kDa, Discs large and Zonula occludens 1) domains (PDZ1 and PDZ2) that participate in substrate binding as well as compartmentalization. In the resting state, *E. coli* DegP assembles into hexamers consisting of two trimeric units forming a small cage, in which the catalytic triad faces inward and the PDZ domains restrict access to the proteolytic chamber [63, 64]. Upon binding of substrates, such as unfolded or misfolded globular proteins, *E. coli* DegP switches from inactive hexamers to larger active oligomers (12-mers and 24-mers). Substrate is then refolded or degraded within the spherical chamber created by these higher order forms [65]. After release of the degraded peptide, or refolded substrate, DegP returns to the inactive hexamer. Substrate binding, oligomerization and formation of the active conformation are therefore directly coupled preventing uncontrolled proteolytic and refolding activities (for a recent review on HtrA see [66]).

It has been proposed that DegP requires almost complete unfolding of substrates before degradation can proceed [64] and recently it was shown that DegP degrades substrates progressively, using a molecular ruler comprising the PDZ1 domain and the proteolytic site of DegP [67]. The crystal structure of the 24-mer reveals an unusual, large spherical shell with an internal cavity (~110 Å in diameter) with wide pores giving access to outer membrane proteins (OMPs). It has been proposed that DegP can protect OMP intermediates during their transit across the periplasm [65]. The fate of the encapsulated substrate, within these large oligomers, may be determined by the temperature or activity of the complex [63, 64] or by the intrinsic folding properties of the substrate [68] resulting in either degradation, refolding or alternatively just protection of the substrate, in the extra-cytoplasmic environment.

In both Gram-negative and Gram-positive bacteria, HtrA participates in folding and degradation of damaged extra-cytoplasmic proteins as well as maintaining the solubility of OMP intermediates and secreted proteins; all of which potentially affect virulence of pathogenic bacteria. In the following section, we will focus on the role of HtrA in virulence of pathogenic bacteria including, where relevant, its contribution to stress tolerance.

HtrA/DegP Contributes to Stress Tolerance and Affects Host Cell Interaction in Gram-Negative Pathogens

More than 20 years ago, HtrA was proposed to be a potential virulence factor in *S. typhimurium* [69]. Ten years later it was shown, that although a *S. typhimurium htrA* mutant was unable to kill mice and could not translocate from the Peyer's patches to other organs after oral infection [70], intravenous infection of mice with the *S. typhimurium htrA* mutant, still resulted in full colonization of livers and spleens [70]. These results suggest that HtrA is not required for infection *per se,* but is required to survive the stressful environments associated with oral uptake. Specifically, an *htrA* mutant of *S. typhimurium* is sensitive to oxidative stress and shows reduced survival in the oxidative environment of macrophages [71]. Similarly, another facultative intracellular Gram-negative pathogen such as *Brucella abortus* show reduced tolerance to oxidative stress *in vitro* as well as reduced survival in macrophages [72]. Furthermore, HtrA contributes to survival in macrophages in *Yersinia enterocolitica* that relies on its ability to resist phagocytosis, when causing diarrhoea [73]. Hence, HtrA may play an important role in the degradation of damaged proteins created, during infection, from the oxidative stressful conditions in phagocytes (Fig. 7.3a). Recently, the role of degradation and refolding/sequestering activities of *S. typhimurium* HtrA were examined by the expression of wild type, or a protease inactive mutant of HtrA ($HtrA_{S210A}$), in an *htrA* mutant strain [74]. While the protease activity of HtrA was not required for growth at high temperature *in vitro*, it was needed during infection *in vivo*, suggesting that the degradation of misfolded and/or specific target proteins plays a key role in virulence [74]. Thus, it appears that reduced stress tolerance is responsible for the attenuated virulence of the *Salmonella htrA* mutant.

Campylobacter jejuni is a common cause of human gastroenteritis. Currently there are several studies, which suggest that HtrA plays an important role in the virulence of *C. jejuni*. Firstly, mutation of *htrA* in *C. jejuni*, reduced adherence to human epithelial cells [75]. Secondly, *htrA* mutant cells exhibited reduced invasion [76]. Finally, in an insect infection model, the *htrA* mutant strain killed fewer larvae than the isogenic wild type strain [77]. Interestingly, in *C. jejuni* HtrA chaperone activity is important for efficient binding to epithelial cells, while in contrast, the protease activity of HtrA is required for optimal internalization once the bacteria have adhered to the epithelial cells [78]. *In vitro* HtrA chaperone activity is necessary for growth of *C. jejuni* at high temperature or under oxidative stress; whereas HtrA protease activity is only essential during severe stress conditions [79]. These data suggest that, in contrast to *Salmonella*, HtrA affects virulence of *C. jejuni*

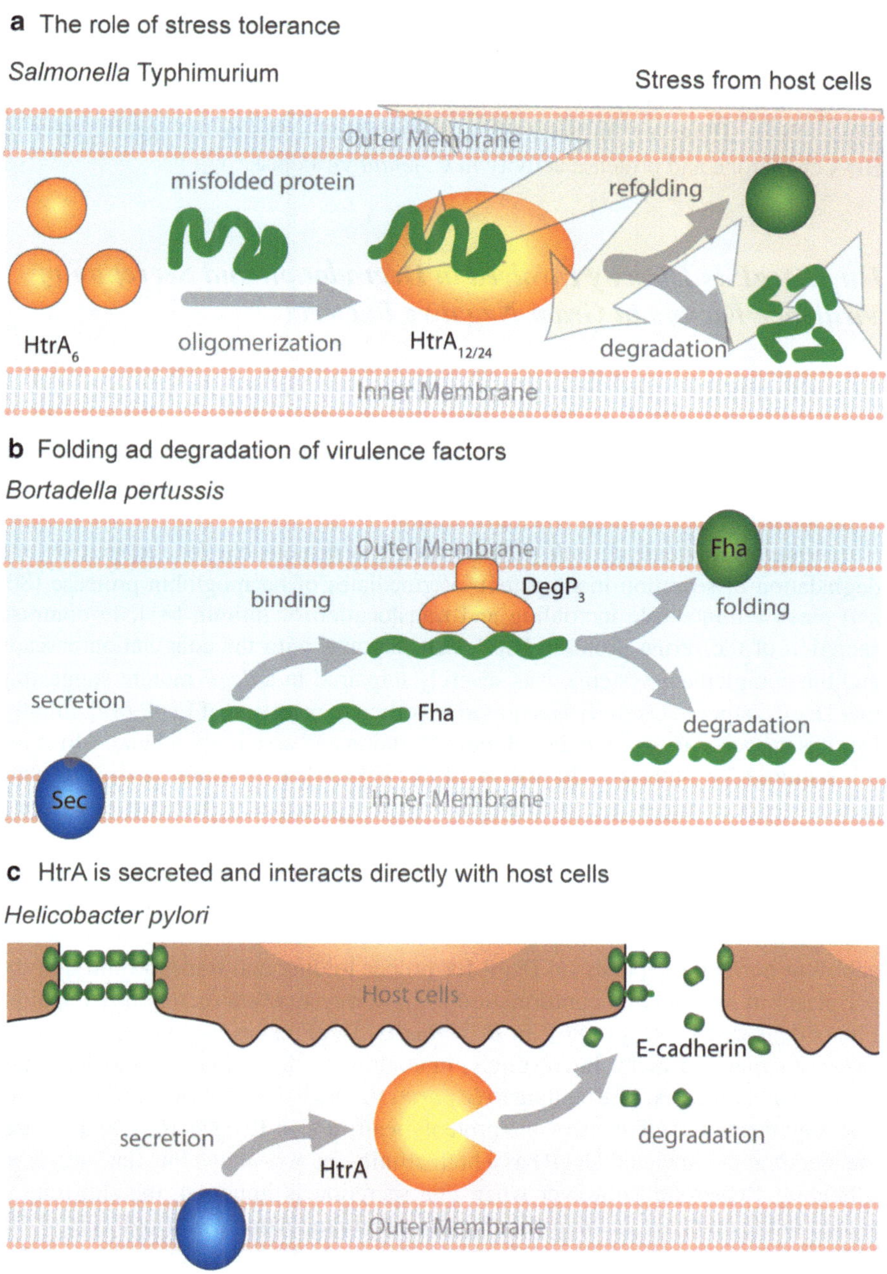

Fig. 7.3 Diverse roles of HtrA/DegP in virulence of Gram-negative pathogens. (**a**) The role of HtrA in stress tolerance in *Salmonella typhimurium*. The stressful environment during infection increases formation of damaged and misfolded proteins that by the binding to hexameric HtrA induces its transition into HtrA oligomers that function to either degrade or refold the proteins. (**b**) Folding and degradation of virulence factors in *Bortadella pertussis*. The virulence factor Fha is translocated across the inner membrane by the Sec system and subsequently binds to membrane associated DegP. Binding to DegP ensures that Fha stays in an extended conformation before folding and insertion into the membrane. Only when secretion is impaired does the DegP trimer degrade Fha. (**c**) Secreted HtrA from Helicobater pylori interacts directly with host cells. HtrA is secreted by an unknown mechanism and degrades E-cadherin in the tight junction of epithelial host cells

through several mechanisms, which may include stress tolerance as well as more direct effects. However, further work is still required to define the specific roles of HtrA chaperone and protease activity in *C. jejuni* virulence.

HtrA/DegP Is Directly Involved in Degradation and Secretion of Virulence Factors in Gram-Negative Bacteria

Secreted virulence factors must temporarily reside in the periplasmic space before they are translocated into the extracellular environment. When secretion is hampered or proteins are damaged, DegP (of Gram-negative pathogenic bacteria) not only degrades a number of secreted virulence factors, but also acts as a chaperone participating in their folding and translocation.

In enteropathogenic *E. coli* (EPEC), DegP is responsible for recognition and degradation of secretion-incompetent intermediates of haemoglobin protease [80] and plays a minor role in folding and translocation of intimin [81]. In contrast, secretion of the serine protease EspP (that interacts with the coagulation cascade and the complement system), was severely impaired in a *degP* mutant suggesting that DegP chaperone activity is required for efficient secretion of EspP [82]. Finally, DegP is proposed to assist in the folding of bundlin and assembly of bundle-forming pili required for fimbrial adhesion in EPEC [83, 84]. In addition, a *degP* mutant showed reduced adherence to epithelial cells in a localized adherence assay compared to the wild type [84], confirming that loss of DegP has a major impact on EPEC virulence.

In the pathogenic bacterium responsible for whooping cough, *Bortadella pertussis*, filamentous haemagglutinin (Fha) is the major adhesin. This long β-helical adhesion normally depends on DegP for proper folding and transport through the periplasm in its extended conformation [85], however, when secretion is impaired Fha is degraded by DegP [86]. In *B. pertussis*, DegP exists as soluble and membrane-associated trimers. Interestingly, in contrast to the binding of globular substrates to *E. coli* HtrA, Fha binding to *B. pertussis* DegP does not trigger the efficient rearrangement of DegP trimers into proteolytically active 12-mers [65]. In this case, the membrane-associated DegP has a high affinity for non-native Fha (and serves as a holding chaperone), however when Fha secretion is impaired, the substrate is stalled in the complex and degraded by the membrane-associated DegP (Fig. 7.3b) [85]. The membrane-associated trimer thus protects unfolded secretory proteins that would be degraded in the large (12-mer and 24-mer) oliogmeric forms of DegP.

Interestingly, in *Shigella flexneri*, DegP has been implicated in the biogenesis of the virulence factor IcsA, a protein that is localized at the pole and involved in polymerization of actin required for cell-to-cell spread. In this case, only the chaperone activity and not the protease activity of DegP was required for IcsA biogenesis [87, 88], most likely ensuring safe transport across the periplasm to the outer membrane. Consistently, less IcsA is localized at the bacterial pole in a *degP* mutant, which may account for the reduced plaque size of a *degP* mutant.

HtrA Is Secreted into Host Cells and Modulates Host Response in Gram-Negative Pathogens

In *Helicobacter pylori*, which colonizes the human gastric epithelium and disrupts the mucosal integrity leading to a severe inflammatory response or gastric cancer, HtrA was found to contribute to virulence by a novel mechanism. HtrA of *H. pylori* is secreted into the extracellular environment [89, 90] and recently, was shown to act as a specific E-cadherin protease that efficiently destroys adherence junctions in polarized epithelial cells (Fig. 7.3c) [91].

Recently is was shown that HtrA of *Clamydia trachomatis* was actively secreted into the host cell cytosol by a process independent of T3SS [92]. Secretion of HtrA into host cell cytosol supports the current model for pathogenesis of chlamydia in which a proteolysis strategy is suggested to manipulate host cell signalling pathways. However, to date, the host cell target of *C. trachomatis* HtrA has not been identified, although most chlamydial species secrete HtrA into host cell cytosol, suggesting that HtrA secretion may be a common virulence mechanism. Nevertheless, it remains to be seen if this virulence mechanism is also employed by other pathogenic bacteria.

HtrA Affects Expression of Virulence Factors and Stress Tolerance in Gram-Positive Pathogens

Although Gram-positive bacteria lack a compartment dedicated to the folding of secreted proteins, in *Streptococcus pyogenes* HtrA was found exclusively at a single microdomain – termed the ExPortal [61]. The ExPortal is located at the hemispherical position distal to either cell pole of *S. pyrogenes* and is dedicated to export of secreted proteins. While HtrA was shown to be necessary for maturation and secretion of the extracellular cysteine protease SpeB [61], the requirement is most likely indirect, as HtrA did not process proSpeB to its mature active form *in vitro* [93]. Streptolysin S is another extracellular virulence factor of *S. pyogenes* that seems to be controlled by HtrA. However, neither of these changes affected the overall pathogenicity of *S. pyogenes*, as an *htrA* mutant was not attenuated in a murine model of subcutaneous infection [94]. The expression of several secreted virulence factors were also altered in a mutant strain of *S. aureus* RN6390 lacking both *htrA* orthologous, and this double mutant showed diminished virulence in a rat model of endocarditis. In contrast, HtrA had minimal effect on exoprotein expression and virulence of the *S. aureus* COL strain [95]. In *S. mutans* the loss of HtrA altered expression of several extracellular proteins thought to play important roles in interaction between the organism and its host, including glucan-binding protein B and glucosyltransferases. The loss of HtrA in *S. mutans* also influences the appearance of biofilm (also known as dental plaque) [96]. These examples show that HtrA control extracellular virulence factors and virulence in Gram-positive pathogens.

HtrA in Gram-positive pathogens also plays a role in stress tolerance and survival in host tissues. An *htrA* mutant of *L. monocytogens* is sensitive to oxidative agents and shows reduced survival in macrophages [97]. Thus, attenuated virulence may be attributed to a decreased ability to persist in macrophages that are fundamental to the ability of this pathogenic bacterium to cause disease. Recently, HtrA was found to be a major virulence determinant of *Bacillus anthracis*, as virulence of the *htrA* mutant strain was severely attenuated in a guinea pig model for anthrax [98]. Indeed the loss of HtrA in *B. anthacis* altered secretion of several proteins, prevented expression of the extracellular protease NprA and reduced tolerance to heat, oxidative, ethanol and osmotic stress. Collectively, these data suggest that the altered secretion of extracellular proteins as well as the increased susceptibility to stress may account for the attenuated virulence in this serious human pathogen.

Extracellular Proteases in Virulence

Proteases are commonly secreted by microorganisms. As we learn more about bacterial pathogens and how they cause disease in humans, it is becoming increasingly clear that extracellular proteases play an important role in infection. In some pathogens such as *S. aureus* and the Group A *Streptococcus* (GAS) *S. pyogenes*, a number of proteases contribute to virulence. These pathogens give rise to infections often involving massive tissue destruction. While in other pathogens, like the intracellular pathogen, *Listeria monocytogene* only a single secreted protease is associated with virulence. The extracellular proteases can be categorized depending of their targets associated with pathogenesis. In some pathogens they are needed for maturation of cell surface or extracellular virulence factors produced by the pathogen itself. Other targets include structural components of host tissue such as fibrinogen or elastin in processes that paves the way for the spreading infection. Even host cell signalling can be disrupted by the extracellular proteases degrading host signal molecules such as cytokines and chemokines. Below we describe, the diverse contributions that extracellular proteases make to pathogenesis with examples from pathogens known to express several extracellular proteases contributing to virulence.

Multiple Roles of Extracellular Proteinases in S. aureus Virulence

Proteases and Their Control

Staphylococcus aureus is known to secrete a number of extracellular enzymes that contribute to the virulence of the organism. Among these are four major proteases, namely the serine glutamyl endopeptidase (SspA or V8 protease), the metalloproteinase (aureolysin, also known as Aur) and two cysteine proteases, staphopain A (ScpA or StpA) and staphopain B (SspB or StbB) [99]. Like other extracellular

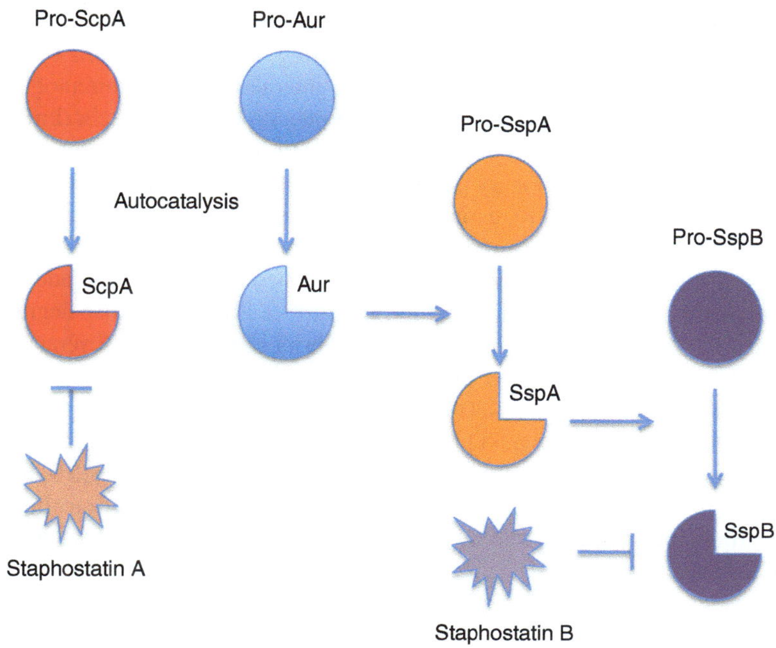

Fig. 7.4 Activation of extracellular proteases. The extracellular proteases are controlled at the level of activity as they are expressed as inactive zymogens that are activated through proteolytic cascades. Aur contributes to the processing of proSspA that in turn matures proSspB. Aur and ScpA appear to be on top of the proteolytic cascade (Modified from [99])

virulence factors in *S. aureus* maximal expression of these proteinases occurs post exponential growth phase and transcription is positively regulated by *agr* (accessory gene regulator) QS system known to control *S. aureus* virulence gene expression [100, 101].

To ensure timely proteolytic activity the extracellular proteases are often expressed as inactive zymogens that are activated by a proteolytic cascade. During export, Aur contributes to the processing of proSspA [101–103] that in turn matures proSspB (Fig. 7.4) [104, 105]. Aur and ScpA appear to be on top of the proteolytic cascade as both proteins are able to undergo autocatalytic activation [106, 107]. To prevent the extracellular proteinases from maturing in the bacterial cytoplasm, protease inhibitors are expressed together with the proteinases. For example, a potent inhibitor of staphopain B, (SspB), called staphostatin B is encoded by the *sspC* gene located in the *ssp* operon [108]. Likewise staophostatin A, encoded by *scpB* inhibits staphopain A, ScpA (Fig. 7.4) [109]. Thus, expression of *S. aureus* extracellular proteases is controlled at multiple levels.

Processing of Virulence Factors

The secreted proteinases are not only involved in their own maturation, but also required for the degradation or processing of other bacterial virulence factors.

Among these are a number of distinct surface proteins, which promote colonization by binding to fibronectin or collagen and interfering with blood clotting by binding fibrinogen or counteract phagocytosis through interactions with immunoglobulins [110, 111]. One example is the cell surface fibronectin-binding protein (FnBP) that is maximally expressed during exponential growth, but disappears rapidly as the culture progresses into stationary phase. The loss of FnBP is dependent on the proteolytic activity of SspA (V8), expressed upon entry into stationary phase and thus, during this growth phase SspA reduces the adhesive properties of the *S. aureus* cells associated with FnBP [112]. Another study confirmed this finding and additionally showed that the immunoglobulin binding, cell surface Protein A is also degraded by SspA [113]. The ability of *S. aureus* to bind fibrinogen is among several proteins mediated by cell wall attached clumping factor, ClfB. Interestingly, Aureolysin cleaves the N-terminal region (~30 kDa) of ClfB, releasing an 120 kDa protein fragment, unable to bind fibrinogen [114]. The proteolytic susceptibility of the proteins attached to the cell surface of *S. aureus*, such as FnBP, Protein A and ClfB, involved in binding to structural host cell components, indicates that these extracellular proteases are important for the release of *S. aureus* from colonization sites and spreading of the infection.

Processing of Host Proteins in Virulence

S. aureus extracellular proteases contribute to virulence, not only through processing of bacterial cell surface virulence factors, but also through proteolysis of host components. One notable feature associated with many types of *S. aureus* infections is the profound tissue destruction observed at the site of infection. While little is known of the processes behind host matrix degradation, the extracellular proteases are known to be involved. ScpA degrades fibres composed of elastin, an elastic protein in connective tissue [106, 115] and both staphopains (ScpA and SspB) contribute to the turnover of collagen, the main component of connective tissue [116]. These findings strongly support an important role for extracellular protease in host matrix degradation and local dissemination of the pathogen.

S. aureus is one of the most common Gram-positive pathogens associated with sepsis, a condition whereby the pathogen, in the blood stream, may spread systemically leading to life-threatening conditions such as infection of the heart valves. In the bloodstream, fibrinogen cleavage by staphopains (SspB and ScpA) interferes with plasma clotability, resulting in a tendency to induce bleeding [116]. Cell wall anchored proteins may contribute to the efficiency of this process, by binding fibrinogen to facilitate close interaction with the secreted proteinases [116].

Neutrophils and monocytes are central in host defences as they phagocytose and kill invading microorganisms [117]. They do so by expressing a variety of receptors such as the CD11b/CD18 integrin that recognize ligands and opsonins present on the surface of microorganisms to be engulfed [118]. On the other hand, pathogens attempt to evade phagocytosis and recently SspB was shown to eliminate phagocytes by selectively cleaving CD11b on phagocytes in a process leading to apoptotic

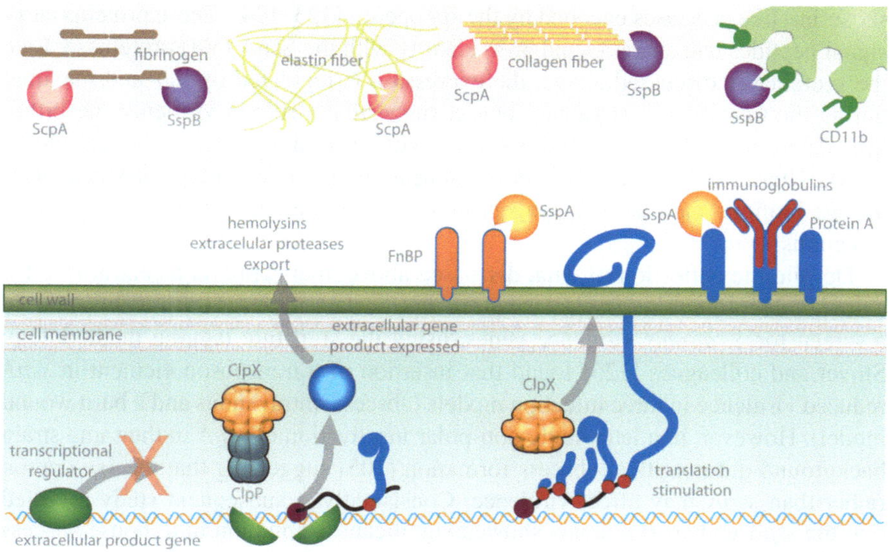

Fig. 7.5 Proteolysis in *S. aureus* virulence. In the cytosol ClpXP stimulates transcription of genes encoding extracellular products such as hemolysins and proteases possibly through the degradation of a transcriptional repressor. ClpX, independently of ClpP stimulates translation of *spa* encoding Protein A that binds immunoglobulins. Outside the *S. aureus* cell the secreted proteases SspA, SspB and ScpA are key players in the turnover of the host matrix components and interfere with host cell signalling

cell death (Fig. 7.5) [119]. In fact SspB seems to have multiple activities in the host environment, as another study revealed that SspB can mimic human proteases and activate chemerin that circulates in blood. Upon proteolytic activation, chemerin becomes a potent chemo-attractant for circulating immature dendritic cells and macrophages [120]. It is speculated that SspB directs the recruitment of specialized immune cells, potentially contributing to the ability of *S. aureus* to elicit and maintain a chronic inflammatory state [120].

Biofilm Formation and Virulence

Biofilm formation is another clinically important property of *S. aureus*. In nosocomial infections, *S. aureus* is commonly encountered in biofilms, which has been related to a number of serious clinical conditions [121]. Staphylococcal biofilm formation is a complicated process involving a number of macromolecules such as extracellular DNA, the polysaccharide intracellular adhesion (PIA) and a number of the specific surface proteins such as Bap, SasG, FnBPs and Protein A (summarized in Marti et al. [122]). The degradation of these surface proteins (Protein A, FnBP and Bap) by SspA and Aur [112, 113] reduces biofilm formation [122]. Another set of extracellular proteases, have also been implicated in biofilm formation, namely six serine-like proteases encoded by the *spl* operon [123, 124]. These proteins carry signal peptides and share 33 and 36% identity with the SspA (V8) proteinase. Like the proteinases discussed above, the expression is regulated by *agr* and maximal during the transition to stationary phase, suggesting a role in virulence, nevertheless, deletion of this operon did not affect virulence in a mouse peritoneal model [124]. Thus, extracellular proteinases appear to promote timely dispersal of *S. aureus* biofilms and as such may promote dissemination of the staphylococcal infections.

Despite the collection of data, discussed above, that supports a central role for extracellular proteinases in virulence little *in vivo* data confirms this involvement. In a screen for *S. aureus* virulence genes, using the signature tagging methodology Stover and colleagues [125] found that insertion of a transposon element in *sspA* reduced virulence in three infection models (abscess, intravenous and a burn wound model). However, in a later study non-polar inactivation of *sspA* in the same strain background did not affect abscess formation [105] suggesting that the *ssp* operon rather than sspA may affect virulence. Consistently, a subsequent study revealed that the *sspA* and *sspB* mutants showed significant attenuation in a mouse abscess model whilst mutations in *aur* or *scpA* did not [101]. Thus the contributions of the extracellular proteases to virulence is difficult to mimic in virulence models perhaps reflecting their complex contribution to virulence.

Immunoglobulins and Chemokines as Targets in Streptococci

Several Streptococci species, including the GAS *S. pyogenes*, are amongst some of the most serious human pathogens. Like *S. aureus*, *S. pyogenes* colonizes a substantial fraction of the healthy human population, where it may cause a range of conditions, from mild such as a soar throat or local skin infections to more serious infections such as blood stream infections and necrotizing fasciitis (the "flesh eating disease") [126]. It was from this organism, more than 60 years ago, that the first cysteine proteinase (SpeB) was isolated [125]. SpeB is secreted as a proenzyme (proSpeB) and the mature protein is responsible for the activation of several human proteins, including immunoglobulins [127]. SpeB specifically cleaves IgG at Gly236 within the flexible hinge region, impairing IgG-mediated opsonization and subsequent phagocytosis [128–130]. Interestingly, SpeB is only secreted during late exponential growth phase of GAS whereas another proteinase (IdeS) that cleaves IgG is secreted during the early exponential growth phase of GAS [131, 132]. Collectively, the activity of both SpeB and IdeS play an important role in the virulence of GAS [133].

Another group of host molecules, targeted by exoproteinases, is the chemokines. In one example, the *S. pyogenes* cell envelope protease (SpyCEP, also called ScpC) degrades the chemokine IL-8, a multifunctional host defense protein involved in activation and recruitment of neutrophils as well as other chemokines [134, 135]. More recently it was shown that SpyCEP is both necessary and sufficient for IL-8 degradation, which decreased IL-8 dependent neutrophil endothelial transmigration and bacterial killing resulting in attenuation in a murine infection model [136]. Interestingly, SpyCEP expression is controlled by cathelicidin (LL-37), an antimicrobial host peptide, which also induces production of the GAS capsule [137]. Similarly, in the Group B *Streptococcus* (GBS), *S. agalactiae* the serine protease cleaves fibrinogen and abolishes the ability of three CXC chemokines to attract and activate neutrophils [138].

Tissue Destruction and Antimicrobial Peptide Resistance Displayed by Gram-Negative Bacteria

P. aeruginosa is an opportunistic human pathogen that is capable of infecting various tissues and organs. The main target of *P. aeruginosa*, is the lung of immunocompromised individuals or individuals with chronic conditions such as cystic fibrosis or acute infections such as severe pneumonia. The major role of proteases in *P. aeruginosa* virulence is thought to be tissue penetration, which is supported by the proteolytic activities of LasB, LasA, protease IV and the alkaline protease encoded by *aprA*. Well characterized is the activity of the *lasB* gene that originally was termed elastase due to its ability to degrade host components including collagen and elastin [139, 140]. LasB stimulates host proteolytic processes such as plasminogen (Plg) activation. Plg is the precursor of the serine protease plasmin, which is primarily responsible for dissolving blood plasma proteins and fibrin clots. Plasmin activation depends activators of Plg (such as, tissue-type Plg activator (tPA) and urokinase-type Plg activator (uPA)), as well as several inhibitors [141]. In this cascade both the LasB metalloproteinase and protease IV, despite displaying quite different substrate specificities, activate pro-uPA, and it has been speculated that this activation allows easier colonization of susceptible host tissue, such as the respiratory tract [142, 143].

The activation of Plg to plasmin is not restricted to *P. aeruginoase*. In *S. typhimurium* and *Yersinia pestis*, surface located proteinases known as omptins, (PgtE and Pla, respectively), are able to activate Plg through cleavage of a specific inhibitor of PA, thereby bypassing normal control of host proteolysis [144]. Both PgtE and Pla have several functions within the human host and the activities of PgtE in *S. typhimurium* is critical for the outcome of an infection, as inactivation of the corresponding gene decreased colonization of organs in an intraperitoneal mouse model [145]. Interestingly, both PgtE and Pla are able to cleave cationic antimicrobial peptides such as the cathelicidin (LL-37) that exert antibacterial activity against both Gram-negative and Gram-positive bacteria [146, 147]. Similarly, in *P. aeruginosa*, *S. pyogenes* and *S. aureus*, proteases such as elastase, target antimicrobial peptides [148, 149]. These findings suggest that the degradation of host produced antimicrobial peptides, by extracellular proteases, may be a common approach used by both Gram-positive and Gram-negative pathogens to evade defences of the human host.

Concluding Remarks

Here we have highlighted examples of how proteases support the infection process in bacterial pathogens. Although the precise mechanism by which they contribute to virulence is not always known some general trends appear. In Gram-positive bacteria the Clp proteolytic complex is a central regulator in controlling the expression of virulence genes. Particularly in *S. aureus* numerous extracellular virulence factors rely on the proteolytic activity of ClpXP, for their expression, and we speculate that ClpXP is also responsible for the degradation of an unidentified repressor of extracellular virulence factor production (Fig. 7.6). Among the secreted enzymes in *S. aureus* that require ClpXP proteolysis for expression, are the extracellular proteases. These proteins play a central part in the dissemination of the infection as they directly degrade host structural components such as fibronectin, collagen and elastin, allowing spreading of the infection. Another important activity of these proteases is the turn-over of proteins involved in host signalling such as processing of host cell ligands or by stimulating proteolysis in the host through the activation of Plg. Thus, there is a close interplay between controlled proteolysis in the bacterial cytosol and the aggressive actions of extracellular proteases in direct contact with host cell proteins.

In Gram-negative bacteria the contribution of proteases to virulence is complex, and timing of virulence factor expression seems to be even more critical than in

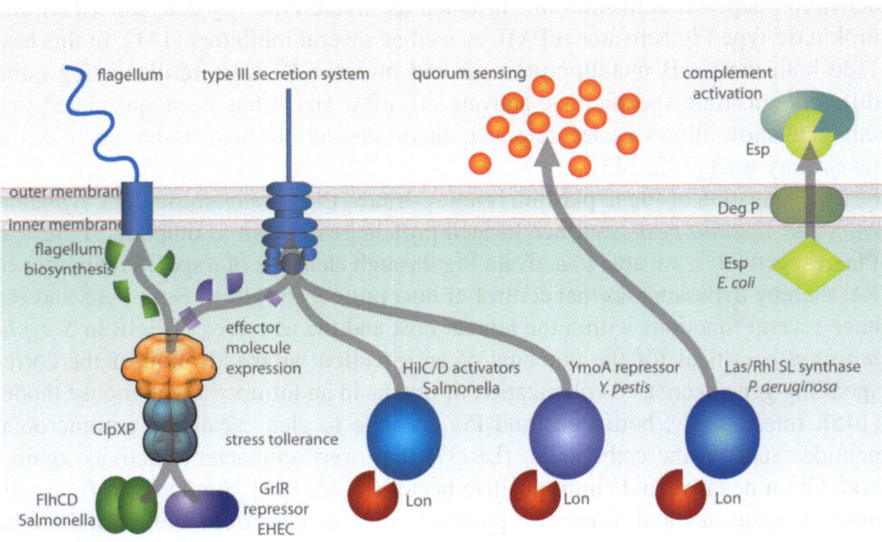

Fig. 7.6 Proteolysis in virulence of Gram-negative bacteria. ClpXP degrades FlhCD (the transcriptional regulator that controls flagella expression in *Salmonella*) and GrlR (a negative regulator of the T3SS effector in EHEC). Lon degrades HilCD and YmoA (which control T3SS production in *Salmonella* and *Y. pestis*, respectively) and the QS systems Las/Rhl in *P. aeruginosa*. In *E. coli*, DegP is involved in the export and activation of Esp

Gram-positive bacteria (Fig. 7.6). For example, in the absence of ClpXP, *S. typhimurium* becomes hyper-flagellated which appears to interfere with host cell contact. Although Lon has also been proposed to contribute to virulence in Gram-negative bacteria, in *Salmonella* it is responsible for the degradation of the transcriptional regulator HilCD (which is needed for production of the T3SS), suggesting that Lon may, in fact, be an anti-virulence factor. In some proteases, that exhibit both chaperone and proteolytic activity, their chaperone activity may play a dominant role in virulence, e.g. the folding of EspP by HtrA in *E. coli* O157:H7. These findings demonstrate the importance of conditional and controlled proteolysis and highlight the need, in future studies, to address the timing of degradation and the environmental signals that trigger this processes in the human host.

Acknowledgement We would like to thank the Danish Council for Independent Research, Natural Sciences for support grant number: 272-08-0371.

References

1. Maurizi MR, Clark WP, Kim SH, Gottesman S (1990) Clp P represents a unique family of serine proteases. J Biol Chem 265(21):12546–12552
2. Wang J, Hartling JA, Flanagan JM (1997) The structure of ClpP at 2.3 A resolution suggests a model for ATP-dependent proteolysis. Cell 91(4):447–456
3. Woo KM, Chung WJ, Ha DB, Goldberg AL et al (1989) Protease Ti from Escherichia coli requires ATP hydrolysis for protein breakdown but not for hydrolysis of small peptides. J Biol Chem 264(4):2088–2091
4. Gur E, Ottofuelling R, Dougan DA (2013) Machines of destruction – AAA+ proteases and the adaptors that control them. In: Dougan DA (ed) Regulated proteolysis in microorganisms. Springer, Subcell Biochem 66:3–33
5. Frees D, Savijoki K, Varmanen P, Ingmer H (2007) Clp ATPases and ClpP proteolytic complexes regulate vital biological processes in low GC, Gram-positive bacteria. Mol Microbiol 63(5):1285–1295
6. Kim YI, Levchenko I, Fraczkowska K, Woodruff RV et al (2001) Molecular determinants of complex formation between Clp/Hsp100 ATPases and the ClpP peptidase. Nat Struct Biol 8(3):230–233
7. Butler SM, Festa RA, Pearce MJ, Darwin KH (2006) Self-compartmentalized bacterial proteases and pathogenesis. Mol Microbiol 60(3):553–562
8. Frees D, Chastanet A, Qazi S, Sorensen K et al (2004) Clp ATPases are required for stress tolerance, intracellular replication and biofilm formation in Staphylococcus aureus. Mol Microbiol 54(5):1445–1462
9. Molière N, Turgay K (2013) General and regulatory proteolysis in Bacillus subtilis. In: Dougan DA (ed) Regulated proteolysis in microorganisms. Springer, Subcell Biochem 66:3–33
10. Frees D, Qazi SN, Hill PJ, Ingmer H (2003) Alternative roles of ClpX and ClpP in Staphylococcus aureus stress tolerance and virulence. Mol Microbiol 48(6):1565–1578
11. Michel A, Agerer F, Hauck CR, Herrmann M et al (2006) Global regulatory impact of ClpP protease of Staphylococcus aureus on regulons involved in virulence, oxidative stress response, autolysis, and DNA repair. J Bacteriol 188(16):5783–5796
12. Bottcher T, Sieber SA (2008) Beta-lactones as specific inhibitors of ClpP attenuate the production of extracellular virulence factors of Staphylococcus aureus. J Am Chem Soc 130(44):14400–14401

13. Jelsbak L, Ingmer H, Valihrach L, Cohn MT et al (2010) The chaperone ClpX stimulates expression of Staphylococcus aureus protein A by Rot dependent and independent pathways. PLoS One 5(9):e12752
14. Gaillot O, Pellegrini E, Bregenholt S, Nair S et al (2000) The ClpP serine protease is essential for the intracellular parasitism and virulence of Listeria monocytogenes. Mol Microbiol 35(6):1286–1294
15. Gaillot O, Bregenholt S, Jaubert F, Di Santo JP et al (2001) Stress-induced ClpP serine protease of Listeria monocytogenes is essential for induction of listeriolysin O-dependent protective immunity. Infect Immun 69(8):4938–4943
16. Nair S, Milohanic E, Berche P (2000) ClpC ATPase is required for cell adhesion and invasion of Listeria monocytogenes. Infect Immun 68(12):7061–7068
17. Borezee E, Pellegrini E, Beretti JL, Berche P (2001) SvpA, a novel surface virulence-associated protein required for intracellular survival of Listeria monocytogenes. Microbiology 147(Pt 11):2913–2923
18. Robertson GT, Ng WL, Foley J, Gilmour R et al (2002) Global transcriptional analysis of clpP mutations of type 2 Streptococcus pneumoniae and their effects on physiology and virulence. J Bacteriol 184(13):3508–3520
19. Kwon HY, Ogunniyi AD, Choi MH, Pyo SN et al (2004) The ClpP protease of Streptococcus pneumoniae modulates virulence gene expression and protects against fatal pneumococcal challenge. Infect Immun 72(10):5646–5653
20. Ibrahim YM, Kerr AR, Silva NA, Mitchell TJ (2005) Contribution of the ATP-dependent protease ClpCP to the autolysis and virulence of Streptococcus pneumoniae. Infect Immun 73(2):730–740
21. Webb C, Moreno M, Wilmes-Riesenberg M, Curtiss R 3rd et al (1999) Effects of DksA and ClpP protease on sigma S production and virulence in Salmonella typhimurium. Mol Microbiol 34(1):112–123
22. Yamamoto T, Sashinami H, Takaya A, Tomoyasu T et al (2001) Disruption of the genes for ClpXP protease in Salmonella enterica serovar Typhimurium results in persistent infection in mice, and development of persistence requires endogenous gamma interferon and tumor necrosis factor alpha. Infect Immun 69(5):3164–3174
23. Thomsen LE, Olsen JE, Foster JW, Ingmer H (2002) ClpP is involved in the stress response and degradation of misfolded proteins in Salmonella enterica serovar Typhimurium. Microbiology 148(Pt 9):2727–2733
24. Fang FC, Libby SJ, Buchmeier NA, Loewen PC et al (1992) The alternative sigma factor katF (rpoS) regulates Salmonella virulence. Proc Natl Acad Sci U S A 89(24):11978–11982
25. Micevski D, Dougan DA (2013) Proteolytic regulation of stress response pathways in Escherichia coli. In: Dougan DA (ed) Regulated proteolysis in microorganisms. Springer, Subcell Biochem 66:105–128
26. Zhou D, Galan J (2001) Salmonella entry into host cells: the work in concert of type III secreted effector proteins. Microbes Infect 3(14–15):1293–1298
27. Ramos HC, Rumbo M, Sirard JC (2004) Bacterial flagellins: mediators of pathogenicity and host immune responses in mucosa. Trends Microbiol 12(11):509–517
28. Tomoyasu T, Ohkishi T, Ukyo Y, Tokumitsu A et al (2002) The ClpXP ATP-dependent protease regulates flagellum synthesis in Salmonella enterica serovar typhimurium. J Bacteriol 184(3):645–653
29. Tomoyasu T, Takaya A, Isogai E, Yamamoto T (2003) Turnover of FlhD and FlhC, master regulator proteins for Salmonella flagellum biogenesis, by the ATP-dependent ClpXP protease. Mol Microbiol 48(2):443–452
30. Lucas RL, Lostroh CP, DiRusso CC, Spector MP et al (2000) Multiple factors independently regulate hilA and invasion gene expression in Salmonella enterica serovar typhimurium. J Bacteriol 182(7):1872–1882
31. Kage H, Takaya A, Ohya M, Yamamoto T (2008) Coordinated regulation of expression of Salmonella pathogenicity island 1 and flagellar type III secretion systems by ATP-dependent ClpXP protease. J Bacteriol 190(7):2470–2478

32. Takaya A, Suzuki M, Matsui H, Tomoyasu T et al (2003) Lon, a stress-induced ATP-dependent protease, is critically important for systemic Salmonella enterica serovar typhimurium infection of mice. Infect Immun 71(2):690–696
33. Iyoda S, Watanabe H (2005) ClpXP protease controls expression of the type III protein secretion system through regulation of RpoS and GrlR levels in enterohemorrhagic Escherichia coli. J Bacteriol 187(12):4086–4094
34. Brotz-Oesterhelt H, Beyer D, Kroll HP, Endermann R et al (2005) Dysregulation of bacterial proteolytic machinery by a new class of antibiotics. Nat Med 11(10):1082–1087
35. Hinzen B, Raddatz S, Paulsen H, Lampe T et al (2006) Medicinal chemistry optimization of acyldepsipeptides of the enopeptin class antibiotics. ChemMedChem 1(7):689–693
36. Kirstein J, Hoffmann A, Lilie H, Schmidt R et al (2009) The antibiotic ADEP reprogrammes ClpP, switching it from a regulated to an uncontrolled protease. EMBO Mol Med 1(1):37–49
37. Sass P, Josten M, Famulla K, Schiffer G et al (2011) Antibiotic acyldepsipeptides activate ClpP peptidase to degrade the cell division protein FtsZ. Proc Natl Acad Sci U S A 108(42):17474–17479
38. Lee BG, Park EY, Lee KE, Jeon H et al (2010) Structures of ClpP in complex with acyldepsipeptide antibiotics reveal its activation mechanism. Nat Struct Mol Biol 17(4):471–478
39. Li DH, Chung YS, Gloyd M, Joseph E et al (2010) Acyldepsipeptide antibiotics induce the formation of a structured axial channel in ClpP: a model for the ClpX/ClpA-bound state of ClpP. Chem Biol 17(9):959–969
40. Gominet M, Seghezzi N, Mazodier P (2011) Acyl depsipeptide (ADEP) resistance in Streptomyces. Microbiology 157(Pt 8):2226–2234
41. Schmitt EK, Riwanto M, Sambandamurthy V, Roggo S et al (2011) The natural product cyclomarin kills Mycobacterium tuberculosis by targeting the ClpC1 subunit of the caseinolytic protease. Angew Chem Int Ed Engl 50(26):5889–5891
42. Morsczeck C, Prokhorova T, Sigh J, Pfeiffer M et al (2008) Streptococcus pneumoniae: proteomics of surface proteins for vaccine development. Clin Microbiol Infect 14(1):74–81
43. Cao J, Li D, Gong Y, Yin N et al (2009) Caseinolytic protease: a protein vaccine which could elicit serotype-independent protection against invasive pneumococcal infection. Clin Exp Immunol 156(1):52–60
44. Wu K, Zhang X, Shi J, Li N et al (2010) Immunization with a combination of three pneumococcal proteins confers additive and broad protection against Streptococcus pneumoniae Infections in mice. Infect Immun 78(3):1276–1283
45. Matsui H, Suzuki M, Isshiki Y, Kodama C et al (2003) Oral immunization with ATP-dependent protease-deficient mutants protects mice against subsequent oral challenge with virulent Salmonella enterica serovar typhimurium. Infect Immun 71(1):30–39
46. Splichal I, Rychlik I, Gregorova D, Sebkova A et al (2007) Susceptibility of germ-free pigs to challenge with protease mutants of Salmonella enterica serovar Typhimurium. Immunobiology 212(7):577–582
47. Rotanova TV, Botos I, Melnikov EE, Rasulova F et al (2006) Slicing a protease: structural features of the ATP-dependent Lon proteases gleaned from investigations of isolated domains. Protein Sci 15(8):1815–1828
48. Cha SS, An YJ, Lee CR, Lee HS et al (2010) Crystal structure of Lon protease: molecular architecture of gated entry to a sequestered degradation chamber. EMBO J 29(20):3520–3530
49. Tsilibaris V, Maenhaut-Michel G, Van Melderen L (2006) Biological roles of the Lon ATP-dependent protease. Res Microbiol 157(8):701–713
50. Takaya A, Kubota Y, Isogai E, Yamamoto T (2005) Degradation of the HilC and HilD regulator proteins by ATP-dependent Lon protease leads to downregulation of Salmonella pathogenicity island 1 gene expression. Mol Microbiol 55(3):839–852
51. Gur E, Sauer RT (2008) Recognition of misfolded proteins by Lon, a AAA(+) protease. Genes Dev 22(16):2267–2277
52. Gur E (2013) The Lon AAA+ protease. In: Dougan DA (ed) Regulated proteolysis in microorganisms. Springer, Subcell Biochem 66:35–51

53. Boddicker JD, Jones BD (2004) Lon protease activity causes down-regulation of Salmonella pathogenicity island 1 invasion gene expression after infection of epithelial cells. Infect Immun 72(4):2002–2013
54. Takaya A, Tomoyasu T, Tokumitsu A, Morioka M et al (2002) The ATP-dependent lon protease of Salmonella enterica serovar Typhimurium regulates invasion and expression of genes carried on Salmonella pathogenicity island 1. J Bacteriol 184(1):224–232
55. Schechter LM, Lee CA (2001) AraC/XylS family members, HilC and HilD, directly bind and derepress the Salmonella typhimurium hilA promoter. Mol Microbiol 40(6):1289–1299
56. Jackson MW, Silva-Herzog E, Plano GV (2004) The ATP-dependent ClpXP and Lon proteases regulate expression of the Yersinia pestis type III secretion system via regulated proteolysis of YmoA, a small histone-like protein. Mol Microbiol 54(5):1364–1378
57. Herbst K, Bujara M, Heroven AK, Opitz W et al (2009) Intrinsic thermal sensing controls proteolysis of Yersinia virulence regulator RovA. PLoS Pathog 5(5):e1000435
58. Takaya A, Tabuchi F, Tsuchiya H, Isogai E et al (2008) Negative regulation of quorum-sensing systems in Pseudomonas aeruginosa by ATP-dependent Lon protease. J Bacteriol 190(12):4181–4188
59. Bertani I, Rampioni G, Leoni L, Venturi V (2007) The Pseudomonas putida Lon protease is involved in N-acyl homoserine lactone quorum sensing regulation. BMC Microbiol 7:71
60. Lipinska B, Fayet O, Baird L, Georgopoulos C (1989) Identification, characterization, and mapping of the Escherichia coli htrA gene, whose product is essential for bacterial growth only at elevated temperatures. J Bacteriol 171(3):1574–1584
61. Rosch JW, Caparon MG (2005) The ExPortal: an organelle dedicated to the biogenesis of secreted proteins in Streptococcus pyogenes. Mol Microbiol 58(4):959–968
62. Lipinska B, Zylicz M, Georgopoulos C (1990) The HtrA (DegP) protein, essential for Escherichia coli survival at high temperatures, is an endopeptidase. J Bacteriol 172(4):1791–1797
63. Krojer T, Garrido-Franco M, Huber R, Ehrmann M et al (2002) Crystal structure of DegP (HtrA) reveals a new protease-chaperone machine. Nature 416(6879):455–459
64. Spiess C, Beil A, Ehrmann M (1999) A temperature-dependent switch from chaperone to protease in a widely conserved heat shock protein. Cell 97(3):339–347
65. Krojer T, Sawa J, Schafer E, Saibil HR et al (2008) Structural basis for the regulated protease and chaperone function of DegP. Nature 453(7197):885–890
66. Clausen T, Kaiser M, Huber R, Ehrmann M (2011) HTRA proteases: regulated proteolysis in protein quality control. Nat Rev Mol Cell Biol 12(3):152–162
67. Krojer T, Pangerl K, Kurt J, Sawa J et al (2008) Interplay of PDZ and protease domain of DegP ensures efficient elimination of misfolded proteins. Proc Natl Acad Sci U S A 105(22):7702–7707
68. Subrini O, Betton JM (2009) Assemblies of DegP underlie its dual chaperone and protease function. FEMS Microbiol Lett 296(2):143–148
69. Johnson K, Charles I, Dougan G, Pickard D et al (1991) The role of a stress-response protein in Salmonella typhimurium virulence. Mol Microbiol 5(2):401–407
70. Humphreys S, Stevenson A, Bacon A, Weinhardt AB et al (1999) The alternative sigma factor, sigmaE, is critically important for the virulence of Salmonella typhimurium. Infect Immun 67(4):1560–1568
71. Baumler AJ, Kusters JG, Stojiljkovic I, Heffron F (1994) Salmonella typhimurium loci involved in survival within macrophages. Infect Immun 62(5):1623–1630
72. Elzer PH, Phillips RW, Kovach ME, Peterson KM et al (1994) Characterization and genetic complementation of a Brucella abortus high-temperature-requirement A (htrA) deletion mutant. Infect Immun 62(10):4135–4139
73. Yamamoto T, Hanawa T, Ogata S, Kamiya S (1996) Identification and characterization of the Yersinia enterocolitica gsrA gene, which protectively responds to intracellular stress induced by macrophage phagocytosis and to extracellular environmental stress. Infect Immun 64(8):2980–2987
74. Lewis C, Skovierova H, Rowley G, Rezuchova B et al (2009) Salmonella enterica Serovar Typhimurium HtrA: regulation of expression and role of the chaperone and protease activities during infection. Microbiology 155(Pt 3):873–881

75. Brondsted L, Andersen MT, Parker M, Jorgensen K et al (2005) The HtrA protease of Campylobacter jejuni is required for heat and oxygen tolerance and for optimal interaction with human epithelial cells. Appl Environ Microbiol 71(6):3205–3212
76. Novik V, Hofreuter D, Galan JE (2010) Identification of Campylobacter jejuni genes involved in its interaction with epithelial cells. Infect Immun 78(8):3540–3553
77. Champion OL, Karlyshev AV, Senior NJ, Woodward M et al (2010) Insect infection model for Campylobacter jejuni reveals that O-methyl phosphoramidate has insecticidal activity. J Infect Dis 201(5):776–782
78. Baek KT, Vegge CS, Brondsted L (2011) HtrA chaperone activity contributes to host cell binding in Campylobacter jejuni. Gut Pathog 3:13
79. Baek KT, Vegge CS, Skorko-Glonek J, Brondsted L (2011) Different contributions of HtrA protease and chaperone activities to Campylobacter jejuni stress tolerance and physiology. Appl Environ Microbiol 77(1):57–66
80. Jong WS, ten Hagen-Jongman CM, Ruijter E, Orru RV et al (2010) YidC is involved in the biogenesis of the secreted autotransporter hemoglobin protease. J Biol Chem 285(51):39682–39690
81. Bodelon G, Marin E, Fernandez LA (2009) Role of periplasmic chaperones and BamA (YaeT/Omp85) in folding and secretion of intimin from enteropathogenic Escherichia coli strains. J Bacteriol 191(16):5169–5179
82. Ruiz-Perez F, Henderson IR, Leyton DL, Rossiter AE et al (2009) Roles of periplasmic chaperone proteins in the biogenesis of serine protease autotransporters of Enterobacteriaceae. J Bacteriol 191(21):6571–6583
83. Humphries RM, Griener TP, Vogt SL, Mulvey GL et al (2010) N-acetyllactosamine-induced retraction of bundle-forming pili regulates virulence-associated gene expression in enteropathogenic Escherichia coli. Mol Microbiol 76(5):1111–1126
84. Vogt SL, Nevesinjac AZ, Humphries RM, Donnenberg MS et al (2010) The Cpx envelope stress response both facilitates and inhibits elaboration of the enteropathogenic Escherichia coli bundle-forming pilus. Mol Microbiol 76(5):1095–1110
85. Baud C, Gutsche I, Willery E, de Paepe D et al (2011) Membrane-associated DegP in Bordetella chaperones a repeat-rich secretory protein. Mol Microbiol 80(6):1625–1636
86. Baud C, Hodak H, Willery E, Drobecq H et al (2009) Role of DegP for two-partner secretion in Bordetella. Mol Microbiol 74(2):315–329
87. Purdy GE, Fisher CR, Payne SM (2007) IcsA surface presentation in Shigella flexneri requires the periplasmic chaperones DegP, Skp, and SurA. J Bacteriol 189(15):5566–5573
88. Purdy GE, Hong M, Payne SM (2002) Shigella flexneri DegP facilitates IcsA surface expression and is required for efficient intercellular spread. Infect Immun 70(11):6355–6364
89. Bumann D, Aksu S, Wendland M, Janek K et al (2002) Proteome analysis of secreted proteins of the gastric pathogen Helicobacter pylori. Infect Immun 70(7):3396–3403
90. Lower M, Weydig C, Metzler D, Reuter A et al (2008) Prediction of extracellular proteases of the human pathogen Helicobacter pylori reveals proteolytic activity of the Hp1018/19 protein HtrA. PLoS One 3(10):e3510
91. Hoy B, Lower M, Weydig C, Carra G et al (2010) Helicobacter pylori HtrA is a new secreted virulence factor that cleaves E-cadherin to disrupt intercellular adhesion. EMBO Rep 11(10):798–804
92. Wu X, Lei L, Gong S, Chen D et al (2011) The chlamydial periplasmic stress response serine protease cHtrA is secreted into host cell cytosol. BMC Microbiol 11:87
93. Cole JN, Aquilina JA, Hains PG, Henningham A et al (2007) Role of group A Streptococcus HtrA in the maturation of SpeB protease. Proteomics 7(24):4488–4498
94. Lyon WR, Caparon MG (2004) Role for serine protease HtrA (DegP) of Streptococcus pyogenes in the biogenesis of virulence factors SpeB and the hemolysin streptolysin S. Infect Immun 72(3):1618–1625
95. Rigoulay C, Entenza JM, Halpern D, Widmer E et al (2005) Comparative analysis of the roles of HtrA-like surface proteases in two virulent Staphylococcus aureus strains. Infect Immun 73(1):563–572
96. Biswas S, Biswas I (2005) Role of HtrA in surface protein expression and biofilm formation by Streptococcus mutans. Infect Immun 73(10):6923–6934

97. Wilson RL, Brown LL, Kirkwood-Watts D, Warren TK et al (2006) Listeria monocytogenes 10403S HtrA is necessary for resistance to cellular stress and virulence. Infect Immun 74(1):765–768
98. Chitlaru T, Zaide G, Ehrlich S, Inbar I et al (2011) HtrA is a major virulence determinant of Bacillus anthracis. Mol Microbiol 81(6):1542–1559
99. Arvidson S (2006) Extracellular enzymes. In: Fischetti VA (ed) Gram-positive pathogens (2nd edn.). ASM Press.
100. Novick RP (2003) Autoinduction and signal transduction in the regulation of staphylococcal virulence. Mol Microbiol 48(6):1429–1449
101. Shaw L, Golonka E, Potempa J, Foster SJ (2004) The role and regulation of the extracellular proteases of Staphylococcus aureus. Microbiology 150(Pt 1):217–228
102. Drapeau GR (1978) Role of metalloprotease in activation of the precursor of staphylococcal protease. J Bacteriol 136(2):607–613
103. Lindsay JA, Foster SJ (1999) Interactive regulatory pathways control virulence determinant production and stability in response to environmental conditions in Staphylococcus aureus. Mol Gen Genet 262(2):323–331
104. Chan PF, Foster SJ (1998) The role of environmental factors in the regulation of virulence-determinant expression in Staphylococcus aureus 8325–4. Microbiology 144(Pt 9):2469–2479
105. Rice K, Peralta R, Bast D, de Azavedo J et al (2001) Description of staphylococcus serine protease (ssp) operon in Staphylococcus aureus and nonpolar inactivation of sspA-encoded serine protease. Infect Immun 69(1):159–169
106. Nickerson N, Ip J, Passos DT, McGavin MJ (2010) Comparison of Staphopain A (ScpA) and B (SspB) precursor activation mechanisms reveals unique secretion kinetics of proSspB (Staphopain B), and a different interaction with its cognate Staphostatin, SspC. Mol Microbiol 75(1):161–177
107. Nickerson NN, Joag V, McGavin MJ (2008) Rapid autocatalytic activation of the M4 metalloprotease aureolysin is controlled by a conserved N-terminal fungalysin-thermolysin-propeptide domain. Mol Microbiol 69(6):1530–1543
108. Massimi I, Park E, Rice K, Muller-Esterl W et al (2002) Identification of a novel maturation mechanism and restricted substrate specificity for the SspB cysteine protease of Staphylococcus aureus. J Biol Chem 277(44):41770–41777
109. Rzychon M, Sabat A, Kosowska K, Potempa J et al (2003) Staphostatins: an expanding new group of proteinase inhibitors with a unique specificity for the regulation of staphopains Staphylococcus spp. cysteine proteinases. Mol Microbiol 49(4):1051–1066
110. Foster TJ (2005) Immune evasion by staphylococci. Nat Rev Microbiol 3(12):948–958
111. Foster TJ, Hook M (1998) Surface protein adhesins of Staphylococcus aureus. Trends Microbiol 6(12):484–488
112. McGavin MJ, Zahradka C, Rice K, Scott JE (1997) Modification of the Staphylococcus aureus fibronectin binding phenotype by V8 protease. Infect Immun 65(7):2621–2628
113. Karlsson A, Saravia-Otten P, Tegmark K, Morfeldt E et al (2001) Decreased amounts of cell wall-associated protein A and fibronectin-binding proteins in Staphylococcus aureus sarA mutants due to up-regulation of extracellular proteases. Infect Immun 69(8):4742–4748
114. McAleese FM, Walsh EJ, Sieprawska M, Potempa J et al (2001) Loss of clumping factor B fibrinogen binding activity by Staphylococcus aureus involves cessation of transcription, shedding and cleavage by metalloprotease. J Biol Chem 276(32):29969–29978
115. Potempa J, Dubin A, Korzus G, Travis J (1988) Degradation of elastin by a cysteine proteinase from Staphylococcus aureus. J Biol Chem 263(6):2664–2667
116. Ohbayashi T, Irie A, Murakami Y, Nowak M et al (2011) Degradation of fibrinogen and collagen by staphopains, cysteine proteases released from Staphylococcus aureus. Microbiology 157(Pt 3):786–792
117. Mayer-Scholl A, Averhoff P, Zychlinsky A (2004) How do neutrophils and pathogens interact? Curr Opin Microbiol 7(1):62–66
118. Greenberg S, Grinstein S (2002) Phagocytosis and innate immunity. Curr Opin Immunol 14(1):136–145

119. Smagur J, Guzik K, Magiera L, Bzowska M et al (2009) A new pathway of staphylococcal pathogenesis: apoptosis-like death induced by Staphopain B in human neutrophils and monocytes. J Innate Immun 1(2):98–108
120. Kulig P, Zabel BA, Dubin G, Allen SJ et al (2007) Staphylococcus aureus-derived staphopain B, a potent cysteine protease activator of plasma chemerin. J Immunol 178(6):3713–3720
121. Otto M (2008) Staphylococcal biofilms. Curr Top Microbiol Immunol 322:207–228
122. Marti M, Trotonda MP, Tormo-Mas MA, Vergara-Irigaray M et al (2010) Extracellular proteases inhibit protein-dependent biofilm formation in Staphylococcus aureus. Microbes Infect 12(1):55–64
123. Boles BR, Horswill AR (2008) Agr-mediated dispersal of Staphylococcus aureus biofilms. PLoS Pathog 4(4):e1000052
124. Reed SB, Wesson CA, Liou LE, Trumble WR et al (2001) Molecular characterization of a novel Staphylococcus aureus serine protease operon. Infect Immun 69(3):1521–1527
125. Coulter SN, Schwan WR, Ng EY, Langhorne MH et al (1998) Staphylococcus aureus genetic loci impacting growth and survival in multiple infection environments. Mol Microbiol 30(2):393–404
126. Bisno AL, Stevens DL (1996) Streptococcal infections of skin and soft tissues. N Engl J Med 334(4):240–245
127. Rasmussen M, Bjorck L (2002) Proteolysis and its regulation at the surface of Streptococcus pyogenes. Mol Microbiol 43(3):537–544
128. Collin M, Olsen A (2001) Effect of SpeB and EndoS from Streptococcus pyogenes on human immunoglobulins. Infect Immun 69(11):7187–7189
129. Collin M, Olsen A (2001) EndoS, a novel secreted protein from Streptococcus pyogenes with endoglycosidase activity on human IgG. EMBO J 20(12):3046–3055
130. Collin M, Svensson MD, Sjoholm AG, Jensenius JC et al (2002) EndoS and SpeB from Streptococcus pyogenes inhibit immunoglobulin-mediated opsonophagocytosis. Infect Immun 70(12):6646–6651
131. Chaussee MS, Phillips ER, Ferretti JJ (1997) Temporal production of streptococcal erythrogenic toxin B (streptococcal cysteine proteinase) in response to nutrient depletion. Infect Immun 65(5):1956–1959
132. von Pawel-Rammingen U, Johansson BP, Bjorck L (2002) IdeS, a novel streptococcal cysteine proteinase with unique specificity for immunoglobulin G. EMBO J 21(7):1607–1615
133. von Pawel-Rammingen U, Bjorck L (2003) IdeS and SpeB: immunoglobulin-degrading cysteine proteinases of Streptococcus pyogenes. Curr Opin Microbiol 6(1):50–55
134. Hidalgo-Grass C, Mishalian I, Dan-Goor M, Belotserkovsky I et al (2006) A streptococcal protease that degrades CXC chemokines and impairs bacterial clearance from infected tissues. EMBO J 25(19):4628–4637
135. Sumby P, Zhang S, Whitney AR, Falugi F et al (2008) A chemokine-degrading extracellular protease made by group A Streptococcus alters pathogenesis by enhancing evasion of the innate immune response. Infect Immun 76(3):978–985
136. Zinkernagel AS, Timmer AM, Pence MA, Locke JB et al (2008) The IL-8 protease SpyCEP/ScpC of group A Streptococcus promotes resistance to neutrophil killing. Cell Host Microbe 4(2):170–178
137. Gryllos I, Tran-Winkler HJ, Cheng MF, Chung H et al (2008) Induction of group A Streptococcus virulence by a human antimicrobial peptide. Proc Natl Acad Sci U S A 105(43):16755–16760
138. Bryan JD, Shelver DW (2009) Streptococcus agalactiae CspA is a serine protease that inactivates chemokines. J Bacteriol 191(6):1847–1854
139. Bever RA, Iglewski BH (1988) Molecular characterization and nucleotide sequence of the Pseudomonas aeruginosa elastase structural gene. J Bacteriol 170(9):4309–4314
140. Heck LW, Morihara K, McRae WB, Miller EJ (1986) Specific cleavage of human type III and IV collagens by Pseudomonas aeruginosa elastase. Infect Immun 51(1):115–118

141. Castellino FJ, Ploplis VA (2005) Structure and function of the plasminogen/plasmin system. Thromb Haemost 93(4):647–654
142. Beaufort N, Seweryn P, de Bentzmann S, Tang A et al (2010) Activation of human pro-urokinase by unrelated proteases secreted by Pseudomonas aeruginosa. Biochem J 428(3):473–482
143. Leduc D, Beaufort N, de Bentzmann S, Rousselle JC et al (2007) The Pseudomonas aeruginosa LasB metalloproteinase regulates the human urokinase-type plasminogen activator receptor through domain-specific endoproteolysis. Infect Immun 75(8):3848–3858
144. Haiko J, Laakkonen L, Juuti K, Kalkkinen N et al (2010) The omptins of Yersinia pestis and Salmonella enterica cleave the reactive center loop of plasminogen activator inhibitor 1. J Bacteriol 192(18):4553–4561
145. Ramu P, Lobo LA, Kukkonen M, Bjur E et al (2008) Activation of pro-matrix metalloproteinase-9 and degradation of gelatin by the surface protease PgtE of Salmonella enterica serovar Typhimurium. Int J Med Microbiol 298(3–4):263–278
146. Galvan EM, Lasaro MA, Schifferli DM (2008) Capsular antigen fraction 1 and Pla modulate the susceptibility of Yersinia pestis to pulmonary antimicrobial peptides such as cathelicidin. Infect Immun 76(4):1456–1464
147. Guina T, Yi EC, Wang H, Hackett M et al (2000) A PhoP-regulated outer membrane protease of Salmonella enterica serovar typhimurium promotes resistance to alpha-helical antimicrobial peptides. J Bacteriol 182(14):4077–4086
148. Schmidtchen A, Frick IM, Andersson E, Tapper H et al (2002) Proteinases of common pathogenic bacteria degrade and inactivate the antibacterial peptide LL-37. Mol Microbiol 46(1):157–168
149. Sieprawska-Lupa M, Mydel P, Krawczyk K, Wojcik K et al (2004) Degradation of human antimicrobial peptide LL-37 by Staphylococcus aureus-derived proteinases. Antimicrob Agents Chemother 48(12):4673–4679

Part III
Regulated Proteolysis in Yeast

Chapter 8
Roles of Cdc48 in Regulated Protein Degradation in Yeast

Alexander Buchberger

Abstract The chaperone-related, ubiquitin-selective AAA (ATPase associated with a variety of cellular activities) protein Cdc48 (also known as TER94, p97 and VCP) is a key regulator of intracellular proteolysis in eukaryotes. It uses the energy derived from ATP hydrolysis to segregate ubiquitylated proteins from stable assemblies with proteins, membranes and chromatin. Originally characterized as essential factor in proteasomal degradation pathways, Cdc48 was recently found to control lysosomal protein degradation as well. Moreover, impaired lysosomal proteolysis due to mutational inactivation of Cdc48 causes protein aggregation diseases in humans. This review introduces the major systems of intracellular proteolysis in eukaryotes and the role of protein ubiquitylation. It then discusses in detail structure, mechanism and cellular functions of Cdc48 with an emphasis on protein degradation pathways in yeast.

Introduction

Regulated intracellular proteolysis is essential for many aspects of cell biology: It takes center stage in protein quality control by eliminating aggregation-prone, potentially toxic conformers including aborted translational products, misfolded and damaged proteins; controls metabolic pathways, signal transduction and cell division through the degradation of key enzymes, transcription factors and signaling proteins; and ensures supply with amino acids and intermediary metabolites, in particular under conditions of nutrient deprivation. Eukaryotic cells possess two

A. Buchberger (✉)
Department of Biochemistry, Biocenter, University of Würzburg,
Am Hubland, 97074 Würzburg, Germany
e-mail: alexander.buchberger@biozentrum.uni-wuerzburg.de

D.A. Dougan (ed.), *Regulated Proteolysis in Microorganisms*, Subcellular Biochemistry 66, DOI 10.1007/978-94-007-5940-4_8,

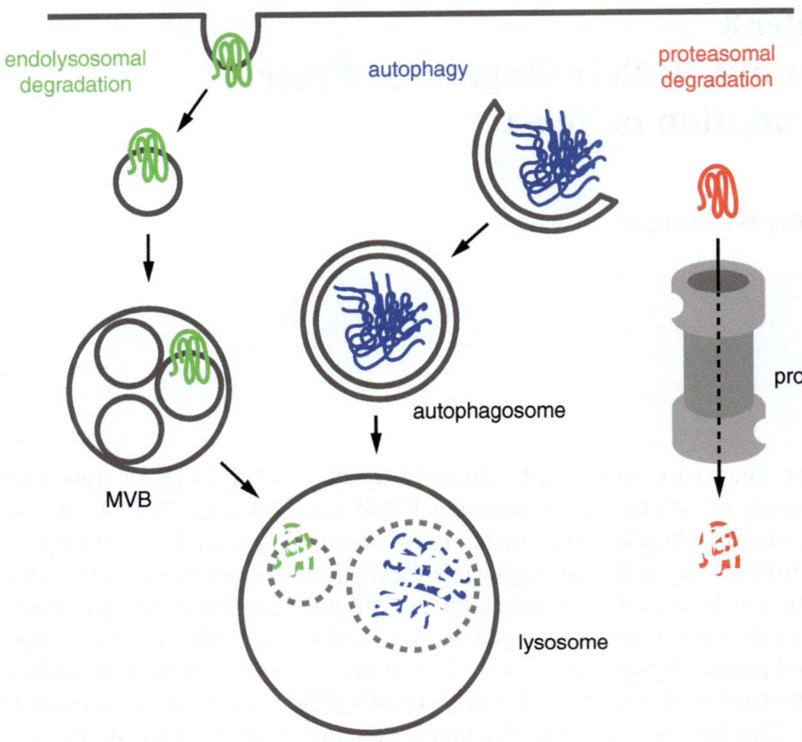

Fig. 8.1 Major pathways of intracellular proteolysis in eukaryotes. Most soluble proteins are degraded by the 26S proteasome (*red pathway*). Cytosolic structures incompatible with proteasomal degradation, including protein aggregates, large stable protein complexes and organelles, are degraded by autophagy (*blue pathway*). They are engulfed by a double membrane forming autophagosomes and delivered to lysosomes. Plasma membrane proteins are eliminated by endolysosomal degradation (*green pathway*). They are endocytosed, sorted into the luminal vesicles of multivesicular bodies (MVBs), and delivered to lysosomes. The acidic lysosomal lumen harbours various hydrolases that degrade cargo delivered by autophagosomes and MVBs

major proteolytic systems: the proteasome and the lysosomal compartment. While most soluble proteins are degraded by the proteasome, the lysosome is able to hydrolyze insoluble protein aggregates, vesicle-embedded plasma membrane proteins, and even entire organelles, viruses and bacteria (Fig. 8.1).

The proteasome is a large, barrel-shaped protease complex consisting of four stacked rings of seven structurally related subunits [1]. Its active sites are deeply buried in a central chamber formed by the two inner rings, and substrate proteins have to pass a central channel gated by the distal rings in order to be degraded. This 20S proteasomal core particle is sufficient for the degradation of polypeptides and some unfolded proteins, but most substrates require the additional presence of a 19S regulatory particle controlling substrate specificity, unfolding and access to the central channel. Together, the 20S and 19S subcomplexes form the eukaryotic 26S proteasome [2].

The lysosome is an acidic organelle harbouring various hydrolases, among them acid-optimal proteases of the cathepsin family [3]. It is the end-point of two proteolytic pathways delivering membrane-confined cargo: the endolysosomal pathway for the degradation of predominantly plasma membrane-derived proteins, and autophagy pathways for the degradation of cytosolic content including soluble proteins, insoluble protein aggregates, and organelles (Fig. 8.1) [4]. In the endolysosomal pathway, endocytosed proteins that evade recycling to the plasma membrane are sorted into luminal vesicles of multivesicular bodies (MVBs), membrane-bounded structures derived from sorting endosomes. MVBs fuse with lysosomes and release their luminal cargo for lysosomal degradation [5]. In autophagy, diverse cytosolic cargo is first engulfed during the *de novo* formation of a double membrane-bound compartment termed the autophagosome. Similar to MVBs, autophagosomes then fuse with lysosomes to form autolysosomes degrading the autophagic cargo [6]. While autophagy of cytosolic content is considered to be relatively non-specific, in particular under starvation conditions, there is clear evidence for the existence of selective autophagy pathways targeting protein complexes/aggregates and organelles [7, 8].

Intriguingly, all three major pathways for regulated proteolysis in eukaryotic cells are controlled by the covalent modification of target proteins with ubiquitin, a small, highly conserved protein of 76 amino acid residues [4]. This process (termed "ubiquitylation", "ubiquitination" or "ubiquitinylation") involves the formation of an isopeptide bond between the carboxy terminus of ubiquitin and the ε amino group of lysine residues on target proteins (Fig. 8.2). Ubiquitylation requires a catalytic cascade of three enzymatic activities E1, E2, and E3 [9]. The E1 (ubiquitin activating enzyme) activates ubiquitin by C-terminal adenylation, followed by formation of an energy-rich thioester with its active site cysteine thiol. Activated ubiquitin is then transferred in a transesterification reaction to the active site cysteine of the E2 (ubiquitin conjugating enzyme), followed by conjugation to target lysine residues mediated by the E3 (ubiquitin protein ligase). The substrate specificity of the system is primarily determined by the large number (dozens to hundreds) of different E3 ligases found in all eukaryotes. Ubiquitin itself can be ubiquitylated on either one of seven lysine residues or its amino terminus, allowing for target protein modification by ubiquitin chains of different length, linkage type and complexity [10–12]. Moreover, ubiquitin chains are also subject to editing by deubiquitylating enzymes (DUBs) [13], further increasing the plasticity of the ubiquitin signal. Importantly, different ubiquitin chains mark substrates for distinct downstream processes (Fig. 8.2) [11, 14]. Chains linked via lysine residue 48 of ubiquitin ("K48-linked chains") are prototypical signals for targeting to, and degradation by, the 26S proteasome [9, 15]. Chains linked via residues K11, K29, and perhaps also K6, K27 and K33, can also constitute proteasomal degradation signals [11, 16]. By contrast, K63-linked chains, "linear" chains linked via the amino terminus of ubiquitin, and single ubiquitin moieties ("mono-ubiquitylation") constitute signals for non-proteasomal fates of substrate proteins, for example in sorting processes, signal transduction, and DNA damage repair [12, 14, 17, 18]. In the endolysosomal pathway, ubiquitylation serves as a crucial sorting signal at two distinct steps.

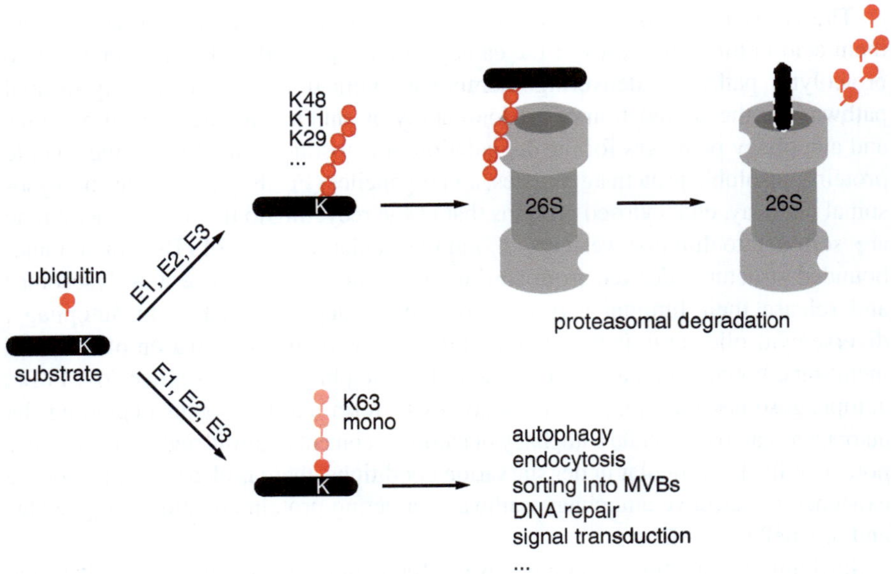

Fig. 8.2 The ubiquitin-proteasome system. Substrate proteins are covalently modified with ubiquitin through the formation of an isopeptide bond with lysine residues (K) in a reaction catalyzed by three enzymatic activities (E1, E2, E3). Polyubiquitin chains linked through ubiquitin residues K48, K11, K29, and perhaps also K6, K27 and K33, target substrate proteins for proteasomal degradation; K63-linked chains and mono-ubiquitin constitute signals in various non-proteasomal processes, among them sorting during autophagy and endolysososmal degradation

Mono-ubiquitylation and K63-linked chains mark plasma membrane proteins for internalization through primary endocytic vesicles [17]. During subsequent endosomal maturation, mono-ubiquitylation acts as a signal for the sorting of cargo into MVB vesicles [5, 19]. In autophagy, selective engulfment of ubiquitylated, cytosolic cargo by autophagosomes is mediated by a conserved family of adaptor proteins [7, 8].

In recent years, the evolutionary conserved protein Cdc48 (also known as TER94 in *Drosophila* and as p97 and VCP in vertebrates) has emerged as an important motor and regulator for a number of proteasomal degradation pathways. Cdc48 is a chaperone-related member of the ATPase associated with various cellular activities (AAA) protein family [20]. It converts the chemical energy released by ATP hydrolysis into mechanical force believed to drive the "segregation" of ubiquitylated substrate proteins from stable protein complexes, membranes and chromatin, thereby facilitating their delivery to and degradation by the 26S proteasome [21–23]. Importantly, Cdc48 has more recently also been shown to be critical for autophagy and endolysosomal protein degradation [24–26]. Even though functional insights into the role of Cdc48 in the latter pathways are just beginning to emerge, it is intriguing to note that Cdc48 is involved in all three major routes of intracellular proteolysis in eukaryotes. Consistent with such a central role in protein degradation,

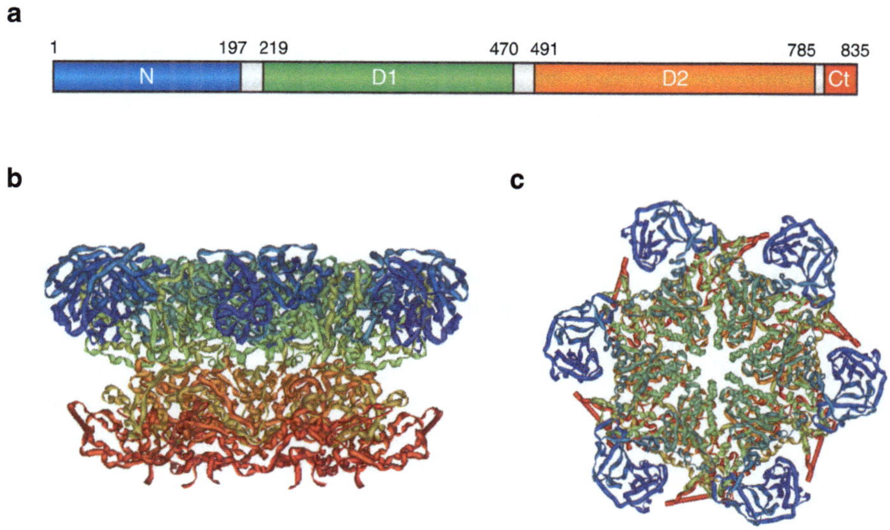

Fig. 8.3 Cdc48 structure. (**a**) Schematic view of Cdc48 domain organisation. The borders of the N, D1 and D2 domains as well of the carboxy-terminal tail (Ct) are indicated by residue numbers. Colours were chosen to closely match the colour scheme in (**b**) and (**c**). (**b**) Side view of the three-dimensional structure of murine p97 (PDB entry 1OZ4 [33]) in ribbon representation. Colour coding is from the amino-terminus (*top, blue*) to the carboxy-terminus (*bottom, red*). (**c**) Top view along the central axis

Cdc48 is essential in all organisms tested, and mutations in the human *VCP* gene cause proteinopathies including neurodegenerative, muscular and motor neuron diseases [24, 27–29].

This review discusses basic structural and functional properties of Cdc48, the complex control of distinct Cdc48 activities by regulatory cofactors, and the role of Cdc48 in various proteolysis pathways. The focus of the review will be on Cdc48-dependent degradation pathways in the Baker's yeast *Saccharomyces cerevisiae*, which serve as a paradigm for Cdc48 functions in higher eukaryotes.

Cdc48 Structure and Mechanism of Action

Three-Dimensional Structure and ATP-Dependent Conformational Changes

Like many other AAA enzymes, Cdc48 forms homohexameric, ring-shaped complexes [30, 31]. The protomer of Cdc48 consists of two Walker-type ATPase domains, D1 and D2, which are flanked by an amino-terminal N domain and a carboxy-terminal, unstructured tail (Fig. 8.3a) Whereas the ATPase domains provide the driving force for Cdc48 function, the N domain and the extreme C-terminus are major docking sites for regulatory cofactors (see below). X-ray crystallography of the mammalian orthologue p97 (65% sequence identity to yeast Cdc48) provided detailed information about the architecture and structural organisation of the Cdc48/p97 hexamer (Fig. 8.3b, c) [32–34]. The D1 and D2 ATPase domains of the six protomers form two stacked rings in head-to-tail orientation, encircling a central, axial pore. The N domains are arranged radial to the D1 domains, whereas structure and orientation of the carboxy-terminal tail could not be solved, probably due to its high conformational flexibility.

Numerous studies employing complementary experimental approaches have provided compelling evidence for nucleotide-dependent conformational changes in Cdc48/p97 (for a detailed review, see [35]). Unfortunately, differences in the proteins studied, i.e. full-length p97 *versus* a ND1 fragment, the methodology, nucleotide and buffer conditions have so far precluded the formulation of a unifying model for the coupling of conformational states to the ATPase cycle [35]. Nevertheless, from a global view, the most pronounced nucleotide-dependent changes are (i) rigid body movements of the N domain relative to the D1 ring and (ii) widening/closing of the central pore of the D2 ring.

The biochemical characterization of Cdc48 orthologues from yeast, worm and mammals consistently revealed differences between the ATPase activities of the D1 and D2 domains [36–40]. ATP hydrolysis by D2 was found to contribute about 90% to the overall ATPase activity of Cdc48, whereas mutations in D1 compromising ATP hydrolysis had only moderate effects. In line with a slow turnover at the D1 active site inferred from these ATPase measurements, preparations of hexameric mammalian p97 contain up to six molecules of ADP stably bound to D1 that are in very slow exchange with free nucleotide (reviewed in [35]). It has therefore been proposed that the D1 nucleotide-binding site has a structural rather than catalytic function [41]. *In vivo*, ATP hydrolysis by both ATPase domains is essential for viability [37, 39], but for different reasons. While the ATPase activity of D2 is believed to be the main driving force for the segregase activity of Cdc48, hydrolysis at the D1 active site appears to merely be required to eliminate a yet ill-defined inhibitory effect of the D1 ATP state on Cdc48 function [39].

Models for Cdc48 Segregase Activity

Despite considerable knowledge about the three-dimensional structure of Cdc48/p97, the mechanism underlying its segregase activity has not been elucidated in molecular detail. Structurally related bacterial AAA+proteins, such as ClpA, ClpB, ClpX, FtsH and HslU from *E. coli* employ a "threading" mechanism of substrate unfolding [20, 42–44]. Instrumental to this mechanism are evolutionary highly conserved loops of the D1 and (when present) D2 domains that face the central pore and contain an aromatic residue next to a large hydrophobic or basic residue. These loops contact substrate proteins and are believed to transform conformational

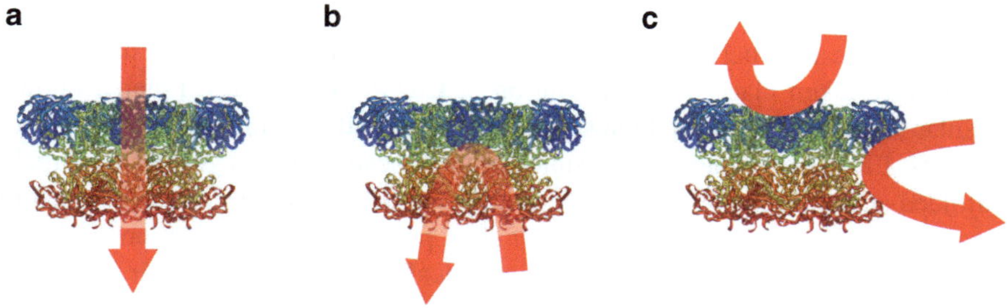

Fig. 8.4 Models for Cdc48 segregase activity. Side view of murine p97 as in Fig. 8.3b. *Red arrows* indicate the routes of substrate proteins according to the different models. (**a**) Threading. (**b**) D2 in – D2 out. (**c**) Surface only. See text for details

changes accompanying ATP binding and hydrolysis into ratchet-like movements, thereby pulling the substrate polypeptide chain through the central pore and unfolding it [42, 43].

In contrast to bacterial unfoldases, there exists no convincing experimental evidence that Cdc48 substrates are threaded through the entire length of the central pore (Fig. 8.4a). In fact, several lines of circumstantial evidence may argue against such a mechanism. First, Cdc48 substrates are typically marked by ubiquitin chains on one or several lysine residues. It is unclear how the resulting branched polypeptide could be efficiently threaded through the narrow central pore of Cdc48, aside from the fact that ubiquitin is remarkably stable thermodynamically and hence difficult to unfold. Based on studies of the Cdc48-dependent ER-associated protein degradation (ERAD) quality control pathway (see below), this conceptual problem was proposed to be solved by the presence of Cdc48-associated DUBs which could initially deubiquitylate substrates, followed by threading through the central pore and re-ubiquitylation for proteasomal targeting by Cdc48-associated E3 ligases [45, 46]. However, in disfavour of a critical role for DUBs in ERAD substrate turnover by Cdc48, siRNA-mediated depletion of Cdc48/p97-associated DUBs in mammalian cells does not, with the exception of the poorly characterized DUB Usp13, result in ERAD defects [47]. Similarly, deletion of known Cdc48-associated DUBs in yeast has not been reported to cause ERAD phenotypes. Second, the presence of the N domain appears to prevent threading of polypeptides. Interestingly, an archaeal homologue of Cdc48, VAT, was found to be capable of unfolding the ClpA and ClpX model substrate GFP-ssrA *in vitro* under certain experimental conditions [48]. Similar to the Clp-type AAA+proteins, this activity required the aromatic side chains in the conserved D1 and D2 loops facing the central pore. In addition, deletion of the VAT N domain enhanced its unfolding activity towards GFP-ssrA. In contrast to VAT and Clp-type AAA+proteins, however, Cdc48 orthologues lack the critical aromatic residue in the D1 loop [49] (P. Zwickl, personal communication). Consistently, wild type mammalian p97 was found to be unable to unfold GFP-ssrA *in vitro* and could only be converted into a GFP-ssrA unfoldase by introducing one or two aromatic D1 loop residues and simultaneously deleting its N domain [49].

Aromatic D1 loop residues, however, render Cdc48 inactive *in vivo* (our unpublished results), suggesting that a counter-selection against a Clp-like threading mechanism might have taken place during evolution of Cdc48 orthologues. Finally, known cellular functions of Cdc48 do not necessarily demand total unfolding of substrate proteins. ERAD substrates are presumably translocated to the cytosol in a largely unfolded state (see below), and other quality control substrates can be assumed to be at least partially unfolded as well. Of note, presentation of just an unstructured region(s) by proteasomal substrates is typically sufficient for their complete unfolding by the AAA subunits of the proteasomal 19S regulatory particle [50, 51], suggesting that the critical function of Cdc48 in protein quality control pathways may not be complete protein unfolding. Furthermore, Cdc48 substrates not destined for degradation by the 26S proteasome (see below) need not or must not be unfolded at all.

As an alternative mechanism to threading through the central pore of Cdc48, it was proposed that substrates enter and exit the pore at the D2 end (Fig. 8.4b) [38]. Interestingly, the D2 pore-facing loops of all Cdc48 orthologues possess the key aromatic residue absent in the D1 loop, allowing for a ratchet-like activity of the D2 ring similar to Clp-type unfoldases [49] (P. Zwickl, personal communication). Consistent with such a "D2 in – D2 out" model of Cdc48 activity, these aromatic residues are critical for Cdc48 function *in vitro* and *in vivo* [38] (M. Esaki and T. Ogura, cited as unpublished results in [39]). Furthermore, it has been hypothesized that the high local concentration of guanidyl groups present at a "denaturation collar" of arginine residues lining the D2 pore provides the denaturant for protein unfolding [38]. Even though direct evidence for such a denaturing property is missing, the "D2 in – D2 out" model is consistent with the hypothesis that local unfolding of weakly structured regions in substrate proteins rather than their complete unfolding is sufficient for subsequent global unfolding and degradation by the 26S proteasome [51].

A third model does not invoke substrate entry into the central pore at all. Instead, the pronounced domain movements of Cdc48 throughout its ATPase cycle, perhaps amplified by cofactor proteins acting as levers, could be sufficient to segregate or extract substrates from their cellular environments (Fig. 8.4c). Even though there is presently no experimental evidence to support such a "surface only" model, it is remarkable that the homotrimeric cofactor p47 (see below) was found to bind centrally on top of the hexameric p97 ring in a cryo-EM study [52]. On the basis of these data, it is difficult to imagine how p47-dependent substrates could enter the central pore of Cdc48 from either top or bottom.

It should be noted that the models depicted above need not be mutually exclusive, but could apply for distinct subsets of Cdc48 substrates. For instance, substrates destined for proteasomal degradation may be (partially) unfolded by threading or by the "D2 in – D2 out" mechanism, whereas p47-dependent substrates in membrane fusion processes may be prevented from entering the central pore and processed when bound to the Cdc48 surface. Clearly, all three models need to be scrutinized on the basis of additional structural and functional studies.

Regulation by Cofactors

The involvement of Cdc48 in fundamentally distinct cellular processes necessitates the tight control of its segregase activity. Indeed, a large number of cofactors regulate central aspects of Cdc48 function, including its subcellular localization and substrate specificity and fate [53]. The critical importance of regulatory cofactors for Cdc48 activity is underscored by the fact that yeast null mutants in major Cdc48 cofactors are either severely sick or non-viable [54–57]. Furthermore, human disease-associated p97 mutant proteins were found to exhibit significantly perturbed cofactor interactions [25, 58, 59]. According to their function, Cdc48 cofactors can be broadly classified as either substrate-recruiting or substrate-processing (Fig. 8.5) [53, 60]. Substrate-recruiting cofactors typically possess ubiquitin binding domains and function as adaptors between Cdc48 and specific substrates. Alternatively, or additionally, they can control Cdc48 recruitment to specific substrates/pathways by means of their subcellular localization, e.g. as integral ER membrane proteins. Substrate-processing cofactors, on the other hand, typically exhibit E3 ligase or DUB activity and modulate the ubiquitin signal on substrates in order to regulate their downstream fate [23, 60]. The two principal docking sites for regulatory cofactors are the N domain and the C-terminal tail of Cdc48 (Fig. 8.5). The majority of cofactors possess one or two defined modules for Cdc48 binding. These include the ubiquitin regulatory X (UBX) and UBX-like domains and the linear sequence motifs binding-site 1 (BS1), VCP-interacting motif (VIM), and VCP-binding motif (VBM), which all bind to partially overlapping regions of the N domain; and the peptide N-glycosidase/ubiquitin-associated (PUB) and PLAA, Ufd3 and Lub1 (PUL) domains interacting with the extreme carboxy-terminus of Cdc48 [23, 53, 61–63].

An efficient regulation of Cdc48 cellular functions requires that different cofactors do not randomly interact with the six N-terminal and six C-terminal cofactor binding sites present per Cdc48 hexamer. In accordance with this consideration, cofactor interactions are organized in a hierarchical manner dominated by the mutually exclusive binding of the major substrate-recruiting cofactors Ufd1-Npl4 and Shp1 (known as p47 in vertebrates) [53, 64]. Broadly speaking, the heterodimeric cofactor Ufd1-Npl4 recruits substrates of proteasomal degradation or processing pathways to Cdc48, whereas Shp1/p47 is believed to act primarily as an adaptor for non-proteasomal substrates of Cdc48. As discussed in the previous section, the two different cofactors may even dictate distinct mechanisms of Cdc48 substrate turnover. Importantly, the activities of the $Cdc48^{Ufd1\text{-}Npl4}$ and $Cdc48^{Shp1}$ complexes can be fine-tuned by additional substrate-recruiting cofactors [53]. For instance, the yeast cofactors Ubx2 and Ubx5 direct the $Cdc48^{Ufd1\text{-}Npl4}$ complex to the ER membrane for ERAD [65, 66] and to nuclear chromatin for DNA damage repair [67], respectively. Consistent with such a mechanism of multi-layered specificity control, only one Ufd1-Npl4 heterodimer binds one Cdc48 hexamer [68], leaving additional binding sites for Ubx2, Ubx5 and related cofactors. Conversely, the latter cofactors can only form stable complexes with Cdc48 in the presence of Ufd1-Npl4 [69].

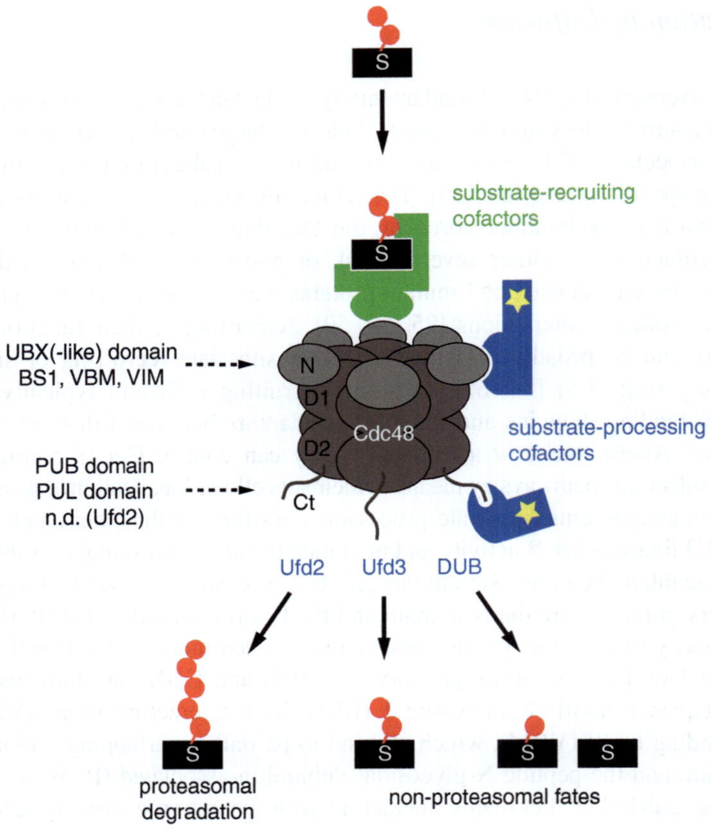

Fig. 8.5 Control of Cdc48 by regulatory cofactors. Substrate proteins (S) marked by short ubiquitin chains (*red circles*) are recognized by substrate-recruiting cofactors (*green*) and segregated from protein complexes, lipid membranes, or chromatin (not shown). Substrate-processing cofactors (*blue*) catalyze ubiquitin chain elongation, targeting the substrate for efficient proteasomal degradation (Ufd2); prevent chain elongation (Ufd3); or catalyze deubiquitylation (DUB, e.g. Otu1). The latter two activities prevent proteasomal targeting. Enzymatic activity of substrate-processing cofactors is indicated by a *yellow asterisk*. The domain organisation of Cdc48 according to Fig. 8.3a (N, D1, D2, Ct) and Cdc48 binding modules of cofactors (*left*) are indicated. Note that the Cdc48 binding site of Ufd2 has not been characterized in detail (n.d.) (Modified from [23])

Interestingly, there is evidence for Ufd1-independent functions of Npl4 in a recently discovered mitochondrial stress response pathway in yeast [70] and in certain mammalian ERAD pathways [71, 72]. While these findings raise the intriguing possibility that different binding partners of Npl4 define yet another layer of specificity control, this hypothesis still awaits a rigorous structural and biochemical characterization of the Cdc48-cofactor complexes involved.

Compared to the $Cdc48^{Ufd1\text{-}Npl4}$ complex, little is known about fine-tuning of the $Cdc48^{Shp1}$ complex by additional cofactors. The only well-established example so far is the role of mammalian VCIP135 in controlling the function of the related

$p97^{p47}$ and $p97^{p37}$ complexes in the homotypic fusion of Golgi and ER membrane vesicles [73–76]. VCIP135 is a large Cdc48/p97 cofactor possessing DUB activity [74], which initially suggested its classification as substrate-processing factor [23, 53]. However, recent results showed that the DUB activity of VCIP135 is exclusively required for the post-mitotic, $p97^{p47}$-mediated fusion of Golgi vesicles, but not for other VCIP135-controlled fusion events [75, 76]. Thus, VCIP135 may also be considered an additional substrate-recruiting factor directing $p97^{p47}$ and $p97^{p37}$ to ER and Golgi vesicles, analogous to the role of Ubx2 in recruiting $Cdc48^{Ufd1\text{-}Npl4}$ to the ER.

Recently, a potential third major Cdc48 cofactor was identified. UBXD1 interacts with both the N domain and the C-terminus of Cdc48/p97, by virtue of its VIM and PUB domains, respectively [77, 78]. Consistent with UBXD1 occupying both major cofactor binding regions of Cdc48 in a proposed stoichiometry of three molecules UBXD1 per hexamer [69], binding of UBXD1 appears to be mutually exclusive with the two other major Cdc48 cofactors, Ufd1-Npl4 and p47 [25, 77, 79]. Homologues of UBXD1 are found in all eukaryotes except fungi, suggesting that it may control a fundamental cellular function of Cdc48. Indeed, a recent study implicated the mammalian $p97^{UBXD1}$ complex in endolysosomal sorting, even though the precise role of UBXD1 in this process remains to be defined [25] (see below).

Role of Cdc48 in Intracellular Proteolysis

Proteasomal Degradation Pathways

The UFD Pathway

Cdc48 and its cofactors were first implicated in proteolysis on the basis of a genetic screen for yeast mutants defective in the proteasomal degradation of the ubiquitin fusion protein, ubiquitin-β-galactosidase (Ub-P-βGal) [54]. Despite the fact that the physiological relevance of this "ubiquitin fusion degradation" (UFD) pathway is still unclear, studies of the UFD pathway revealed important principles of Cdc48-mediated protein degradation. The UFD genetic screen [54] and subsequent candidate approaches [80–83] led to the identification of the E3 ligase specific for Ub-P-βGal (Ufd4) and of all proteins critical for Cdc48-dependent protein degradation by the 26S proteasome. Besides Cdc48 itself, these include Ufd1 and Npl4, forming the heterodimeric substrate-recruiting cofactor Ufd1-Npl4; the substrate-processing cofactor Ufd2; and the proteasomal targeting proteins Rad23 and Dsk2 (Fig. 8.6a). Ufd4 catalyzes the modification of Ub-P-βGal with one or few ubiquitin moieties [54, 84], which are insufficient for proteasomal targeting and degradation [85]. This "oligo-ubiquitylated" substrate is recognized by Ufd1-Npl4 and recruited to the $Cdc48^{Ufd1\text{-}Npl4}$ complex [86]. There, the substrate-processing cofactor Ufd2 by virtue of its "E4" ubiquitin ligase activity extends the oligo-ubiquitin tag to a long K48-linked chain representing an efficient proteasomal degradation signal [84, 87].

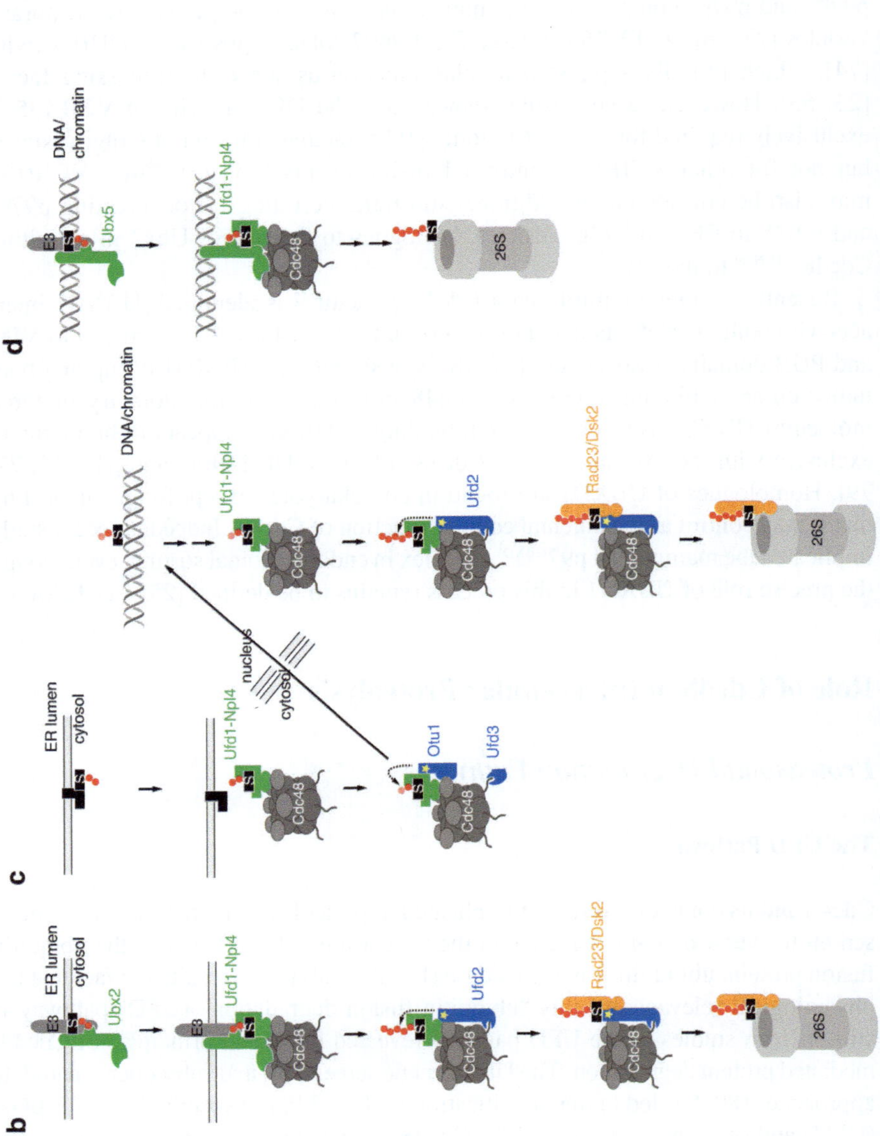

a
b
c
d
S
E3
ER lumen
cytosol
Ubx2
Ufd1-Npl4
Cdc48
Ufd2
Rad23/Dsk2
26S
Otu1
Ufd3
nucleus
DNA/chromatin
Ubx5
DNA/
chromatin

Fig. 8.6 Cdc48-dependent proteasomal degradation pathways. (**a**) The UFD pathway. The substrate (S) Ub-P-βGal is oligo-ubiquitylated by the E3 ligase Ufd4 (not shown) and recruited to Cdc48 by Ufd1-Npl4. Ufd2 catalyzes the ubiquitin chain elongation (*dashed arrow*) and hands over the polyubiquitylated substrate to the proteasomal adaptor proteins Rad23 and Dsk2. (**b**) The ERAD pathway. The Cdc48 machinery is linked to the degradation of ER quality control substrates by Ubx2, an ER membrane-localized substrate-recruiting cofactor coordinating interactions of $Cdc48^{Ufd1\text{-}Npl4}$ with ERAD substrates and E3 ligases. (**c**) The OLE pathway. The processed, oligo-ubiquitylated p90 form of the transcription factor Spt23 is mobilized from stable associations with the unprocessed, ER membrane-tethered p120 precursor (*black*) by $Cdc48^{Ufd1\text{-}Npl4}$. It is unclear how the Cdc48 complex is recruited to the substrate. After mobilization, Ufd3 and Otu1 prevent Ufd2-catalyzed chain elongation and stabilize p90 (*dashed arrow*), enabling nuclear translocation and activation of target gene expression. Subsequently, in a mechanistically poorly understood process, p90 is presumably recruited by Ufd1-Npl4 to Cdc48 and proteasomally degraded in a Ufd2- and Rad23-/Dsk2-dependent manner. (**d**) Rpb1 degradation. In analogy to Ubx2 (panel b), $Cdc48^{Ufd1\text{-}Npl4}$ is recruited to chromatin-associated, ubiquitylated Rpb1 by the substrate-recruiting factor Ubx5 and segregates the substrate from chromatin. The 26S proteasome has been shown to interact with Cdc48, but a potential involvement of Ufd2 and Rad23/Dsk2 as well as the ubiquitylation state of the substrate have not been investigated

Polyubiquitylated Ub-P-βGal is then delivered to the 26S proteasome by the related proteasomal adaptor proteins Rad23 and Dsk2 [81, 83], which possess binding sites both for polyubiquitin chains and the proteasomal 19S regulatory particle [88].

Notably, not only Cdc48 and its cofactors, but also Ufd2 and Rad23/Dsk2 are linked by physical interactions [86, 89] (Fig. 8.6a), which led to the proposal that Cdc48 substrates are "escorted" to the 26S proteasome in order to ensure efficient degradation [86]. The importance of a substrate hand-over between Ufd2 and Rad23/Dsk2 was recently underscored by a study analyzing the $Cdc48^{Ufd1\text{-}Npl4}$-dependent degradation of a linear tetra-ubiquitin fusion protein of β-galactosidase (Ub_4-βGal) [90]. Consistent with tetra-ubiquitin being the minimal signal for proteasomal degradation [85], this longer ubiquitin fusion tag obliterated the need for polyubiquitylation by the combined activities of Ufd4 and Ufd2. However, a catalytically inactive truncated form of Ufd2 was still required for efficient proteasomal degradation of Ub_4-βGal, presumably in its capacity to escort the substrate from the Cdc48 complex to the proteasomal adaptors Rad23/Dsk2. Because binding of Rad23/Dsk2 to Ufd2 and the 19S complex is mutually exclusive [89], the relatively stable interaction between Ufd2 and Rad23 [91] must be abrogated after substrate hand-over to Rad23. Intriguingly, this process was recently found to require Cdc48 itself, raising the interesting possibility that the Ufd2-Rad23 complex is a (pseudo-)substrate for the Cdc48 segregase activity [92]. Thus, Cdc48 could be involved in three different steps of UFD substrate turnover: in (i) local destabilization/unfolding of the substrate [51]; (ii) Ufd2-catalyzed ubiquitin chain elongation [84, 86]; and (iii) breaking the Ufd2-Rad23 interaction [92]. Further studies are needed to establish the relative importance of these three functions for the UFD and other degradation pathways.

ERAD

The UFD pathway can be considered to represent a conserved "module" of $Cdc48^{Ufd1\text{-}Npl4}$ activity that can be integrated into various proteasomal degradation pathways. In ERAD, this module is placed downstream of a sophisticated quality control network of ER chaperones and lectins. This quality control network identifies irreparably damaged or misfolded ER proteins and directs them to dedicated E3 ligases at the ER membrane for ubiquitylation and subsequent proteasomal degradation [93] (Fig. 8.6b). In *S. cerevisisae*, the two principal ERAD E3s are the HRD (HMG-CoA reductase degradation) complex and Doa10 [94]. The HRD complex consists of a core and ancillary components. The core, composed of the catalytic subunit Hrd1 (also known as Der3) and the luminal substrate receptor Hrd3, is required for the degradation of all substrates, while the ancillary factors Der1, Usa1 and Yos9 are only involved in the degradation of certain subsets of HRD substrates [93, 94]. HRD directs the degradation of all luminal ERAD substrates and of ER membrane proteins exposing degradation signals in their luminal or transmembrane regions [95]. Doa10 is required for the degradation of ER membrane proteins exposing cytosolic degradation signals [95], as well as for the degradation of several soluble non-ERAD substrates [96–98]. In order to be ubiquitylated by Hrd1, luminal substrates recognized by Hrd3 must be at least partially retro-translocated to the cytosolic face of the ER membrane *via* still intensely debated channel proteins and mechanisms (discussed in [99, 100]). In contrast, transmembrane substrates of HRD and Doa10 can presumably be ubiquitylated *in situ*.

Downstream of HRD- and Doa10-catalyzed substrate ubiquitylation, the different ERAD routes converge, as all substrates must be dislocated from the ER membrane by Cdc48 for subsequent delivery to the 26S proteasome (Fig. 8.6b). These steps require the same components as those in the UFD pathway, i.e. the $Cdc48^{Ufd1\text{-}Npl4}$ complex [101–105], Ufd2 [86, 106] and Rad23/Dsk2 [86, 107]. The essential function of $Cdc48^{Ufd1\text{-}Npl4}$ in ERAD is the ATP-dependent retro-translocation and dislocation of ubiquitylated substrates that are stably associated with or inserted into the ER membrane. Ufd2 and Rad23/Dsk2 then escort the dislocated substrates to the 26S proteasome, perhaps helping to suppress unwanted aggregation of non-native substrates in the cytosol. Importantly, the link between the "UFD module" and the ERAD pathways is provided by the substrate-recruiting cofactor Ubx2. Ubx2 is an integral ER membrane protein that serves as ER anchor for the $Cdc48^{Ufd1\text{-}Npl4}$ complex and coordinates interactions between $Cdc48^{Ufd1\text{-}Npl4}$, ERAD substrates, and the E3 ligases [65, 66] (Fig. 8.6b). In addition to Ubx2, which is critical for the degradation of most ERAD substrates, the Cdc48 cofactors Ubx4 [108] and Vms1 [109] play auxiliary roles in the degradation of certain ERAD substrates. Both appear to act independently of Ubx2 and probably downstream of Cdc48-driven substrate dislocation, but their precise role in ERAD remains to be defined. Of note, Ubx4 and Vms1 have also been implicated in DNA damage repair and mitochondrial stress response, respectively (see below), raising the possibility that they perform a more general function in Cdc48-mediated degradation pathways. Besides Ubx4 and Vms1, cytosolic chaperones, mainly from the Hsp70 and Hsp40 families, have also been demonstrated to assist in the degradation of some ERAD substrates [106, 110–112], presumably by preventing their aggregation after dislocation from the ER membrane.

In addition to its central role in protein quality control, the ERAD pathway is also used for regulatory proteolysis in the control of lipid metabolism [113, 114]. In yeast, the HMG-CoA reductase isoenzyme Hmg2 is degraded in a sterol-dependent manner via the canonical pathway for transmembrane substrates of the HRD complex [65, 105, 115–117]. In a negative feedback loop, high metabolic flux through Hmg2 is believed to result in altered fluidity of the ER membrane, causing conformational changes in the transmembrane domain of Hmg2 that lead to the exposure of degradation signals recognized by the HRD E3 ligase [114, 118, 119].

The OLE Pathway

The mobilization and degradation of the transcription factor Spt23 is another $Cdc48^{Ufd1\text{-}Npl4}$-mediated degradation pathway initiated at the ER membrane, but is distinct from the canonical ERAD pathways (Fig. 8.6c). In this so-called OLE pathway, the inactive ER membrane-anchored form of Spt23 (p120) is mono- or oligo-ubiquitylated by the E3 ligase Rsp5 and processed by the 26S proteasome into its active form, p90 [120]. Ubiquitylated p90 is segregated from its stable association with unprocessed, ER membrane-anchored p120 precursors by the $Cdc48^{Ufd1\text{-}Npl4}$ complex [82] (Fig. 8.6c). In contrast to ERAD, it is still unclear how $Cdc48^{Ufd1\text{-}Npl4}$ is recruited to the ER membrane for the mobilization of p90, as this process does not appear to involve Ubx2 (our unpublished data). One possibility is that $Cdc48^{Ufd1\text{-}Npl4}$ is recruited directly by the 26S proteasome during proteasomal processing of p120, because Cdc48 has been identified as a proteasome interacting protein [121, 122]. Following its mobilization, p90 is transported to the nucleus, most likely in complex with $Cdc48^{Ufd1\text{-}Npl4}$, where it can activate expression of the key target gene, *OLE1* encoding Δ9 fatty acid desaturase [86]. The transcription factor activity of p90 is terminated by Ufd2-mediated polyubiquitylation, Rad23/Dsk2-mediated delivery to the 26S proteasome, and proteasomal degradation [86] (Fig. 8.6c). Notably, the homologous, functionally overlapping transcription factor Mga2 is controlled in a similar manner, even though mechanistic details appear to differ from Spt23 processing and mobilization [123, 124]. The OLE pathway provides another example for an involvement of Cdc48 in a negative feedback loop of lipid metabolism. Ole1 controls the concentration of unsaturated fatty acids which in turn influences ER membrane fluidity and thickness. These parameters, through regulation of Rsp5 access, are believed to control the balance between proteasomal processing of p120 into p90 and processive proteasomal degradation of p120, in a complex scenario involving casein kinase 2 and the prolyl isomerase Ess1 (Pin1 in mammals) [125].

Two aspects of the $Cdc48^{Ufd1\text{-}Npl4}$ activity in the OLE pathway are worth mentioning. Firstly, this pathway provides a clear example of a temporal and spatial separation of Cdc48 segregase function at the ER membrane from Ufd2-triggered proteasomal degradation in the nucleus [82, 86] (Fig. 8.6c). This raises the interesting, but entirely speculative possibility that UFD and ERAD substrates are, at least partially, degraded in the nucleus as well by the significant nuclear subpopulation of 26S proteasomes [126, 127]. Secondly, in order to ensure proper expression of *OLE1*, it is vital for the OLE pathway to prevent the premature, Ufd2-catalyzed polyubiquitylation and degradation of p90. This is achieved by the combined action of two Cdc48 substrate-processing factors, Otu1 and Ufd3, which antagonize Ufd2 function by virtue of its DUB activity and by competing for Cdc48 binding, respectively [60, 78] (Fig. 8.6c). Importantly, the antagonistic effects of Ufd2 *versus* Ufd3 and Otu1 in the degradation of Cdc48 substrates led to the current model that the ubiquitylation status and, thus, the ultimate fate of Cdc48 substrates can be actively modulated and edited by substrate-processing cofactors of Cdc48 [23, 60, 128, 129] (Fig. 8.5, bottom).

Nuclear Substrates

The $Cdc48^{Ufd1\text{-}Npl4}$ complex is involved in the proteasomal degradation of a number of additional cellular proteins [67]. Recently, $Cdc48^{Ufd1\text{-}Npl4}$ was shown to be critical for the degradation of the largest subunit of RNA polymerase II, Rpb1 [67]. Upon induction of DNA damage by UV radiation, RNA polymerase II becomes irreversibly stalled at the sites of DNA lesions and must be removed by proteasomal degradation in order to allow transcription to recommence [130, 131]. UV-induced Rpb1 degradation was found to depend on the $Cdc48^{Ufd1\text{-}Npl4}$ complex and, in addition, on the substrate-recruiting cofactor Ubx5 [67]. Moreover, ubiquitylated Rpb1 accumulated on chromatin in a *cdc48* mutant, and binding of Ubx5 to Rpb1 was greatly increased upon UV treatment [67]. Together, these data strongly suggest that Ubx5 recruits $Cdc48^{Ufd1\text{-}Npl4}$ to stalled, ubiquitylated Rpb1, for its extraction from chromatin and delivery to the proteasome (Fig. 8.6d). Thus, the function of Ubx5 bears analogy to the role of Ubx2 in recruiting $Cdc48^{Ufd1\text{-}Npl4}$ to the ER membrane in ERAD. Because potential roles of Ufd2 and Rad23/Dsk2 in Rpb1 degradation are yet to be investigated, it remains unclear if Rpb1 is escorted to the proteasome in a manner similar to UFD, ERAD and OLE pathway substrates. Contrary to this possibility, the finding that ubiquitylated Rpb1 interacts with the proteasome in a *cdc48* mutant was interpreted to indicate that Cdc48 does not act upstream of the proteasome, but rather exerts its activity on proteasome-bound substrates resistant to unfolding by the AAA subunits of the 19S complex. However, while further experimental work is clearly needed to elucidate the details of Rpb1 degradation, it should be noted that the specific molecular defect(s) caused by the conditional *cdc48* allele used are currently unknown and could, for instance, lead to the delivery of ubiquitylated substrates in a conformation incompatible with proteasomal degradation. Interestingly, Ubx4 is also involved in Rpb1 degradation, apparently in parallel to the pathway involving Ubx5 [67], but its specific role in this and other degradation pathways has yet to be identified.

Another transcription-related function of $Cdc48^{Ufd1\text{-}Npl4}$ is the extraction of the transcriptional repressor Matα2 from its DNA target sites during yeast mating type switching [132]. In this case however, in contrast to the proteasomal degradation of Spt23 p90 and Rpb1, the segregase activity of $Cdc48^{Ufd1\text{-}Npl4}$ appears to be uncoupled from proteasomal degradation. This finding raises the interesting possibility that $Cdc48^{Ufd1\text{-}Npl4}$ plays a more general role in transcriptional regulation by remodeling transcription factor complexes.

CDC48 was first identified in a yeast genetic screen for cell cycle defects, and conditional *cdc48* mutants arrest as large budded cells with 2n DNA content in metaphase [57, 133]. Studies in yeast and in higher eukaryotes revealed that Cdc48 acts on a number of substrates in different phases of the cell cycle. Even though the involvement of Ufd1-Npl4 is best established for cell cycle functions in metazoans, it is likely that Cdc48-dependent degradation of cell cycle substrates in yeast requires Ufd1-Npl4 as well. In *S. cerevisiae*, Cdc48 has been shown to be critical for the degradation of the G1 cyclin-dependent kinase inhibitor Far1 at the G1/S transition [134], and of the G1 cyclin Cln2 after entry into S phase [135]. During mitotic exit, Cdc48 has been implicated in spindle disassembly in yeast and in *Xenopus* egg extracts [136], although conflicting results exist at least for the latter system [137]. In yeast, the microtubule-binding protein Ase1 and the Polo-like kinase homologue Cdc5 were postulated to be critical targets of $Cdc48^{Ufd1\text{-}Npl4}$ that must either be degraded or sequestered from microtubules in order to allow spindle disassembly to occur [136]. No further information on mechanisms and degradation pathways exists for any of the cell cycle substrates of Cdc48 in yeast. In addition, $Cdc48^{Ufd1-Npl4}$ has recently been implicated in cell wall integrity signaling and G1 progression [138], but Cdc48 substrate(s) critical for this function have yet to be identified.

Other Proteasomal Pathways

An interesting example for an involvement of $Cdc48^{Ufd1-Npl4}$ in the regulation of cellular metabolism is the degradation of Fructose-1,6,-bisphosphatase (FBPase), a key enzyme of gluconeogenesis. Upon shift from ethanol to glucose as carbon source, yeast cells switch metabolism from gluconeogenesis to glycolysis. This switching requires the rapid inactivation of FBPase, which is accomplished, in part, by proteasomal degradation [139, 140]. FBPase degradation depends on $Cdc48^{Ufd1-Npl4}$, Rad23/Dsk2, but not Ufd2 [141]. Ubx4 again appears to play an ancillary role in this process. Interestingly, the glucose-induced degradation of pyruvate carboxykinase, another gluconeogenetic enzyme, is also mediated by $Cdc48^{Ufd1-Npl4}$ [141], suggesting that Cdc48 may play a more general role in metabolic control than previously appreciated.

Cdc48-mediated proteasomal degradation was recently also implicated in a mitochondrial stress-responsive system, but here it appears to employ the substrate-recruiting cofactor Vms1 together with Npl4, not the canonical heterodimer Ufd1-Npl4 [70]. Upon mitochondrial stress, Vms1 relocates from the cytosol to mitochondria and recruits Cdc48 and Npl4, but not Ufd1. The $Cdc48^{Vms1-Npl4}$ complex is required for the degradation of the mitochondrial outer membrane protein Fzo1 and probably further mitochondrial substrates [70]. In metazoans, Cdc48/p97 is involved in the proteasomal degradation of the mitofusins Mfn1 and Mfn2 and of the anti-apoptotic Bcl-2 family member Mcl1 [142, 143]. While Vms1 is conserved from yeast to humans, a direct role in the Cdc48/p97-dependent degradation of mitochondrial proteins in metazoans has yet to be shown. Mechanistically, the role of Cdc48 in the degradation of outer mitochondrial membrane proteins may resemble the Cdc48-mediated dislocation of ERAD substrates, but this hypothesis requires further experimental support. Regardless of the exact mechanism of Cdc48 activity, the identified mitochondrial targets link Cdc48/p97 to mitophagy of damaged mitochondria and to apoptosis, and thus to central aspects of cellular stress response and survival [144, 145].

Autophagy

Besides its role in a growing number of proteasomal degradation pathways, Cdc48 is also involved in different forms of autophagy, both in yeast and in higher eukaryotes. In yeast, Cdc48 and its substrate-recruiting factor Shp1 were found to be required for general starvation-induced macroautophagy and for selective piecemeal microautophagy of the nucleus [146]. Intriguingly, these processes do not appear to involve the ubiquitin-proteasome system. Instead, the ubiquitin-like modifier Atg8 was identified as a critical target of Shp1 in autophagy. Atg8 (LC3 in mammals) is anchored to membranes via reversible lipidation of its C-terminus. This is required for the formation and extension of the autophagosomal isolation membrane and for the incorporation of cargo into autophagosomes [6, 147]. In starved *shp1* mutant cells, neither GFP-Atg8-positive autophagosomes in the cytosol nor autophagic bodies in the vacuole (the yeast lysosome) could be detected, suggesting that Shp1 is critical for autophagosome biogenesis [146], but its exact role in this process and the significance of its physical interaction with Atg8 remain to be determined.

Nitrogen starvation induces the degradation of the yeast ribosomal 60S subunit in a selective autophagy pathway termed ribophagy [148]. Ribophagy depends on Cdc48, its substrate-processing cofactor Ufd3 (also known as Doa1) and the DUB complex Ubp3-Bre5 in a process which is still poorly understood [148, 149]. It has been proposed that ribophagy involves the dynamic ubiquitylation of ribosomes and/or the ribophagy machinery [26, 148]. Notably, in contrast to the study by Krick et al. [146], the study by Ossareh-Nazari et al. reported that general macroautophagy of GFP-Atg8 proceeds normally in a *cdc48* mutant [149]. Even though differences in the *cdc48* alleles and particular starvation conditions used may contribute to these conflicting results, additional experiments are required to clarify this critical point.

Mammalian cells depleted of Cdc48/p97 or expressing mutant forms of Cdc48/p97 associated with the familial proteinopathy termed Inclusion Body Myopathy, Paget's disease of the bone and Fronto-temporal Dementia (IBMPFD) accumulate nondegrading autophagosomes that fail to mature into autolysosomes in order to eliminate their cargo [150–152]. This finding suggests that impaired autophagy could be the critical cellular defect underlying the pathogenesis of IBMPFD and causing the characteristic accumulation of insoluble cytoplasmic inclusions positive for TDP-43 and ubiquitin. Of note, Cdc48/p97 appears to be required for a late step during autophagy in mammals, i.e. the fusion of autophagosomes and lysosomes into proteolytically active autolysosomes, whereas $Cdc48^{Shp1}$ in yeast has been implicated in some early step during autophagosome biogenesis [146]. It is presently unknown if yeast and mammalian autophagy have fundamentally different requirements for Cdc48 activity, or if Cdc48 possesses two distinct functions in autophagy that are evolutionary conserved from yeast to humans.

Endolysosomal Degradation

Cdc48 was very recently also implicated in the third major intracellular proteolytic pathway, endolysososmal degradation. Mammalian p97 and its cofactor UBXD1 were found to target mono-ubiquitylated forms of the caveolar scaffolding protein caveolin-1 on its endocytic route to endolysosomes [25]. Intriguingly, overexpression of IBMPFD-causing variants of Cdc48/p97 blocked endolysososmal sorting, suggesting that impaired endolysosomal degradation may contribute to, or aggravate, the autophagy defects observed in IBMPFD.

UBXD1 is not conserved in yeast, and there exists no direct evidence for an involvement of yeast Cdc48 in endolysosomal degradation so far. Interestingly, however, the substrate-processing cofactor Ufd3 has been shown to be critical for sorting ubiquitylated membrane proteins into MVBs on their biosynthetic route from the *trans*-Golgi network to the vacuole [153]. The same study also provided evidence for a role of Ufd3 in the sorting of endocytic cargo into MVBs [153], but this possibility has not been directly addressed yet. While results obtained using a Ufd3-binding deficient *cdc48* mutant strain suggested that the function of Ufd3 in MVB sorting is independent of Cdc48 [154], this issue deserves further attention in light of the new findings for mammalian Cdc48/p97.

Other Functions of Cdc48

While the focus of this review is the role of yeast Cdc48 in intracellular proteolysis, a number of additional proteolytic substrates and pathways involving Cdc48/p97 have been identified in metazoans. Exciting examples include the proteasomal degradation of the hypoxia-inducible transcription factor 1α [79], the myosin chaperone Unc45 [155], and the replication licensing factor Cdt1 [156, 157], as well as roles of Cdc48/p97 in DNA double strand break repair [158, 159] and lifespan control [129]. For a recent review focusing on Cdc48 functions in metazoans, see [160, 161].

In addition to its activities in various proteolytic pathways, Cdc48 also possesses functions in several non-proteolytic processes that are beyond the scope of this review. For example, the substrate-recruiting cofactor Shp1/p47 controls Cdc48/p97 activities in membrane fusion processes that do not appear to involve the degradation of Cdc48 substrates (reviewed in [53, 162–164]). In yeast, *shp1* mutants exhibit vacuolar fragmentation as well as defective V-ATPase targeting and/or activity [165, 166]. It is presently unclear if and how these phenotypes may be related to the autophagy function of Shp1 discussed above. Interestingly, Shp1 was identified as a positive regulator of the catalytic subunit of yeast protein phosphatase 1, Glc7 [55], although the underlying regulatory mechanism is unknown. In agreement with such a function, Cdc48 and Shp1 were recently shown to be important for the balance between the antagonistic Glc7 phosphatase and Ipl1 (Aurora B) kinase activities on elusive nuclear substrates [57]. Consistently, loss of Shp1 function causes hyperphosphorylation of Ipl1 substrates, defective bipolar attachment of kinetochores to spindle microtubules, and chromosomal mispositioning [57]. However, these phenotypes are most likely secondary to a partial loss of mitotic Glc7 function(s), and so far direct mitotic substrates of $Cdc48^{Shp1}$ have not been identified. In metazoans, by contrast, Cdc48/p97 together with Ufd1-Npl4 has an established function as a direct negative regulator of Aurora B [167, 168]. In *Xenopus* egg extracts, $p97^{Ufd1\text{-}Npl4}$ was shown to remove ubiquitylated Aurora B from chromosomes at the end of mitosis, in order to allow for chromosome decondensation and nucleus formation [167]. In HeLa cells, $p97^{Ufd1\text{-}Npl4}$ negatively regulates Aurora B earlier in mitosis, at the metaphase/anaphase transition, to allow chromosome segregation to occur properly [168]. Interestingly, after its extraction from chromatin, Aurora B is stable in *Xenopus* egg extracts, while it is apparently subject to proteasomal degradation in HeLa cells [168]. It is presently unclear if the respective activities of $Cdc48^{Shp1}$ and $p97^{Ufd1\text{-}Npl4}$ in the negative regulation of yeast and metazoan Aurora B are mechanistically related, or if they rather reflect independent solutions for the general requirement of tight spatiotemporal regulation of Aurora B during mitosis.

Outlook

The Baker's yeast *S. cerevisiae* has proven a valuable model organism for the genetic, biochemical and cellular analysis of Cdc48. A wealth of studies performed in the yeast system formed the basis for our detailed understanding of Cdc48 cellular functions, in particular in proteasomal protein degradation. While the yeast system will continue to deepen our knowledge on fundamental cellular functions of Cdc48, metazoan models including worm, fly and mouse are rapidly closing the gap and provide valuable insights into exciting additional Cdc48 functions in the complexity of multicellular organisms. Ultimately, these complementary approaches and models should help to better understand the defects caused by Cdc48 malfunction in human disease in the near future.

References

1. Baumeister W, Walz J, Zuhl F, Seemuller E (1998) The proteasome: paradigm of a self-compartmentalizing protease. Cell 92(3):367–380
2. Lasker K, Förster F, Bohn S, Walzthoeni T (2012) Molecular architecture of the 26S proteasome holocomplex determined by an integrative approach. Proc Natl Acad Sci USA 109(5):1380–1387
3. Kornfeld S, Mellman I (1989) The biogenesis of lysosomes. Annu Rev Cell Biol 5:483–525
4. Clague MJ, Urbe S (2010) Ubiquitin: same molecule, different degradation pathways. Cell 143(5):682–685
5. Katzmann DJ, Odorizzi G, Emr SD (2002) Receptor downregulation and multivesicular-body sorting. Nat Rev Mol Cell Biol 3(12):893–905
6. Nakatogawa H, Suzuki K, Kamada Y, Ohsumi Y (2009) Dynamics and diversity in autophagy mechanisms: lessons from yeast. Nat Rev Mol Cell Biol 10(7):458–467
7. Kirkin V, McEwan DG, Novak I, Dikic I (2009) A role for ubiquitin in selective autophagy. Mol Cell 34(3):259–269
8. Kraft C, Peter M, Hofmann K (2010) Selective autophagy: ubiquitin-mediated recognition and beyond. Nat Cell Biol 12(9):836–841
9. Hershko A, Ciechanover A (1998) The ubiquitin system. Annu Rev Biochem 67:425–479
10. Ikeda F, Dikic I (2008) Atypical ubiquitin chains: new molecular signals. 'Protein modifications: beyond the usual suspects' review series. EMBO Rep 9(6):536–542
11. Komander D (2009) The emerging complexity of protein ubiquitination. Biochem Soc Trans 37(Pt 5):937–953

12. Behrends C, Harper JW (2011) Constructing and decoding unconventional ubiquitin chains. Nat Struct Mol Biol 18(5):520–528
13. Komander D, Clague MJ, Urbe S (2009) Breaking the chains: structure and function of the deubiquitinases. Nat Rev Mol Cell Biol 10(8):550–563
14. Pickart CM, Fushman D (2004) Polyubiquitin chains: polymeric protein signals. Curr Opin Chem Biol 8(6):610–616
15. Hochstrasser M (1996) Ubiquitin-dependent protein degradation. Annu Rev Genet 30:405–439
16. Xu P, Duong DM, Seyfried NT, Cheng D et al (2009) Quantitative proteomics reveals the function of unconventional ubiquitin chains in proteasomal degradation. Cell 137(1):133–145
17. Hicke L, Dunn R (2003) Regulation of membrane protein transport by ubiquitin and ubiquitin-binding proteins. Annu Rev Cell Dev Biol 19:141–172
18. Chen ZJ (2005) Ubiquitin signalling in the NF-[kappa]B pathway. Nat Cell Biol 7(8):758–765
19. Piper RC, Katzmann DJ (2007) Biogenesis and function of multivesicular bodies. Annu Rev Cell Dev Biol 23:519–547
20. Hanson PI, Whiteheart SW (2005) AAA+ proteins: have engine, will work. Nat Rev Mol Cell Biol 6(7):519–529
21. Bays NW, Hampton RY (2002) Cdc48-Ufd1-Npl4: stuck in the middle with Ub. Curr Biol 12(10):R366–R371
22. Elsasser S, Finley D (2005) Delivery of ubiquitinated substrates to protein-unfolding machines. Nat Cell Biol 7(8):742–749
23. Buchberger A (2010) Control of ubiquitin conjugation by Cdc48 and its cofactors. Subcell Biochem 54:17–30
24. Ju JS, Weihl CC (2010) Inclusion body myopathy, Paget's disease of the bone and fronto-temporal dementia: a disorder of autophagy. Hum Mol Genet 19(R1):R38–R45
25. Ritz D, Vuk M, Kirchner P, Bug M et al (2011) Endolysosomal sorting of ubiquitylated caveolin-1 is regulated by VCP and UBXD1 and impaired by VCP disease mutations. Nat Cell Biol 13(9):1116–1123
26. Dargemont C, Ossareh-Nazari B (2012) Cdc48/p97, a key actor in the interplay between autophagy and ubiquitin/proteasome catabolic pathways. Biochim Biophys Acta 1823(1): 138–144
27. Watts GD, Wymer J, Kovach MJ, Mehta SG et al (2004) Inclusion body myopathy associated with Paget disease of bone and frontotemporal dementia is caused by mutant valosin-containing protein. Nat Genet 36(4):377–381
28. Weihl CC, Pestronk A, Kimonis VE (2009) Valosin-containing protein disease: inclusion body myopathy with Paget's disease of the bone and fronto-temporal dementia. Neuromuscul Disord 19(5):308–315
29. Johnson JO, Mandrioli J, Benatar M, Abramzon Y et al (2010) Exome sequencing reveals VCP mutations as a cause of familial ALS. Neuron 68(5):857–864
30. Rouiller I, Butel VM, Latterich M, Milligan RA et al (2000) A major conformational change in p97 AAA ATPase upon ATP binding. Mol Cell 6(6):1485–1490
31. Zhang X, Shaw A, Bates PA, Newman RH et al (2000) Structure of the AAA ATPase p97. Mol Cell 6(6):1473–1484
32. DeLaBarre B, Brunger AT (2005) Nucleotide dependent motion and mechanism of action of p97/VCP. J Mol Biol 347(2):437–452
33. DeLaBarre B, Brunger AT (2003) Complete structure of p97/valosin-containing protein reveals communication between nucleotide domains. Nat Struct Biol 10(10):856–863
34. Beuron F, Flynn TC, Ma J, Kondo H et al (2003) Motions and negative cooperativity between p97 domains revealed by cryo-electron microscopy and quantised elastic deformational model. J Mol Biol 327(3):619–629
35. Pye VE, Dreveny I, Briggs LC, Sands C et al (2006) Going through the motions: the ATPase cycle of p97. J Struct Biol 156(1):12–28

36. Song C, Wang Q, Li CC (2003) ATPase activity of p97-valosin-containing protein (VCP). D2 mediates the major enzyme activity, and D1 contributes to the heat-induced activity. J Biol Chem 278(6):3648–3655
37. Ye Y, Meyer HH, Rapoport TA (2003) Function of the p97-Ufd1-Npl4 complex in retrotranslocation from the ER to the cytosol: dual recognition of nonubiquitinated polypeptide segments and polyubiquitin chains. J Cell Biol 162(1):71–84
38. DeLaBarre B, Christianson JC, Kopito RR, Brunger AT (2006) Central pore residues mediate the p97/VCP activity required for ERAD. Mol Cell 22(4):451–462
39. Esaki M, Ogura T (2010) ATP-bound form of the D1 AAA domain inhibits an essential function of Cdc48p/p97. Biochem Cell Biol 88(1):109–117
40. Nishikori S, Esaki M, Yamanaka K, Sugimoto S et al (2011) Positive cooperativity of the p97 AAA ATPase is critical for essential functions. J Biol Chem 286(18):15815–15820
41. Wang Q, Song C, Li CC (2003) Hexamerization of p97-VCP is promoted by ATP binding to the D1 domain and required for ATPase and biological activities. Biochem Biophys Res Commun 300(2):253–260
42. Zolkiewski M (2006) A camel passes through the eye of a needle: protein unfolding activity of Clp ATPases. Mol Microbiol 61(5):1094–1100
43. Sauer RT, Baker TA (2011) AAA+ proteases: ATP-fueled machines of protein destruction. Annu Rev Biochem 80:587–612
44. Gur E, Ottofuelling R, Dougan DA (2013) Machines of destruction – AAA+proteases and the adaptors that control them. In: Dougan DA (ed) Regulated proteolysis in microorganisms. Springer, Subcell Biochem 66:3–33
45. Ernst R, Claessen JH, Mueller B, Sanyal S et al (2011) Enzymatic blockade of the ubiquitin-proteasome pathway. PLoS Biol 8(3):e1000605
46. Claessen JH, Kundrat L, Ploegh HL (2011) Protein quality control in the ER: balancing the ubiquitin checkbook. Trends Cell Biol 22(1):22–32
47. Sowa ME, Bennett EJ, Gygi SP, Harper JW (2009) Defining the human deubiquitinating enzyme interaction landscape. Cell 138(2):389–403
48. Gerega A, Rockel B, Peters J, Tamura T et al (2005) VAT, the thermoplasma homolog of mammalian p97/VCP, is an N domain-regulated protein unfoldase. J Biol Chem 280(52): 42856–42862
49. Rothballer A, Tzvetkov N, Zwickl P (2007) Mutations in p97/VCP induce unfolding activity. FEBS Lett 581(6):1197–1201
50. Prakash S, Tian L, Ratliff KS, Lehotzky RE et al (2004) An unstructured initiation site is required for efficient proteasome-mediated degradation. Nat Struct Mol Biol 11(9):830–837
51. Beskow A, Grimberg KB, Bott LC, Salomons FA et al (2009) A conserved unfoldase activity for the p97 AAA-ATPase in proteasomal degradation. J Mol Biol 394(4):732–746
52. Beuron F, Dreveny I, Yuan X, Pye VE et al (2006) Conformational changes in the AAA ATPase p97-p47 adaptor complex. EMBO J 25(9):1967–1976
53. Schuberth C, Buchberger A (2008) UBX domain proteins: major regulators of the AAA ATPase Cdc48/p97. Cell Mol Life Sci 65(15):2360–2371
54. Johnson ES, Ma PC, Ota IM, Varshavsky A (1995) A proteolytic pathway that recognizes ubiquitin as a degradation signal. J Biol Chem 270(29):17442–17456
55. Zhang S, Guha S, Volkert FC (1995) The *Saccharomyces* SHP1 gene, which encodes a regulator of phosphoprotein phosphatase 1 with differential effects on glycogen metabolism, meiotic differentiation, and mitotic cell cycle progression. Mol Cell Biol 15(4): 2037–2050
56. Schuberth C, Richly H, Rumpf S, Buchberger A (2004) Shp1 and Ubx2 are adaptors of Cdc48 involved in ubiquitin-dependent protein degradation. EMBO Rep 5(8):818–824
57. Cheng YL, Chen RH (2010) The AAA-ATPase Cdc48 and cofactor Shp1 promote chromosome bi-orientation by balancing Aurora B activity. J Cell Sci 123(Pt 12):2025–2034
58. Fernandez-Saiz V, Buchberger A (2010) Imbalances in p97 co-factor interactions in human proteinopathy. EMBO Rep 11(6):479–485

59. Manno A, Noguchi M, Fukushi J, Motohashi Y et al (2010) Enhanced ATPase activities as a primary defect of mutant valosin-containing proteins that cause inclusion body myopathy associated with Paget disease of bone and frontotemporal dementia. Genes Cells 15(8):911–922
60. Rumpf S, Jentsch S (2006) Functional division of substrate processing cofactors of the ubiquitin-selective Cdc48 chaperone. Mol Cell 21(2):261–269
61. Stolz A, Hilt W, Buchberger A, Wolf DH (2011) Cdc48: a power machine in protein degradation. Trends Biochem Sci 36(10):515–523
62. Bohm S, Lamberti G, Fernandez-Saiz V, Stapf C et al (2011) Cellular functions of Ufd2 and Ufd3 in proteasomal protein degradation depend on Cdc48 binding. Mol Cell Biol 31(7):1528–1539
63. Yeung HO, Kloppsteck P, Niwa H, Isaacson RL et al (2008) Insights into adaptor binding to the AAA protein p97. Biochem Soc Trans 36(Pt 1):62–67
64. Meyer HH, Shorter JG, Seemann J, Pappin D et al (2000) A complex of mammalian ufd1 and npl4 links the AAA-ATPase, p97, to ubiquitin and nuclear transport pathways. EMBO J 19(10):2181–2192
65. Schuberth C, Buchberger A (2005) Membrane-bound Ubx2 recruits Cdc48 to ubiquitin ligases and their substrates to ensure efficient ER-associated protein degradation. Nat Cell Biol 7(10):999–1006
66. Neuber O, Jarosch E, Volkwein C, Walter J et al (2005) Ubx2 links the Cdc48 complex to ER-associated protein degradation. Nat Cell Biol 7(10):993–998
67. Verma R, Oania R, Fang R, Smith GT et al (2011) Cdc48/p97 mediates UV-dependent turnover of RNA Pol II. Mol Cell 41(1):82–92
68. Pye VE, Beuron F, Keetch CA, McKeown C et al (2007) Structural insights into the p97-Ufd1-Npl4 complex. Proc Natl Acad Sci USA 104(2):467–472
69. Hanzelmann P, Buchberger A, Schindelin H (2011) Hierarchical binding of cofactors to the AAA ATPase p97. Structure 19(6):833–843
70. Heo JM, Livnat-Levanon N, Taylor EB, Jones KT et al (2010) A stress-responsive system for mitochondrial protein degradation. Mol Cell 40(3):465–480
71. Ballar P, Pabuccuoglu A, Kose FA (2011) Different p97/VCP complexes function in retrotranslocation step of mammalian ER-associated degradation (ERAD). Int J Biochem Cell Biol 43(4):613–621
72. Ballar P, Shen Y, Yang H, Fang S (2006) The role of a novel p97/valosin-containing protein-interacting motif of gp78 in endoplasmic reticulum-associated degradation. J Biol Chem 281(46):35359–35368
73. Uchiyama K, Jokitalo E, Kano F, Murata M et al (2002) VCIP135, a novel essential factor for p97/p47-mediated membrane fusion, is required for Golgi and ER assembly in vivo. J Cell Biol 159(5):855–866
74. Wang Y, Satoh A, Warren G, Meyer HH (2004) VCIP135 acts as a deubiquitinating enzyme during p97-p47-mediated reassembly of mitotic Golgi fragments. J Cell Biol 164(7):973–978
75. Uchiyama K, Totsukawa G, Puhka M, Kaneko Y et al (2006) p37 is a p97 adaptor required for Golgi and ER biogenesis in interphase and at the end of mitosis. Dev Cell 11(6):803–816
76. Totsukawa G, Kaneko Y, Uchiyama K, Toh H et al (2011) VCIP135 deubiquitinase and its binding protein, WAC, in p97ATPase-mediated membrane fusion. EMBO J 30(17): 3581–3593
77. Kern M, Fernandez-Saiz V, Schafer Z, Buchberger A (2009) UBXD1 binds p97 through two independent binding sites. Biochem Biophys Res Commun 380(2):303–307
78. Stapf C, Cartwright E, Bycroft M, Hofmann K et al (2011) The general definition of the p97/valosin-containing protein (VCP)-interacting motif (VIM) delineates a new family of p97 cofactors. J Biol Chem 286(44):38670–38678
79. Alexandru G, Graumann J, Smith GT, Kolawa NJ et al (2008) UBXD7 binds multiple ubiquitin ligases and implicates p97 in HIF1alpha turnover. Cell 134(5):804–816
80. Ghislain M, Dohmen RJ, Levy F, Varshavsky A (1996) Cdc48p interacts with Ufd3p, a WD repeat protein required for ubiquitin-mediated proteolysis in *Saccharomyces cerevisiae*. EMBO J 15(18):4884–4899

81. Lambertson D, Chen L, Madura K (1999) Pleiotropic defects caused by loss of the proteasome-interacting factors Rad23 and Rpn10 of Saccharomyces cerevisiae. Genetics 153(1):69–79
82. Rape M, Hoppe T, Gorr I, Kalocay M et al (2001) Mobilization of processed, membrane-tethered SPT23 transcription factor by CDC48(UFD1/NPL4), a ubiquitin-selective chaperone. Cell 107(5):667–677
83. Rao H, Sastry A (2002) Recognition of specific ubiquitin conjugates is important for the proteolytic functions of the ubiquitin-associated domain proteins Dsk2 and Rad23. J Biol Chem 277(14):11691–11695
84. Koegl M, Hoppe T, Schlenker S, Ulrich HD et al (1999) A novel ubiquitination factor, E4, is involved in multiubiquitin chain assembly. Cell 96(5):635–644
85. Thrower JS, Hoffman L, Rechsteiner M, Pickart CM (2000) Recognition of the polyubiquitin proteolytic signal. EMBO J 19(1):94–102
86. Richly H, Rape M, Braun S, Rumpf S et al (2005) A series of ubiquitin binding factors connects CDC48/p97 to substrate multiubiquitylation and proteasomal targeting. Cell 120(1):73–84
87. Saeki Y, Tayama Y, Toh-e A, Yokosawa H (2004) Definitive evidence for Ufd2-catalyzed elongation of the ubiquitin chain through Lys48 linkage. Biochem Biophys Res Commun 320(3):840–845
88. Buchberger A (2002) From UBA to UBX: new words in the ubiquitin vocabulary. Trends Cell Biol 12(5):216–221
89. Kim I, Mi K, Rao H (2004) Multiple interactions of rad23 suggest a mechanism for ubiquitylated substrate delivery important in proteolysis. Mol Biol Cell 15(7):3357–3365
90. Zhao S, Ulrich HD (2010) Distinct consequences of posttranslational modification by linear versus K63-linked polyubiquitin chains. Proc Natl Acad Sci U S A 107(17): 7704–7709
91. Hanzelmann P, Stingele J, Hofmann K, Schindelin H et al (2010) The yeast E4 ubiquitin ligase Ufd2 interacts with the ubiquitin-like domains of Rad23 and Dsk2 via a novel and distinct ubiquitin-like binding domain. J Biol Chem 285(26):20390–20398
92. Baek GH, Kim I, Rao H (2011) The Cdc48 ATPase modulates the interaction between two proteolytic factors Ufd2 and Rad23. Proc Natl Acad Sci USA 108(33):13558–13563
93. Buchberger A, Bukau B, Sommer T (2010) Protein quality control in the cytosol and the endoplasmic reticulum: brothers in arms. Mol Cell 40(2):238–252
94. Ismail N, Ng DT (2006) Have you HRD? Understanding ERAD is DOAble! Cell 126(2):237–239
95. Vashist S, Ng DT (2004) Misfolded proteins are sorted by a sequential checkpoint mechanism of ER quality control. J Cell Biol 165(1):41–52
96. Ravid T, Kreft SG, Hochstrasser M (2006) Membrane and soluble substrates of the Doa10 ubiquitin ligase are degraded by distinct pathways. EMBO J 25(3):533–543
97. Swanson R, Locher M, Hochstrasser M (2001) A conserved ubiquitin ligase of the nuclear envelope/endoplasmic reticulum that functions in both ER-associated and Matalpha2 repressor degradation. Genes Dev 15(20):2660–2674
98. Hwang CS, Shemorry A, Varshavsky A (2010) N-terminal acetylation of cellular proteins creates specific degradation signals. Science 327(5968):973–977
99. Bagola K, Mehnert M, Jarosch E, Sommer T (2011) Protein dislocation from the ER. Biochim Biophys Acta 1808(3):925–936
100. Claessen JH, Kundrat L, Ploegh HL (2012) Protein quality control in the ER: balancing the ubiquitin checkbook. Trends Cell Biol 22(1):22–32
101. Bays NW, Gardner RG, Seelig LP, Joazeiro CA et al (2001) Hrd1p/Der3p is a membrane-anchored ubiquitin ligase required for ER-associated degradation. Nat Cell Biol 3(1):24–29
102. Ye Y, Meyer HH, Rapoport TA (2001) The AAA ATPase Cdc48/p97 and its partners transport proteins from the ER into the cytosol. Nature 414(6864):652–656
103. Braun S, Matuschewski K, Rape M, Thoms S et al (2002) Role of the ubiquitin-selective CDC48(UFD1/NPL4) chaperone (segregase) in ERAD of OLE1 and other substrates. EMBO J 21(4):615–621

104. Jarosch E, Taxis C, Volkwein C, Bordallo J et al (2002) Protein dislocation from the ER requires polyubiquitination and the AAA-ATPase Cdc48. Nat Cell Biol 4(2):134–139
105. Rabinovich E, Kerem A, Frohlich KU, Diamant N et al (2002) AAA-ATPase p97/Cdc48p, a cytosolic chaperone required for endoplasmic reticulum-associated protein degradation. Mol Cell Biol 22(2):626–634
106. Nakatsukasa K, Huyer G, Michaelis S, Brodsky JL (2008) Dissecting the ER-associated degradation of a misfolded polytopic membrane protein. Cell 132(1):101–112
107. Medicherla B, Kostova Z, Schaefer A, Wolf DH (2004) A genomic screen identifies Dsk2p and Rad23p as essential components of ER-associated degradation. EMBO Rep 5(7):692–697
108. Alberts SM, Sonntag C, Schafer A, Wolf DH (2009) Ubx4 modulates cdc48 activity and influences degradation of misfolded proteins of the endoplasmic reticulum. J Biol Chem 284(24):16082–16089
109. Tran JR, Tomsic LR, Brodsky JL (2011) A Cdc48p-associated factor modulates endoplasmic reticulum-associated degradation, cell stress, and ubiquitinated protein homeostasis. J Biol Chem 286(7):5744–5755
110. Zhang Y, Nijbroek G, Sullivan ML, McCracken AA et al (2001) Hsp70 molecular chaperone facilitates endoplasmic reticulum-associated protein degradation of cystic fibrosis transmembrane conductance regulator in yeast. Mol Biol Cell 12(5):1303–1314
111. Han S, Liu Y, Chang A (2007) Cytoplasmic Hsp70 promotes ubiquitination for endoplasmic reticulum-associated degradation of a misfolded mutant of the yeast plasma membrane ATPase, PMA1. J Biol Chem 282(36):26140–26149
112. Taxis C, Hitt R, Park SH, Deak PM et al (2003) Use of modular substrates demonstrates mechanistic diversity and reveals differences in chaperone requirement of ERAD. J Biol Chem 278(38):35903–35913
113. Hampton RY (2002) Proteolysis and sterol regulation. Annu Rev Cell Dev Biol 18:345–378
114. Hampton RY, Garza RM (2009) Protein quality control as a strategy for cellular regulation: lessons from ubiquitin-mediated regulation of the sterol pathway. Chem Rev 109(4):1561–1574
115. Hampton RY, Gardner RG, Rine J (1996) Role of 26S proteasome and HRD genes in the degradation of 3-hydroxy-3-methylglutaryl-CoA reductase, an integral endoplasmic reticulum membrane protein. Mol Biol Cell 7(12):2029–2044
116. Bays NW, Wilhovsky SK, Goradia A, Hodgkiss-Harlow K et al (2001) HRD4/NPL4 is required for the proteasomal processing of ubiquitinated ER proteins. Mol Biol Cell 12(12):4114–4128
117. Carroll SM, Hampton RY (2010) Usa1p is required for optimal function and regulation of the Hrd1p endoplasmic reticulum-associated degradation ubiquitin ligase. J Biol Chem 285(8):5146–5156
118. Gardner RG, Shearer AG, Hampton RY (2001) In vivo action of the HRD ubiquitin ligase complex: mechanisms of endoplasmic reticulum quality control and sterol regulation. Mol Cell Biol 21(13):4276–4291
119. Sato BK, Schulz D, Do PH, Hampton RY (2009) Misfolded membrane proteins are specifically recognized by the transmembrane domain of the Hrd1p ubiquitin ligase. Mol Cell 34(2):212–222
120. Hoppe T, Matuschewski K, Rape M, Schlenker S et al (2000) Activation of a membrane-bound transcription factor by regulated ubiquitin/proteasome-dependent processing. Cell 102(5):577–586
121. Verma R, Chen S, Feldman R, Schieltz D et al (2000) Proteasomal proteomics: identification of nucleotide-sensitive proteasome-interacting proteins by mass spectrometric analysis of affinity-purified proteasomes. Mol Biol Cell 11(10):3425–3439
122. Isakov E, Stanhill A (2011) Stalled proteasomes are directly relieved by P97 recruitment. J Biol Chem 286(35):30274–30283
123. Shcherbik N, Zoladek T, Nickels JT, Haines DS (2003) Rsp5p is required for ER bound Mga2p120 polyubiquitination and release of the processed/tethered transactivator Mga2p90. Curr Biol 13(14):1227–1233
124. Shcherbik N, Haines DS (2007) Cdc48p(Npl4p/Ufd1p) binds and segregates membrane-anchored/tethered complexes via a polyubiquitin signal present on the anchors. Mol Cell 25(3):385–397

125. Siepe D, Jentsch S (2009) Prolyl isomerase Pin1 acts as a switch to control the degree of substrate ubiquitylation. Nat Cell Biol 11(8):967–972
126. Russell SJ, Steger KA, Johnston SA (1999) Subcellular localization, stoichiometry, and protein levels of 26 S proteasome subunits in yeast. J Biol Chem 274(31):21943–21952
127. Laporte D, Salin B, Daignan-Fornier B, Sagot I (2008) Reversible cytoplasmic localization of the proteasome in quiescent yeast cells. J Cell Biol 181(5):737–745
128. Jentsch S, Rumpf S (2007) Cdc48 (p97): a "molecular gearbox" in the ubiquitin pathway? Trends Biochem Sci 32(1):6–11
129. Kuhlbrodt K, Janiesch PC, Kevei E, Segref A et al (2011) The Machado-Joseph disease deubiquitylase ATX-3 couples longevity and proteostasis. Nat Cell Biol 13(3):273–281
130. Beaudenon SL, Huacani MR, Wang G, McDonnell DP et al (1999) Rsp5 ubiquitin-protein ligase mediates DNA damage-induced degradation of the large subunit of RNA polymerase II in *Saccharomyces cerevisiae*. Mol Cell Biol 19(10):6972–6979
131. Woudstra EC, Gilbert C, Fellows J, Jansen L et al (2002) A Rad26-Def1 complex coordinates repair and RNA pol II proteolysis in response to DNA damage. Nature 415(6874):929–933
132. Wilcox AJ, Laney JD (2009) A ubiquitin-selective AAA-ATPase mediates transcriptional switching by remodelling a repressor-promoter DNA complex. Nat Cell Biol 11(12): 1481–1486
133. Moir D, Stewart SE, Osmond BC, Botstein D (1982) Cold-sensitive cell-division-cycle mutants of yeast: isolation, properties, and pseudoreversion studies. Genetics 100(4): 547–563
134. Fu X, Ng C, Feng D, Liang C (2003) Cdc48p is required for the cell cycle commitment point at start via degradation of the G1-CDK inhibitor Far1p. J Cell Biol 163(1):21–26
135. Archambault V, Chang EJ, Drapkin BJ, Cross FR et al (2004) Targeted proteomic study of the cyclin-Cdk module. Mol Cell 14(6):699–711
136. Cao K, Nakajima R, Meyer HH, Zheng Y (2003) The AAA-ATPase Cdc48/p97 regulates spindle disassembly at the end of mitosis. Cell 115(3):355–367
137. Heubes S, Stemmann O (2007) The AAA-ATPase p97-Ufd1-Npl4 is required for ERAD but not for spindle disassembly in Xenopus egg extracts. J Cell Sci 120(Pt 8):1325–1329
138. Hsieh MT, Chen RH (2011) Cdc48 and cofactors Npl4-Ufd1 are important for G1 progression during heat stress by maintaining cell wall integrity in Saccharomyces cerevisiae. PLoS One 6(4):e18988
139. Schork SM, Bee G, Thumm M, Wolf DH (1994) Catabolite inactivation of fructose-1, 6-bisphosphatase in yeast is mediated by the proteasome. FEBS Lett 349(2):270–274
140. Santt O, Pfirrmann T, Braun B, Juretschke J et al (2008) The yeast GID complex, a novel ubiquitin ligase (E3) involved in the regulation of carbohydrate metabolism. Mol Biol Cell 19(8):3323–3333
141. Barbin L, Eisele F, Santt O, Wolf DH (2010) The Cdc48-Ufd1-Npl4 complex is central in ubiquitin-proteasome triggered catabolite degradation of fructose-1,6-bisphosphatase. Biochem Biophys Res Commun 394(2):335–341
142. Tanaka A, Cleland MM, Xu S, Narendra DP et al (2010) Proteasome and p97 mediate mitophagy and degradation of mitofusins induced by Parkin. J Cell Biol 191(7):1367–1380
143. Xu S, Peng G, Wang Y, Fang S et al (2011) The AAA-ATPase p97 is essential for outer mitochondrial membrane protein turnover. Mol Biol Cell 22(3):291–300
144. Karbowski M, Youle RJ (2011) Regulating mitochondrial outer membrane proteins by ubiquitination and proteasomal degradation. Curr Opin Cell Biol 23(4):476–482
145. Braun RJ, Zischka H (2008) Mechanisms of Cdc48/VCP-mediated cell death: from yeast apoptosis to human disease. Biochim Biophys Acta 1783(7):1418–1435
146. Krick R, Bremer S, Welter E, Schlotterhose P et al (2010) Cdc48/p97 and Shp1/p47 regulate autophagosome biogenesis in concert with ubiquitin-like Atg8. J Cell Biol 190(6):965–973
147. Ohsumi Y (2001) Molecular dissection of autophagy: two ubiquitin-like systems. Nat Rev Mol Cell Biol 2(3):211–216
148. Kraft C, Deplazes A, Sohrmann M, Peter M (2008) Mature ribosomes are selectively degraded upon starvation by an autophagy pathway requiring the Ubp3p/Bre5p ubiquitin protease. Nat Cell Biol 10(5):602–610

149. Ossareh-Nazari B, Bonizec M, Cohen M, Dokudovskaya S et al (2010) Cdc48 and Ufd3, new partners of the ubiquitin protease Ubp3, are required for ribophagy. EMBO Rep 11(7):548–554
150. Ju JS, Fuentealba RA, Miller SE, Jackson E et al (2009) Valosin-containing protein (VCP) is required for autophagy and is disrupted in VCP disease. J Cell Biol 187(6):875–888
151. Tresse E, Salomons FA, Vesa J, Bott LC et al (2010) VCP/p97 is essential for maturation of ubiquitin-containing autophagosomes and this function is impaired by mutations that cause IBMPFD. Autophagy 6(2)
152. Vesa J, Su H, Watts GD, Krause S et al (2009) Valosin containing protein associated inclusion body myopathy: abnormal vacuolization, autophagy and cell fusion in myoblasts. Neuromuscul Disord 19(11):766–772
153. Ren J, Pashkova N, Winistorfer S, Piper RC (2008) DOA1/UFD3 plays a role in sorting ubiquitinated membrane proteins into multivesicular bodies. J Biol Chem 283(31):21599–21611
154. Qiu L, Pashkova N, Walker JR, Winistorfer S et al (2010) Structure and function of the PLAA/Ufd3-p97/Cdc48 complex. J Biol Chem 285(1):365–372
155. Janiesch PC, Kim J, Mouysset J, Barikbin R et al (2007) The ubiquitin-selective chaperone CDC-48/p97 links myosin assembly to human myopathy. Nat Cell Biol 9(4):379–390
156. Raman M, Havens CG, Walter JC, Harper JW (2011) A genome-wide screen identifies p97 as an essential regulator of DNA damage-dependent CDT1 destruction. Mol Cell 44(1):72–84
157. Franz A, Orth M, Pirson PA, Sonneville R et al (2011) CDC-48/p97 coordinates CDT-1 degradation with GINS chromatin dissociation to ensure faithful DNA replication. Mol Cell 44(1):85–96
158. Meerang M, Ritz D, Paliwal S, Garajova Z et al (2011) The ubiquitin-selective segregase VCP/p97 orchestrates the response to DNA double-strand breaks. Nat Cell Biol 13(11):1376–1382
159. Acs K, Luijsterburg MS, Ackermann L, Salomons FA et al (2011) The AAA-ATPase VCP/p97 promotes 53BP1 recruitment by removing L3MBTL1 from DNA double-strand breaks. Nat Struct Mol Biol 18(12):1345–1350
160. Yamanaka K, Sasagawa Y, Ogura T (2012) Recent advances in p97/VCP/Cdc48 cellular functions. Biochim Biophys Acta 1823(1):130–137
161. Meyer H, Bug M, Bremer S (2012) Emerging functions of the VCP/p97 AAA-ATPase in the ubiquitin system. Nat Cell Biol 14(2):117–123
162. Uchiyama K, Kondo H (2005) p97/p47-mediated biogenesis of Golgi and ER. J Biochem 137(2):115–119
163. Meyer HH (2005) Golgi reassembly after mitosis: the AAA family meets the ubiquitin family. Biochim Biophys Acta 1744(3):481–492
164. Buchberger A (2006) Cdc48 (p97) and its cofactors. In: Mayer RJ, Ciechanover AJ, Rechsteiner M (eds) Protein degradation, vol 3: cell biology of the ubiquitin-proteasome system. Wiley-VCH, Weinheim, pp 194–211
165. Seeley ES, Kato M, Margolis N, Wickner W et al (2002) Genomic analysis of homotypic vacuole fusion. Mol Biol Cell 13(3):782–794
166. Sambade M, Alba M, Smardon AM, West RW et al (2005) A genomic screen for yeast vacuolar membrane ATPase mutants. Genetics 170(4):1539–1551
167. Ramadan K, Bruderer R, Spiga FM, Popp O et al (2007) Cdc48/p97 promotes reformation of the nucleus by extracting the kinase Aurora B from chromatin. Nature 450(7173):1258–1262
168. Dobrynin G, Popp O, Romer T, Bremer S et al (2011) Cdc48/p97-Ufd1-Npl4 antagonizes Aurora B during chromosome segregation in HeLa cells. J Cell Sci 124(Pt 9):1571–1580

Chapter 9
The Role of AAA+ Proteases in Mitochondrial Protein Biogenesis, Homeostasis and Activity Control

Wolfgang Voos, Linda A. Ward, and Kaye N. Truscott

Abstract Mitochondria are specialised organelles that are structurally and functionally integrated into cells in the vast majority of eukaryotes. They are the site of numerous enzymatic reactions, some of which are essential for life. The double lipid membrane of the mitochondrion, that spatially defines the organelle and is necessary for some functions, also creates a physical but semi-permeable barrier to the rest of the cell. Thus to ensure the biogenesis, regulation and maintenance of a functional population of proteins, an autonomous protein handling network within mitochondria is required. This includes resident mitochondrial protein translocation machinery, processing peptidases, molecular chaperones and proteases. This review highlights the contribution of proteases of the AAA+ superfamily to protein quality and activity control within the mitochondrion. Here they are responsible for the degradation of unfolded, unassembled and oxidatively damaged proteins as well as the activity control of some enzymes. Since most knowledge about these proteases has been gained from studies in the eukaryotic microorganism *Saccharomyces cerevisiae*, much of the discussion here centres on their role in this organism. However, reference is made to mitochondrial AAA+ proteases in other organisms, particularly in cases where they play a unique role such as the mitochondrial unfolded protein response. As these proteases influence mitochondrial function in both health and disease in humans, an understanding of their regulation and diverse activities is necessary.

W. Voos (✉)
Institut für Biochemie und Molekularbiologie (IBMB), Universität Bonn,
Nussallee 11, 53115 Bonn, Germany
e-mail: wolfgang.voos@uni-bonn.de

L.A. Ward • K.N. Truscott (✉)
Department of Biochemistry, La Trobe Institute for Molecular Science (LIMS),
La Trobe University, Melbourne, VIC 3086, Australia
e-mail: k.truscott@latrobe.edu.au

D.A. Dougan (ed.), *Regulated Proteolysis in Microorganisms*, Subcellular Biochemistry 66,
DOI 10.1007/978-94-007-5940-4_9,

Introduction

The highly compartmentalised nature of a eukaryotic cell brings about increased complexity relative to a prokaryotic cell, with respect to the biogenesis and regulation of its proteome. With the vast majority of eukaryotic cellular proteins encoded by the nuclear genome and synthesised on cytosolic ribosomes, many steps are required for their successful trafficking from the cytosol to the correct subcellular location and their subsequent maturation and folding. Furthermore, proteins regardless of their subcellular location, need to be removed in a regulated manner to allow their replenishment or to alter their concentration in the cell. The balance between protein biogenesis and degradation determines the concentration of a particular type of protein in the cell at any particular time. This is often referred to as the steady state level of a protein. An ability to regulate protein concentration in a post-translational manner permits a rapid and/or graded response to physiological and environmental stimuli. It can also ensure the correct subunit stoichiometry of protein complexes and the maintenance of functional populations of proteins through the removal of unfolded, misfolded or otherwise damaged proteins. But how can the concentration and integrity of intracellular proteins, located in various parts of the cell, be adequately regulated in a post-translational manner? In cells, protein degradation is achieved by the enzyme-mediated hydrolysis of peptide bonds between amino acid residues. This end-point in the life of a protein is attained by two main systems in eukaryotic cells; the lysosome/vacuole-mediated "acid cocktail" mechanism and the AAA+ protease-mediated "chamber of doom" mechanism. In the former case, portions of cellular content, organelles and individual proteins (in mammals) are delivered to the lysosome (vacuole in plants and fungi) for degradation; a process known as autophagy. In the lumen of this organelle, internalised cellular proteins are hydrolysed to small peptides and amino acids via a cocktail of exo- and endo-proteases optimally active in the acid environment of this compartment. The major autophagic pathways are macro-, micro- and chaperone-mediated autophagy and are reviewed elsewhere [1–3]. The second major mechanism of protein degradation is via the action of ATP-dependent proteases. In this case individual proteins are selectively delivered to large oligomeric proteases, unfolded (if required) and translocated into an internal catalytic chamber where hydrolysis of the proteins into short peptides takes place. The 26S proteasome is an example of an ATP-dependent protease in which substrates are tagged with ubiquitin for delivery and degradation [4]. Many cellular proteins located in the cytoplasm, nucleus, endoplasmic reticulum and the mitochondrial outer membrane, are degraded by the ubiquitin-proteasome pathway. However, this is not the only ATP-dependent mechanism of protein degradation operational in eukaryotic cells. Unique systems are also present in chloroplast, peroxisomes and mitochondria [5, 6]. The ATP-dependent proteases of mitochondria are the topic of this chapter.

Early research on mitochondrial protein degradation examined if mitochondrial proteins were turned over with the same or different half-lives. A fixed half-life of degradation for all proteins would be suggestive of a lysosomal mode of mitochondrial protein degradation. On the other hand, vastly different degradation half-lives would indicate a compartment based intrinsic capacity to degrade proteins selectively. Studies on protein turnover in rat liver mitochondria revealed that there was considerable protein-dependent variation in half-lives from hours to days [7, 8]. Also, it was observed that the products of mitochondrial translation in isolated rat liver mitochondria were rapidly degraded in the absence of cytoplasmic synthesised partner proteins [9]. Such findings gave support for an intrinsic protease activity in mitochondria, providing the impetus to search for the factor or factors responsible. Classical biochemical approaches using a range of additives (e.g. ATP) and protease inhibitors (e.g. phenylmethylsulfonyl fluoride, *o*-phenanthroline) were used to dissect the protease activity of mitochondria. In these studies, both endogenous (e.g. incompletely synthesised or uncomplexed mitochondrial DNA (mtDNA)-encoded protein subunits of the respiratory chain) and exogenous (e.g. casein) proteins were examined as substrates. Collectively, the research identified ATP-dependent metallo- and serine protease activities in both animal and fungal mitochondria [10–13]. Further biochemical and genetic analysis identified the distinct proteases responsible for these different activities within mitochondria [14–22].

This chapter highlights current knowledge regarding the role of ATP-dependent proteases in the biogenesis, quality control and regulation of the mitochondrial proteome largely but not exclusively in the single-cell eukaryotic organism *Saccharomyces cerevisiae*. The intrinsic ATP-dependent proteases of yeast mitochondria are known as *i*-AAA protease (intermembrane space active-ATPase associated with a variety of activities), *m*-AAA protease (matrix active-AAA) and Pim1p (proteolysis in mitochondria) also known as Lon. In higher eukaryotes the casein lytic protease XP (ClpXP) is also present in mitochondria [15, 23, 24]. Although not covered in this chapter, it should be noted that the cytoplasmic ubiquitin proteasome system contributes to the regulation of the mitochondrial outer membrane proteome [25, 26]. Also described in this chapter are instances where mitochondrial ATP-dependent proteases perform unique roles in animals (i.e. not present in yeast), for example *Caenorhabditis elegans* ClpXP, in the mitochondrial unfolded protein response (section "The Mitochondrial Unfolded Protein Response (UPR^{mt})"). We begin with an overview of mitochondrial function and protein biogenesis including translocation of nuclear-encoded and mtDNA-encoded proteins, their maturation by processing peptidases and their folding and assembly into active units.

Structure and Function of Mitochondria

Structurally, mitochondria are distinct double membrane bound organelles. They are dynamic however, undergoing constant fusion and division [27]. Functionally, many enzymes reside in mitochondria that are important for processes such as ATP production, fatty acid, amino acid and carbohydrate metabolism, and the synthesis of Fe-S clusters and heme. Depending on the organism, mitochondria also play key roles in cell signalling pathways such as programmed cell death, calcium homeostasis and innate immunity [28–30]. While mitochondria contain their own DNA, they are not self-sufficient because the organellar genome is generally very small. The vast majority of mitochondrial proteins are nuclear-encoded. Thus, full mitochondrial function requires the co-ordinated synthesis of its proteome expressed from two genomes. The retention of a small mitochondrial genome, a remnant of the ancient α-proteobacterium endosymbiont from which mitochondria evolved [31], means that the organelle must retain an ability (and all the necessary components) to carry out tasks such as DNA replication and repair, transcription, processing of RNA and protein synthesis. Many other proteins and complexes are required for the integrity of this organelle and its cooperation and communication with other cellular entities [32–38]. Around 1,000 different proteins are expected to perform all the necessary functions in *S. cerevisiae* mitochondria. Using a mass spectrometry approach, 750 different yeast mitochondrial proteins were identified following growth in YPG medium with an estimated 90% coverage of the proteome [39]. Further yeast mitochondrial proteins may exist under different growth or environmental conditions.

The Biogenesis of Mitochondrial Proteins

Synthesis and Transport of Mitochondrial Proteins

Nuclear-encoded mitochondrial proteins are synthesised on ribosomes in the cytosol and transported to the organelle; directed by targeting information contained within the protein. The nascent polypeptide may be assisted on its journey by molecular chaperones that act to maintain the protein in an import competent state, protecting against aggregation and premature degradation and in some cases delivering the protein directly to the import machinery [40]. The types of targeting signals that direct mitochondrial proteins to the correct location within the organelle can take a number of forms. However, the most common mechanism involves that of an N-terminal targeting signal, also known as a presequence. Around 70% of nuclear-encoded mitochondrial proteins in yeast possess an N-terminal targeting signal [41]. These generally range in size from 15 to 65 amino acids, carry a net positive charge and have a propensity to form amphipathic α-helices [41, 42]. Such signals direct the proteins to the mitochondrial matrix or, when combined with sorting signals, assist in directing proteins to the mitochondrial inner membrane or inter membrane space (IMS). The remaining ~30% of mitochondrial proteins contain internal targeting signals. In these cases the targeting information may take the form of a signal motif or be spread throughout the protein [43]. For many proteins containing internal mitochondrial localisation signals, the precise nature of the targeting elements is not known.

There are several different pathways of protein import into mitochondria involving specialised protein translocation machineries. However, most nuclear-encoded preproteins engage the central translocase of the outer membrane (TOM) complex to cross the outer lipid bilayer of mitochondria. Following import across the outer membrane, the preprotein interacts with pathway specific translocases and cofactors

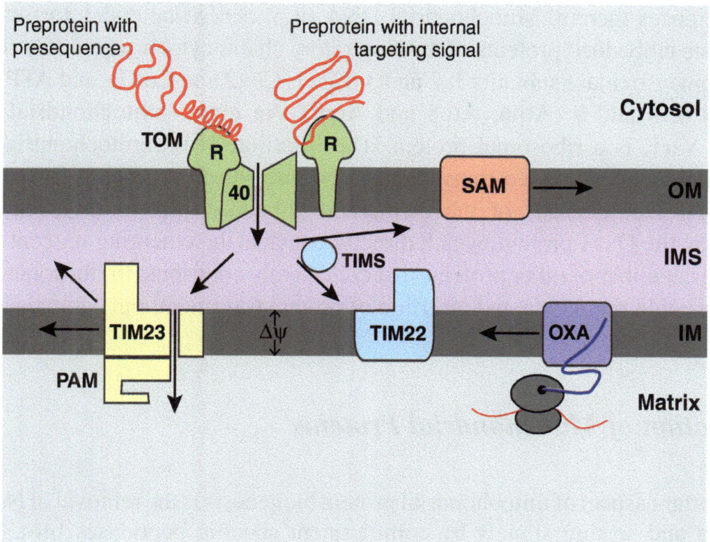

Fig. 9.1 Major import and export pathways for the biogenesis of mitochondrial proteins. Preproteins synthesised in the cytosol are recognised by receptors exposed on the surface of mitochondria directing them to the outer membrane translocation channel formed by Tom40. As the import channel is narrow, the preproteins are imported in an unfolded state as linear polypeptides or loop structures. Once the lipid bilayer has been traversed the preprotein import pathways diverge. Presequence-containing preproteins interact sequentially with the IMS exposed receptors of the TOM and TIM23 complexes. They are then translocated across the inner membrane in a membrane potential-dependent manner via the voltage-activated cation-selective channel formed by Tim23. In the absence of a sorting signal, the emerging presequence interacts with the presequence associated motor (PAM) complex that mediates import via an ATP-driven binding and release mechanism. Some preproteins are translocated into the inner membrane (IM) and IMS via the TIM23 complex directed by a hydrophobic sorting signal situated directly behind the presequence. The presequence directs the preprotein to the TIM23 complex via the standard presequence pathway however a stop transfer signal leads to the arrest of the preprotein in the TIM23 complex. It then laterally transfers into the IM. A variation on this pathway is when laterally sorted presequence-containing proteins contain a protease cleavage site following the stop transfer signal. In such cases, endoprotease(s) cleave these preproteins near to their membrane anchor releasing the mature protein into the IMS. With respect to other pathways, both outer membrane β-barrel proteins and polytopic inner membrane proteins engage holder chaperones (small TIM complexes) in the IMS to assist their transition through this aqueous environment. While the β-barrel preproteins insert into the outer membrane (OM) aided by the sorting and assembly machinery (SAM) complex, polytopic inner membrane proteins are imported via the TIM22 complex in a membrane potential-dependent manner. For non-presequence containing preproteins destined for the IMS, import into this compartment occurs directly via the TOM complex. It seems however that a requisite factor for successful or efficient import, in these cases, is the presence of a binding partner, either a folding catalyst (e.g. Mia40) or a functional partner protein. Finally, mitochondrial inner membrane proteins that are encoded by mtDNA are imported in a co-translational manner mediated by the OXA complex

(for a brief overview see Fig. 9.1). For example, preproteins containing N-terminal targeting sequences transition to the translocase of the inner membrane 23 (TIM23) complex for import into the matrix, inner membrane or IMS. For a detailed description of the mechanisms and models of protein import *see* recent reviews by [43, 44] and references therein. Mitochondrial DNA of *S. cerevisiae* codes for seven inner membrane embedded proteins of the respiratory chain; cytochrome *b* (Cyt*b* or Cob), cytochrome *c* oxidase subunits 1, 2 and 3 (Cox1, Cox2 and Cox3), and ATP synthase subunits 6, 8 and 9 (Atp6, Atp8 and Atp9). An eighth mitochondrial-encoded protein, Var1, is a ribosomal protein. The insertion of the mitochondria-encoded respiratory subunits is co-translational and mediated by the OXA translocase (Fig. 9.1). This consists of the membrane-embedded protein insertion complex formed by the Oxa1 protein, which directly contacts the emerging nascent polypeptide, and a number of other proteins that collectively contribute to ribosome positioning, translation regulation and insertion of protein transmembrane domains [45].

Maturation of Mitochondrial Proteins

An important aspect of mitochondrial protein biogenesis is the removal of N-terminal targeting and sorting signals by compartment specific peptidases [46]. The vast majority of presequence-containing mitochondrial matrix proteins are cleaved by the mitochondrial processing peptidase (MPP) [41]. This enzyme is a dimeric metalloendopeptidase consisting of α and β subunits [47]. The recent determination of the N-terminal sequences of yeast mitochondrial proteins (N-proteome) has clarified the substrate specificity of MPP [41]. The substrate specificity of MPP is characterised by a very high frequency (>75%) of Arg in the P2 position [48], as deduced from the N-proteome data (Fig. 9.2). This MPP substrate specificity is often referred to as the R-2 motif [42]. Residues most commonly found in the P1' position of MPP substrates are Tyr, Leu, Phe and Ala while Ser and Ala are most commonly found in the P2' position. For many mitochondrial proteins with a presequence, cleavage by MPP is the only matrix proteolytic maturation step (Fig. 9.2, Path I). However, for a subset of proteins, further proteolytic processing is required (Fig. 9.2, Path II and III). For example, in yeast mitochondria, 14 proteins have been identified as substrates of the peptidase octapeptidyl aminopeptidase 1 (Oct1) [41, 42, 49]. Following removal of the presequence by MPP, Oct1 cleaves eight amino acid residues from the N-terminus of some matrix proteins. The substrate specificity of Oct1 extends to P8 with Phe found in this position in the vast majority of cases (Fig. 9.2, Path II). The P1' residue in the MPP cleavage pattern becomes the P8 site for Oct1 recognition. The N-proteome study lead to the identification of a new player in mitochondrial matrix protein maturation, the metalloaminopeptidase termed intermediate cleaving peptidase of 55 kDa (Icp55). For a subset of mitochondrial matrix proteins, Icp55 cleaves a single amino acid residue from the N-terminus of proteins following cleavage of the mitochondrial targeting sequence by MPP (Fig. 9.2, Path III). Specifically, the P1' position in the MPP substrate recognition pattern becomes the P1 site for Icp55 recognition. With respect to substrate specificity, Icp55 appears to favour bulky hydrophobic amino acids Tyr, Phe and Leu in the P1 position with Tyr occurring most frequently. Ser is most frequently found in the P1' position, and to a lesser extent Ala and Thr. A quite different specificity for Icp55 has been reported [50] where the scissile bond sits three amino acid residues from the MPP

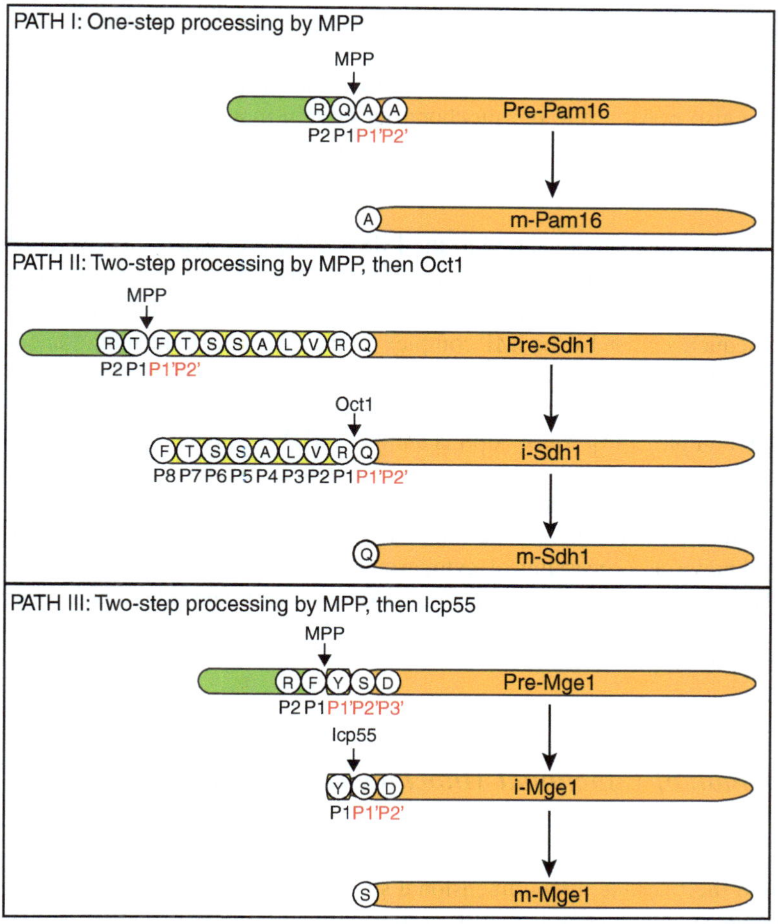

Fig. 9.2 MPP-dependent pathways of mitochondrial protein maturation. Schematic indicating primary sequences of Pam16, Sdh1 and Mge1 precursors around the cleavage site (scissile bond) for the peptidase MPP and where relevant Oct1 and Icp55. Amino acid residues N-terminal of the scissile bond on the substrate are nominated P1, P2, P3 etc. and those residues C-terminal are nominated P1', P2', P3' etc. In *Path I*, cleavage of the presequence by MPP directly generates the mature protein without any further proteolytic processing as exemplified by the substrate Pam16. In *Path II*, MPP cleavage generates an immature intermediate species. This first cleavage event is necessary for subsequent recognition of the substrate by the peptidase Oct1. Cleavage of the substrate by Oct1 removes eight residues from the N-terminus of the protein to generate the mature form. The primary sequence of Sdh1 around the respective cleavage sites is shown in this example. Finally in *Path III*, MPP cleavage again generates an immature intermediate species however in this case it only differs from the mature protein by one residue. The final maturation step is mediated by the aminopeptidase Icp55. The sequence of Mge1 is shown in this example. Note the presequences and mature domains are not to scale. The length of presequences and mature protein are protein dependent and vary considerably

cleavage site. Further analysis is required to fully understand the precise substrate specificity of Icp55. In the case where processing by either Icp55 or Oct1 is prevented (i.e. in gene deletion strains), the resulting intermediate proteins are moderately unstable relative to the mature protein [41, 49]. The intermediates resemble bacterial

N-degrons; degradation signals which mediate the turnover of proteins by the N-end rule pathway [51–53]. This is discussed further in *section* "Activity Control of Mitochondrial Proteins by AAA+ Proteases". Finally, the ATP-dependent *m*-AAA protease also plays a role in mitochondrial protein maturation [54–56]. Normally responsible for the complete degradation of proteins into peptides (*see later*), this maturation activity is unique and dependent on specific features of the substrate protein.

Some proteins that are sorted to the IMS via the presequence pathway are processed by the IMS peptidase (IMP) complex [43, 46]. Following lateral transfer into the inner membrane such preproteins are processed by MPP in the matrix to remove the presequence, then by the IMP complex on the IMS side of the membrane, releasing the protein from its transmembrane anchor and hence into the aqueous compartment. The signal sequence of the mtDNA-encoded protein Cox2 is also cleaved by the IMP complex [43]. Yeast Atp6 is another mtDNA-encoded protein synthesised with an N-terminal signal sequence, however, in this case, it is cleaved by a different protein, the metallopeptidase ATP synthase 23 (Atp23) [57, 58]. Yeast mitochondria also possess an inner membrane integrated serine endopeptidase of the rhomboid family known as processing of cytochrome *c* peroxidase (Pcp1). Two substrates have been described; cytochrome *c* peroxidase (Ccp1), and mitochondrial genome maintenance protein 1 (Mgm1), a mitochondrial outer membrane dynamin-like GTPase [59–63].

Formation of Functional Mitochondrial Proteins

Once proteins are imported into the mitochondrial matrix and processed they must fold into their unique three-dimensional structure to be functional. Such processes can occur spontaneously but in the crowded environment of cells or organelles, protein folding may be assisted by molecular chaperones. Chaperones capture folding intermediates preventing non-productive aggregation or rescuing kinetically trapped species [64]. The major chaperones of yeast mitochondria belong to the Hsp70 and Hsp60 protein families. Both types of chaperones are intrinsically tied to the import and folding reaction of mitochondrial precursor proteins [65, 66]. Mitochondrial Hsp70 (mtHsp70), encoded by the yeast gene *SSC1* is a component of the mitochondrial import machinery for proteins of the matrix compartment, driving the full membrane translocation of these precursors and assisting their folding reaction. These folding reactions are performed in a close collaboration with the Hsp60 chaperone complex in the matrix. In addition, yeast mitochondria contain a member of the Hsp100/ClpB family, called Hsp78, which is missing in metazoan cells. Like its relatives in bacterial cells, Hsp78 is able to recover polypeptides from the aggregated state, promoting the recovery of mitochondrial functions after temperature stress [67]. Another special feature of yeast mitochondria is the presence of the chaperone component of the ClpXP protease (see below), called Mcx1 (for mitochondrial ClpX), although the corresponding protease component is missing

[68, 69]. The function of this chaperone in yeast mitochondria is unknown. Finally, once the imported protein subunits are folded, the final steps in the biogenesis of proteins may be their oligomerisation, binding to prosthetic groups such as flavin, Fe-S clusters or heme, or assembly into multi-subunit structures such as the respiratory chain complexes. It should be noted that in this context, yeast mitochondria contain a specialised Hsp70 system consisting of the chaperone Ssq1 and its cofactors for the transfer of Fe-S cluster cofactors from their assembly scaffold protein to the respective apoenzymes [70]. Further post-translation modifications such as acetylation and phosphorylation may contribute to the activity of mitochondrial proteins [71, 72].

Mitochondrial Protein Quality Control (PQC) by Compartment Specific AAA+ Proteases

While many sophisticated proteinaceous molecular components and pathways mediate the synthesis, transport, folding and assembly of mitochondrial proteins, these are not fail-safe processes. ATP-dependent proteases therefore serve as ever-present organelle protein quality control (PQC) systems that constantly survey protein integrity. Collectively, they have the capacity to degrade unfolded/oxidatively damaged, immature and unassembled proteins. Together with molecular chaperones, these proteases ensure that a functional population of proteins is maintained in mitochondria. Mitochondria, as endosymbiotic organelles, contain many chaperones and proteases homologous to those found in bacteria. For example, major ATP-dependent proteases are represented with homologs of Lon, ClpXP and FtsH families found in mitochondria, while bacterial HslUV and ClpAP families are absent. There are however significant differences in the composition of mitochondrial proteases between eukaryotic organisms. While metazoan mitochondria contain a Lon protease, a ClpXP protease and at least two distinct AAA proteases of the FtsH family, the situation in yeast mitochondria [67] is somewhat different. Yeast mitochondria lack a Clp protease homolog, leaving Lon/Pim1 as the only ATP-dependent protease in the matrix. The main components of the yeast mitochondrial PQC system and their location are depicted in Fig. 9.3. The main functional distinction between these types of proteases is their substrate specificity – the membrane-integrated proteases preferentially degrade membrane proteins while the matrix proteases are responsible for the degradation of water soluble proteins.

Like their bacterial counterparts, mitochondrial proteases are ATP-dependent self-compartmentalising complexes. That is, the catalytic residues responsible for the hydrolysis of peptide bonds line the internal walls of an enclosed aqueous degradation channel or chamber formed by a multi-subunit protease. Recognition of the protein targets is mediated by the ATPase component of the proteases and the degradation signals (degrons) displayed by the target protein. Also, the ATPase domain is responsible for the unfolding and translocation of the substrates into the internal proteolytic cavity. This chambered mechanism of protein degradation prevents the

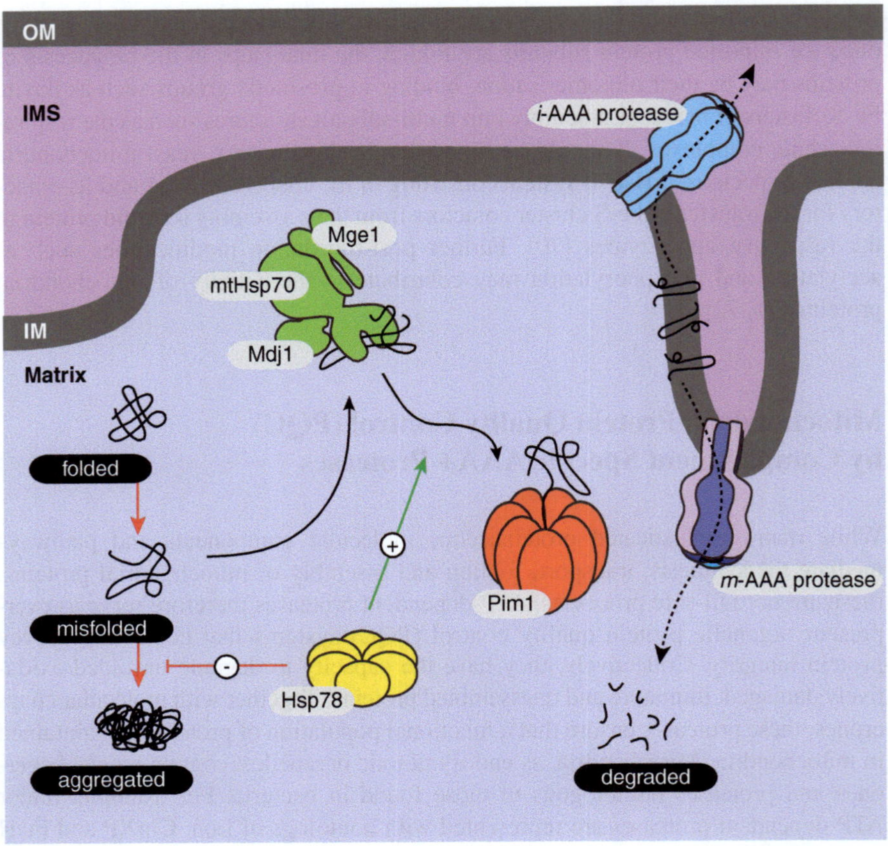

Fig. 9.3 Schematic overview of the PQC system in the matrix of yeast mitochondria. Misfolded proteins are recognized and bound primarily by the mitochondrial Hsp70 (mtHsp70) system that works together with its partner proteins Mge1 (nucleotide exchange factor) and Mdj1 (co-chaperone of the J-protein family). If unfolding efforts fail, the misfolded proteins are transferred to the matrix protease Pim1 that degrades the substrates to small oligopeptides in an ATP-dependent reaction. Pim1-dependent degradation is assisted by the Hsp100 chaperone Hsp78. If the maximum capacity of the PQC system is exceeded, misfolded polypeptides accumulate and eventually aggregate. Aggregated polypeptides may be resolubilised in a concerted process by Hsp78 and mtHsp70. Disaggregated polypeptides are either refolded or directly degraded by the Pim1/Lon protease. Membrane proteins of the IM are degraded by either of the two AAA proteases. The *m*-AAA protease, composed of the subunits Yta10 and Yta12 faces the matrix compartment while the *i*-AAA protease (a Yme1 homo-oligomer) faces the IMS

inadvertent unregulated hydrolysis of native proteins. The ATPase domains of these processive proteases are very similar to each other and to many other ATPases with diverse functions in cells including cellular regulation and PQC. These related proteins are all members of the ATPases associated with various cellular activities (AAA+) superfamily of proteins [73, 74]. As the mitochondrial ATP-dependent proteases Pim1, *i*-AAA, *m*-AAA and ClpXP are all members of the AAA+ superfamily of proteins, this group of proteases will herein be referred to as AAA+ proteases.

Pim1, the Soluble AAA+ Protease of the Yeast Mitochondrial Matrix

The most prominent member of the AAA+ proteases is the Lon protease from *Escherichia coli*, which gives its name to the protein family [75]. Pim1 in yeast mitochondria and the related Lon proteases from other eukaryotic organisms such as LONM in human mitochondria (also known as mitochondrial Lon peptidase 1 (LonP1)) and Lon1 and Lon4 from plant mitochondria [5] are highly conserved and, as far as it has been studied, also exhibit similar functions. Hence, Pim1 from the model organism *S. cerevisiae* commonly serves as the typical representative for the Lon-A type ATP-dependent proteases found in mitochondria.

Identification of Pim1 and Phenotype of *pim1* Deletion Strain

Due to their endosymbiotic origin it had been assumed that mitochondria, similar to bacteria, contain proteases for the specific degradation of their endogenous proteins. Indeed, the first identification of an ATP-dependent proteolytic activity in mitochondria with properties similar to the bacterial protease Lon reach back to the early 1990s [16, 76]. The yeast gene responsible for this activity was cloned independently by two groups and the encoded protease was named Pim1 or Lon based on its high amino acid homology to the *E. coli* Lon protease [77, 78].

A deletion of the *PIM1* gene in yeast resulted in severe consequences for the function of mitochondria. The main phenotype is a growth defect on non-fermentable carbon sources like glycerol or lactate [77]. The inability of Δ*pim1* mitochondria to respire correlates with a non-functional mitochondrial genome (*see section* "Association with Mitochondrial DNA (mtDNA)"). Further examination demonstrated the lack of any ATP-dependent proteolysis activity in soluble mitochondrial extracts [78]. These experiments indicated that Pim1 represents the main proteolytic activity in the mitochondrial matrix. First direct demonstrations of its protease function were based on its ability to degrade radioactively labelled reporter proteins that were imported into isolated mitochondria [79]. Pim1-dependent degradation was shown to be restricted to soluble protein substrates located in the matrix compartment. In contrast, membrane associated substrates, even those facing the matrix compartment were preferentially degraded by the membrane-integrated *m*-AAA protease [80].

Structure-Function Relationships

The *PIM1* gene of *S. cerevisiae* encodes a 1133-amino acid long protein with an overall sequence identity of 30–35% to bacterial family members. Pim1 is encoded in the nuclear genome and has to be imported into the mitochondria after it is synthesised on cytosolic ribosomes. Hence, it is characterised by a long N-terminal amino acid extension that serves as a mitochondrial targeting signal that is missing in its bacterial relatives. Similar to the other ATP-dependent proteases, Pim1 belongs to the AAA+ protein superfamily [81]. AAA+ protease complexes generally consist of a proteolytic subunit and a regulatory subunit that is responsible for substrate recognition and unfolding [82, 83]. Due to the high sequence conservation, most information on the catalytic mechanism of Pim1 has so far been deduced from the properties of its bacterial relative Lon. Similar to other AAA+ proteases, Pim1 consists of two major domains and also activities: a protease with broad specificity and an ATPase responsible for the recognition and/or unfolding of polypeptide substrates. A unique property of Lon-type proteases is that the ATPase domain and the proteolytic domain are located on a single polypeptide chain. An additional third N-terminal domain (N domain) essentially interacts with the potential substrate proteins in concert with the ATPase domain. The C-terminal peptidase contains the catalytic dyad consisting of a conserved serine and lysine residue [84]. The main difference to the bacterial proteases, apart from the targeting sequence, is the presence of three rather large hydrophilic inserts, one in the N-terminal part after the targeting sequence, one between amino acids 300 and 380 and the last between amino acids 840 and 900. Pim1 shares the presence of these inserts with its other eukaryotic homologs, although the sequence conservation in these regions is rather low. So far no data on the significance of these inserts is available but it could be speculated that they are connected with some eukaryotic-specific function.

The assembly of the Pim1 oligomeric protease complex exhibits a special feature. After translocation of the polypeptide chain across the mitochondrial membranes, the presequence is first cleaved in a standard process after position 37 by MPP. However, functional activity of the protease requires the autocatalytic removal of an additional 61 amino acid long pro-region. This second endoproteolytic cleavage is performed by an intermolecular (and likely also intramolecular) reaction by the protease itself and closely coupled to the ATP-dependent assembly of the protease subunits [85].

Lon-type proteases are organized into large homo-oligomeric, ring-shaped protein complexes consisting of several subunits of ~100 kDa each. While bacterial family members are usually composed of six subunits, the data available for Pim1 indicate a heptameric structure [86]. Up to now, no direct ultrastructural studies have been performed on Pim1 but it can be assumed that it is structurally organized in a similar way to its bacterial homologs. As described earlier, a general feature of ATP-dependent proteases is that they form a cylindrical complex with a rather large inner cavity. The active sites are oriented to the interior of this cavity thus forming a proteolytic chamber that is shielded from the environment. Substrates can only enter the proteolytic chamber by pore structures situated at the ends of the cylindrical structure. Substrate entry is usually tightly regulated. Recent structural studies of bacterial proteases related to Lon confirmed this overall structural arrangement [87].

Protease Classification and Cleavage Specificity

The presence of a proteolytic active site characterized by the amino acid Ser at position 1015 confirms that Pim1 as well as Lon are serine proteases with chymotrypsin-like cleavage specificity (proteolysis at regions rich in hydrophobic amino acids) [88]. Similar to other PQC proteases, the regulatory ATPase domain of Lon also seems to exhibit chaperone activity since overexpression of a proteolytically inactive mutant of Pim1 supported the assembly of mitochondrial respiratory complexes [89]. The structurally separated enzymatic domains of Pim1 permitted the functional relationship of the two different domains to be addressed by the recombinant generation of individual mutant constructs [90]. Single point mutations abolishing either ATPase or protease activity both destroyed proteolytic activity. Neither of these mutants were able to support respiration-dependent growth of the cells. When the ATPase or protease domains were expressed separately as individual proteins, neither allowed growth on non-fermentable carbon sources. Interestingly, co-expression of both domains as separate proteins restored full proteolytic and respiratory activities, indicating that both domains are able to functionally interact even as physically separate polypeptides, in order to fulfil the wild-type functions.

General Function of Pim1 in PQC

It was noted relatively early that proteins imported into the mitochondrial matrix compartment under *in vitro* conditions were susceptible to degradation by Pim1 [79, 80]. However, the preproteins used in these experiments represented artificial reporter constructs, which had only a minor structural or functional relationship to the native proteins present in mitochondria. Although ATP-dependent proteases like Pim1 are supposed to have a broad substrate specificity, the number of identified endogenous mitochondrial substrate proteins remained very small for a long time. Due to the restricted availability of antisera against mitochondrial proteins, the initial publications identified a subunit of the matrix processing peptidase (MPP-β) and the β-subunit of the mitochondrial ATP-synthase ($F_1\beta$) as potential substrates [77]. Further progress was achieved by the first proteomic characterisation of mitochondrial protein turnover. Although the mitochondrial proteome is remarkably stable at least under normal conditions, these experiments revealed a set of novel protease substrates [91]. Depending on their behaviour these proteins could be sorted into three different groups: (i) proteins that were protease substrates under all conditions tested, (ii) proteins that remained stable at normal temperatures but became degraded at elevated temperatures, and (iii) proteins that were degraded only at normal temperatures. Proteins in the second group probably represent classical Pim1 substrates, conformationally labile polypeptides that become (partially) unfolded at elevated temperatures and need to be removed. However, the behaviour of group (iii) was counterintuitive at first glance. Later, experiments revealed that proteins of this group had a high tendency to aggregate at elevated temperatures. Typically, under stress conditions a competition between degradation and aggregation takes place (see below). Since aggregated polypeptides are essentially resistant to Pim1-dependent proteolysis, certain aggregation-prone polypeptides are apparently stabilized at higher temperatures although they have of course lost their activity. Interestingly, this hypothesis correlates well with the observation that the accumulation of electron-dense inclusions in the matrix compartment of Δ*pim1* mitochondria [77] most likely represent aggregated damaged polypeptides that could not be removed due to the absence of the protease.

Role of Pim1 Under Oxidative Stress

Apart from elevated temperatures, mitochondria are also vunerable to another stress condition: oxidative stress. Mitochondria themselves are major generators of reactive oxygen species (ROS). In particular, ROS are produced by non-specific electron transfer reactions from components of the mitochondrial respiratory chain to molecular oxygen. Typical examples are superoxide radicals or hydrogen peroxide and its related reaction products [92, 93]. Mitochondrial ROS production is a normal side-reaction of ATP-production by oxidative phosphorylation but can be drastically elevated in certain pathological situations. Due to their high non-specific reactivity, ROS molecules tend to modify and thereby inactivate any type of macromolecule; nucleic acids, lipids and polypeptides. Hence, a major question concerning the role of the mitochondrial PQC system with its core component Pim1 is the impact of its protective function under oxidative stress conditions. This is of particular importance with regard to the multitude of human pathologies that involve some form of mitochondrial oxidative stress, most prominent examples are neurodegenerative diseases [94].

The consequences of oxidative stress conditions on the mitochondrial proteolytic system were studied using a quantitative proteomic approach, identifying proteins that were degraded after treatment of isolated mitochondria with different forms of ROS [95]. Here, three conditions were distinguished: elevated superoxide concentrations, treatment with external H_2O_2 and the artificial inhibition of the respiratory chain. An important observation of this study, was that the sensitivity of the various proteins to the different types of ROS stress was relatively specific. This implies that each of the various conditions leading to oxidative stress can have very different consequences on cellular functions. One class of mitochondrial proteins that have been identified as prominent substrates of ROS-induced proteolysis are enzymes that contain an Fe-S cluster as a cofactor. The Fe-S cluster seems to be very sensitive to modifications by ROS molecules and is likely to be destroyed under these conditions. As a consequence of the loss of their cofactor, the affected polypeptides become conformationally destabilized and will be recognized by the quality control proteases like Pim1. For example, the soluble and Fe-S containing enzyme aconitase (Aco1 in yeast), a component of the citrate cycle, was found as one of the most ROS-sensitive and degradation-prone proteins in the mitochondrial matrix [95, 96]. However, one has to take into account that the presence of a Fe-S cluster does not automatically turn a protein into a protease substrate. This is exemplified by subunit 2 of the succinate dehydrogenase complex (Sdh2), which did not show any alterations in protein levels after ROS stress.

ROS modifications represent covalent, and in most cases irreversible alterations in the affected molecules. Since refolding of ROS-damaged polypeptides, even with the help of chaperones, would not be possible, ATP-dependent proteases like Pim1 have to play a decisive role in the removal of ROS-modified proteins. Their removal is a key response of the cellular PQC system to many different pathological conditions [97, 98]. Interestingly, Pim1 is the only ATP-dependent protease in yeast that is significantly up-regulated under oxidative stress conditions [95]. In contrast, the main chaperones like Hsp70 and Hsp60 show only slightly elevated levels under heat stress but not after ROS treatment. The important protective function of Pim1 under oxidative stress is corroborated by an enhanced sensitivity of Δ*pim1* cells to elevated levels of H_2O_2 in the growth medium.

Pim1 Cooperates with Molecular Chaperones

Substrate Capture

In general, protein degradation in the cellular context is governed by a fine-tuned interplay or cooperation of protease enzymes and molecular chaperones. Instead of focusing on the functions of the individual protein components, it is rather more appropriate to view this system as a functional network for PQC [67, 99]. Chaperones play important roles in recognising and binding to damaged (at least partially unfolded) polypeptide chains. The interaction with a chaperone component is therefore often the first step in preparation of the substrate for recognition by the protease. Cooperation of proteases with chaperones can be either very direct, i.e. the ClpP protease complexes with chaperones (e.g. ClpA or ClpX) to generate the ATP-dependent proteolytic machine or more indirect, when chaperones and proteases act in an independent but coordinated fashion. The latter case applies to the Pim1 protease. It has to be noted that the intrinsic chaperone-like activity of Pim1 is disregarded in this context (see also below). It has been demonstrated that Pim1 closely cooperates with the mitochondrial Hsp70 system in the matrix compartment [79, 80]. Efficient Pim1-dependent degradation of imported reporter proteins that tend to misfold is assisted by mtHsp70. The Hsp70-cochaperone Mdj1, a member of the DnaJ-like protein family, was shown to be closely involved in this function. The primary function of chaperones in this context seems to be the maintenance of the misfolded non-aggregated state of the substrate increasing the time available for safe removal.

Role in Aggregation Prevention

The matrix compartment of fungal mitochondria, like yeast, contains a chaperone that is typical for bacteria but not found in higher eukaryotic species [100]. This chaperone, named Hsp78, is a homolog of the bacterial ClpB protein that has been implicated in the cellular repair of protein aggregates [101]. Enzymes of the ClpB family are able to resolubilise aggregated polypeptides in a reaction that requires ATP and a close cooperation with the respective Hsp70 chaperone system [102]. Hsp78 has been shown to perform a similar role in the mitochondrial matrix [103, 104]. However, when analysing the degradation rates of imported reporter proteins it was observed that Δ*hsp78* deletion mutants exhibited a significant defect in Pim1-

dependent protein degradation [105]. This unprecedented cooperation between Hsp78 and Pim1 during degradation takes place independent of the aggregation state of the substrates and was also observed with destabilised mutants of an abundant endogenous substrate protein named acetohydroxyacid reductoisomerase (Ilv5) [106]. In support of this conclusion it was observed that the yield of disaggregated polypeptides was significantly increased in Δ*pim1* mutant mitochondria [103], again indicating a close functional relationship between Hsp78 and Pim1. This relationship between a ClpB-type chaperone and a Lon-type protease seems to be a special feature of fungal mitochondria. It could be speculated that single cell eukaryotes are subjected to more extreme environmental temperature changes than multicellular eukaryotes and hence require a more efficient removal system for heat-denatured polypeptides.

All the intricate activities of the chaperone-protease network forming cellular PQC systems have two ultimate goals. One is the refolding of damaged or misfolded polypeptide chains in order to regain their enzymatic activities. A second equally important goal is the prevention of the accumulation of misfolded protein species that would otherwise form insoluble toxic aggregates [107]. In a recent study, the aggregation propensity of mitochondrial proteins under heat stress conditions was tested on a proteome scale [108]. In this context it is of particular interest that Pim1 was identified as a major mediator of mitochondrial aggregation reactions, exhibiting an even more significant protective impact than the chaperone proteins from the Hsp70 family. It was shown that the amount of aggregated polypeptides under heat stress was significantly increased in Δ*pim1* mitochondria. Overexpression of Pim1, on the other hand, exhibited a pronounced protective role under these conditions. In the absence of a ClpP protease in yeast mitochondria, Pim1 is the only ATP-dependent protease that is able to degrade soluble substrate proteins in the mitochondrial matrix. Taken together, Pim1 fulfils a key role in the mitochondrial PQC system by removing misfolded proteins in the matrix compartment.

Recognition Motifs and Mechanism of Targeted Degradation

A major general question concerning the degradation of substrate polypeptides by ATP-dependent proteases is the mechanism of substrate selectivity. In principle, the substrate specificities of these proteases are very broad, enabling the proteolysis of any kind of protein substrate. However, in this case it is required that the proteolysis is restricted to short-lived or damaged proteins. As already described, substrate degradation is mainly regulated by the selective entry of polypeptide chains into the proteolytic chamber of the protease complex. The main proteolytic enzyme of the cell, the proteasome recognises a molecular tag on the substrate, a polyubiquitin chain, which mediates engagement of the target substrate with the protease and hence its subsequent degradation [109]. A related tagging system exists in Mycobacteria, where proteins can be tagged for degradation by an ubiquitin-like molecule known as prokaryotic ubiquitin like protein (pup) (for a recent review see [110]). Also, a different type of tagging system exists in bacteria whereby stalled translation products are tagged with an 11 amino acid sequence known as the SsrA tag that targets the abnormal protein for degradation by the ClpXP protease system [53, 111, 112]. Interestingly, a molecular tag system, as a means of substrate recognition, has not been identified for proteolytic reactions in mitochondria. In contrast, current evidence points to a mechanism where the conformational state of a polypeptide chain serves as the main criterion for its selection as a protease substrate. Since it was well established that newly imported reporter proteins containing a dihydrofolate reductase (DHFR) domain are "standard" substrates of the protease Pim1 in the matrix compartment, it was surprising that imported DHFR fusions with only a small N-terminal extension failed to be degraded [113]. A detailed analysis revealed that Pim1 requires an unstructured N-terminal segment of 50–60 amino acid residues in front of the DHFR domain to commence degradation. Pim1 is also able to recognize internal segments as long as they are exposed on the surface of the substrate protein [88]. In addition, it was observed that Pim1 was unable to degrade a folded DHFR domain despite its relatively low thermodynamic stability. Full degradation of DHFR fusion proteins was only possible if the folding state of DHFR was compromised by higher ambient temperatures or destabilizing mutations [113]. Hence, the substrate selectivity of Pim1 (Fig. 9.4) is based on the combination of two properties, (i) the requirement for an unstructured segment in the substrate for initiation of proteolysis and (ii) the low or even absent intrinsic unfolding capacity of Pim1. Taken together, these properties restrict proteolysis to unfolded or damaged polypeptide chains while folded and active enzymes remain largely resistant. Similar properties have been observed in the case of the bacterial protease FtsH [114] and also for the proteasome [115], suggesting that this mechanism represents a basic and probably evolutionary ancient process for substrate selectivity.

This principle of substrate selectivity has been confirmed by the identification of endogenous mitochondrial substrate proteins [91]. A proteomic screen identified proteins containing cofactors or prosthetic groups as main degradation-prone targets of Pim1. In particular, enzymes with Fe-S cluster cofactors, like aconitase and its relatives, were susceptible to degradation. It is conceivable that an absence or the loss of a cofactor would lead to a conformational destabilization of either a part of, or the whole apoenzyme, which could then be recognized by the protease as a potential substrate. Another group of proteins found as potential degradation substrates were single subunits of larger oligomeric complexes. Best examples were the homo-oligomeric chaperone Hsp60 or the dimeric citrate cycle enzyme succinate thiolase. In this case the argument that applies is, that subunits without the respective partners most likely expose unstructured segments that become recognition sites for Pim1.

Additional Cellular Processes Involving Pim1 Function

Association with Mitochondrial DNA (mtDNA)

One of the main effects of a Pim1 mutation in yeast is the loss of mtDNA integrity resulting in a rho$^-$, respiratory deficient phenotype. Elucidation of the biochemical mechanism underlying the lack of respiratory activity in Δ*pim1* cells remains one of

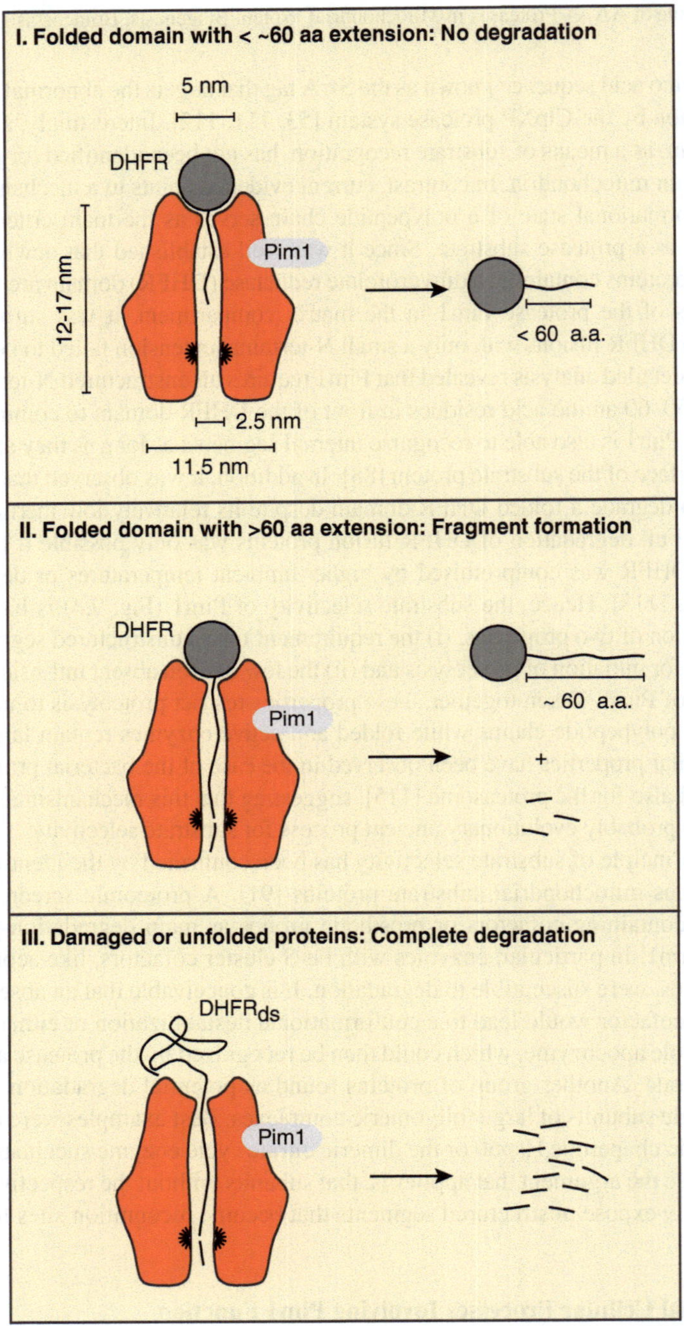

Fig. 9.4 Model for the mechanism of Pim1 substrate selectivity. Tightly folded substrates with no or short unstructured extensions are not cleaved by Pim1 (*I*). The substrate requires a 56-60 amino acid (aa) long unstructured segment for Pim1 to commence its processive degradation reaction. Due to its weak intrinsic unfolding capacity the tightly folded domains of the substrate cannot enter the proteolytic chamber, resulting in clipping of the protein to generate a few specific degradation fragments (*II*). Only damaged or unfolded polypeptide chains are completely degraded by Pim1 (*III*). The Pim1 complex and a DHFR fusion protein as substrate are roughly drawn to scale (based on EM pictures published in [86]) to visualize the dimensions of the protease system. ds, destabilised

the more enigmatic phenomena concerning the cellular function of the protease Pim1. Interestingly, the general effect could be separated into two independent aspects by the identification of a suppressor mutation that retained the integrity of the mtDNA but still rendered the mitochondria respiratory deficient [116]. Using this tool, it could be shown that the proteolytic activity of Pim1 is required for the function of mRNA maturases that are required for the splicing of mitochondrially encoded transcripts, in particular for the expression of the respiratory chain components CoxI and Cob [116]. The precise involvement of Pim1 in this process has not been defined so far but it is conceivable that Pim1 might be required in a processing step during biogenesis of the respective enzymes. Another aspect of Pim1 function, contributing to the respiratory deficient phenotype, was the observation that Pim1 also assists the assembly of respiratory chain components through its chaperone activity, independent of its role in proteolysis [89].

Although these functions would readily explain the respiratory deficient phenotypes of *pim1* mutants, Pim1 has also been shown to play a direct role in mtDNA maintenance. Several lines of evidence show a direct physical interaction of human LONM with the mitochondrial genome [117]. The accumulation of mtDNA lesions was also correlated with a role of LONM under oxidative stress conditions [118], although the details remain unclear. In this context it should be noted that many identified substrate proteins of Pim1 were also shown to be components of the mitochondrial nucleoid, Aco1 being one of the most prominent examples [119]. Integrity of the mitochondrial genome is also maintained if *E. coli* Lon is expressed in mitochondria instead of Pim1 at normal temperature, but not under mild heat stress conditions [120], suggesting that the role in mtDNA maintenance is connected to its proteolytic activity. Although many open questions remain, it can be stated that Pim1/Lon is a multifunctional protein contributing more to mitochondrial maintenance than just PQC.

Lifespan Regulation in Fungi

An important consequence of a successive accumulation of mitochondrial defects seems to be cellular ageing. There is an ongoing discussion on the molecular processes underlying the mitochondrial contribution to ageing processes, summarized under the key terms "mitochondrial free radical theory of aging" [121, 122]. This hypothesis states that during the lifespan of an organism, mitochondria-generated ROS leads to accumulated mitochondrial and cellular damage, resulting in a gradual decline in important functions like ATP synthesis, which usually contribute to cellular survival. Elevated ROS levels lead to various covalent modifications of polypeptide chains, negatively influencing enzyme activities or conformational states. Hence, it can be postulated that the effectiveness of the mitochondrial PQC system may have a significant influence on the lifespan determination of a cell. A direct connection between PQC and aging could be demonstrated in the fungal organism *Podospora anserina* [123], which is used as a model system for studying cellular aging processes. Here, overexpression of Lon resulted in a decreased amount of ROS-modified proteins, consequently a higher resistance against oxidative stress, essentially correlating with a longer lifespan of the cells. Also in *P. anserina* mitochondria, one of the major mitochondrial proteins affected by ROS-related proteolysis was the citrate cycle enzyme aconitase. Although there are many supporting observations for an involvement of mitochondrial PQC in aging processes, the cause-and-effect relationships in particular at a molecular level remain to be clarified.

AAA+ Proteases of the Yeast Mitochondrial Inner Membrane

Whereas the quality of yeast mitochondrial matrix proteins is controlled by Pim1, a different type of proteolytic machine is required for the quality control of inner membrane proteins. In fact, this function is performed by two structurally related inner membrane anchored ATP-dependent proteases, with catalytic domains on opposite sides of the membrane (Fig. 9.3) [14, 18–21, 124, 125]. These are known as *i*-AAA and *m*-AAA proteases, and belong to the M41 (FtsH family) of ATP-dependent metalloendopeptidases found in a wide range of bacteria, and chloroplast and mitochondria of eukaryotes [126]. *E. coli* FtsH and yeast mitochondrial *i*-AAA and *m*-AAA proteases serve as the typical representatives of the bacterial and eukaryotic members of this family. There are however unique oligomeric combinations of *m*-AAA protease in mammalian mitochondria and multiple FtsH proteases in plants [5, 127].

Structure-Function Relationships

All subunits of FtsH family proteins have the same core modular structure. They contain at least one transmembrane anchor in their N-terminal region, a conserved AAA domain followed by a C-terminal protease domain. The AAA domain, conserved in the AAA+ superfamily of proteins, contains the Walker A and Walker B motifs for ATP binding and hydrolysis respectively [14, 19, 21, 74, 125, 128–131]. They are also characterised by a consensus zinc-binding motif HEXXH in the protease domain where Glu is the predicted catalytic residue of the peptidase and the two His residues are expected to coordinate zinc. Indeed, mutation of the conserved Glu in subunits of *i*-AAA and *m*-AAA proteases abolished peptidase activity [14, 125, 128]. The active proteases are ring-shaped oligomeric assemblies of six subunits. The oligomeric arrangement creates a central translocation pore in the ATPase domain, which provides an aqueous path to the internal proteolytic chamber. Like all ATP-dependent proteases the ATPase activity of the protein is not required for peptidase activity *per se*, but rather to present the substrate to the peptidase domain via energy-dependent translocation into the internal proteolytic cavity where hydrolysis of peptide bonds occurs.

Despite a number of common features, the *i*-AAA and *m*-AAA proteases have unique quaternary and topological structures with respect to each other. The *i*-AAA protease is a homo-oligomeric ring-shaped protein anchored to the mitochondrial inner membrane by a single span transmembrane domain situated near its N-terminus with its catalytic domains (ATPase and peptidase) projecting into the mitochondrial IMS (Fig. 9.3) [18, 19, 125, 130]. The protein subunit that makes up the oligomer is known as yeast mitochondrial escape 1 (Yme1) [130, 132]. The *m*-AAA protease on the other hand is a hexameric hetero-oligomer of polypeptides, composed of yeast tat-binding analogs 10 and 12 (Yta10 and Yta12) [14, 129, 133] also known as ATPase family gene 3 (AFG3) and respiratory chain assembly 1 (Rca1) respectively [14, 128, 131]. It is anchored to the inner membrane by two transmembrane domains near the N-terminus of each subunit with the catalytic domains protruding into the matrix space [14, 20]. The topology and modular structure of *i*-AAA and *m*-AAA proteases are well adapted for a role in membrane PQC (*see below*). Due to the lipid-phase anchor, the ATPase domain of these proteins sit close to the membrane, followed by the peptidase domain including a C-terminal coiled-coil region [133]. Thus, the molecular motor (ATPase) of these proteases is in close proximity to target proteins in the inner membrane where they can extract them for delivery to the respective peptidase chambers [134]. Recent low resolution structural studies revealed a space between the ATPase domain and the transmembrane domain of the *m*-AAA protease, which is proposed to provide sufficient room for an unfolded substrate to dock onto a surface recognition site within the ATPase domain, allowing engagement with the protease for subsequent degradation [133]. A detailed description of the known structure-function relationships of these ATP-dependent membrane proteases can be found in [134].

General Functions of *i*-AAA and *m*-AAA Proteases in Inner Membrane PQC

The discovery of energy-dependent degradation of mtDNA-encoded inner membrane proteins [9, 12, 13] was suggestive of a membrane associated ATP-dependent protease [19]. This indeed turned out to be the case with both *i*-AAA and *m*-AAA proteases mediating the degradation of incompletely synthesised and unassembled mtDNA-encoded subunits of respiratory complexes [14, 19–21, 128]. For example, the first substrate of *i*-AAA protease to be identified was Cox2. When this integral inner membrane protein fails to assemble with nuclear-encoded co-subunits it is rapidly degraded [19, 135]. Likewise, *m*-AAA protease can degrade unassembled mtDNA encoded subunits Cox1, Cox3, Cob, Atp6, Atp8 and Atp9 [14, 124]. This activity of the FtsH family of mitochondrial inner membrane proteases means they help regulate the required stoichiometry of inner membrane protein complex subunits (the respiratory complexes). Thus, post-translation regulation of protein concentration ensures the inner membrane remains free of superfluous proteins thereby contributing an important PQC function. The *i*-AAA proteases Yme1 and PaIAP of *S. cerevisiae* and *P. anserina* respectively, both play temperature-related roles in mitochondria, which may be linked to their PQC functions. Mutations in yeast *yme1* cause a heat-sensitive respiratory growth defect [130] while a *palAP* deletion strain of *P. anserina* displays heat-sensitive developmental defects and a reduced lifespan at elevated growth temperatures [136]. A detailed description of the role of these proteases in protein activity control is provided in the following section.

Activity Control of Mitochondrial Proteins by AAA+ Proteases

Another important aspect of AAA+ protease function is their contribution to protein activity control. This can be the direct activation of proteins or the conditional degradation of non-damaged native proteins in order to positively or negatively modulate cellular or organellar pathways. This section describes interconnected processes and regulatory pathways in mitochondria in which AAA+ proteases play a role.

A Unique and Critical Role of m-AAA Protease in Mitochondrial Ribosome Biogenesis

As discussed earlier (*section* "Maturation of Mitochondrial Proteins"), most matrix-destined preproteins that are directed to this compartment are processed by MPP to remove the N-terminal presequence. However, a sub-population of yeast mitochondrial matrix proteins are processed at the N-terminus but do not appear to possess an MPP recognition motif [41, 54]. One such protein is mitochondrial ribosomal protein L32 (MrpL32). As the name indicates, MrpL32 is a component of the mitochondrial ribosome, specifically the large subunit. Its proteolytic maturation, which is necessary for its function, is highly specialised and mediated by the *m*-AAA protease [54, 56]. Thus, *m*-AAA protease can mediate the complete degradation of proteins into peptides and the selective maturation of a preprotein to generate the functionally active mature form. But how the *m*-AAA protease performs multiple and seemingly conflicting functions was a conundrum for a while. The answer however lies with the biogenesis pathway and structural properties of MrpL32. Yeast MrpL32 is synthesised as a 183 amino acid preprotein and directed to the matrix compartment by an N-terminal presequence. Once it is fully imported, it folds into its unique three-dimensional structure in a mechanism that requires the presence of the N-terminal region of the protein [54]. Based on comparison to the bacterial homolog, yeast MrpL32 is expected to form a globular domain at its C-terminus mediated by the metal binding twin Cys motif CxxC-x9-CxxC [54]. Like the bacterial counterpart, the N-terminal region is expected to be in an extended conformation. It is the folded state of the mitochondrial protein that is processed to its mature form by the *m*-AAA protease, which results in the removal of 71 amino acids from the N-terminus [54, 137, 138]. It is anticipated that the extended N-terminal region of MrpL32 is recognised by the *m*-AAA protease and feeds into the inner proteolytic chamber where the catalytic residues act on the polypeptide chain, hydrolysing it into peptides. The folded domain of MrpL32 blocks any further processing of the protein. Thus, the

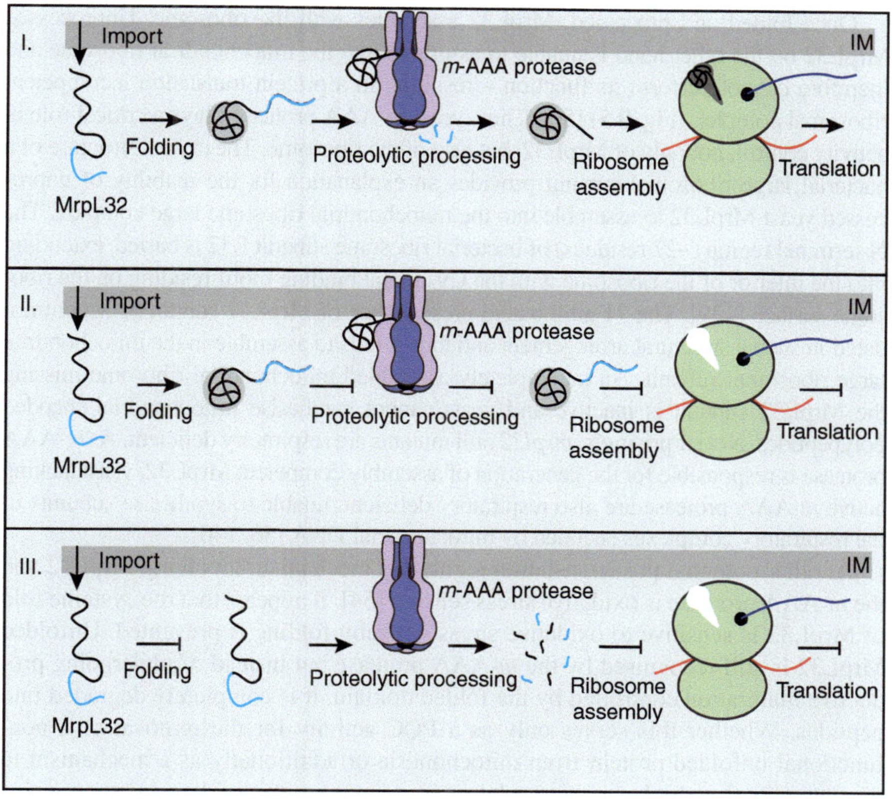

Fig. 9.5 A specialised function of *m*-AAA protease in protein maturation required for mitochondrial ribosome biogenesis. Under normal conditions (*Path I*) the *m*-AAA protease cleaves an N-terminal presequence from folded ribosomal subunit MrpL32 via a partial degradation mechanism. The folded cysteine knot blocks complete processing of this protein and upon release from the protease it assembles into the large subunit of the ribosome. To date the route of entry of the substrate presequence into the proteolytic chamber has not been determined. In situations where the activity of *m*-AAA protease is compromised (e.g. gene deletion in yeast) or genetic mutation in humans, MrpL32 processing is inhibited and thus ribosome assembly and function is impaired (*Path II*). In the case that the folding of MrpL32 is disrupted (*Path III*) due to oxidative stress, the unfolded substrate is fully degraded and thus ribosome assembly and function is also impaired

folding of MrpL32 permits maturation of the protein while averting complete degradation (Fig. 9.5). In such a mechanism, the middle region of MrpL32 is believed to act as a linker bridging the gap between the folded domain butting the external surface of the protease and the proteolytic residues of the inner chamber. Such a scenario is supported by experiments in which the N-terminal region of MrpL32 was extended in length (20 and 40 amino acids respectively) between the presequence and the middle region. Still these longer variants were cleaved to produce proteins of similar length to wild type MrpL32 [54].

Once folded and processed, MrpL32 assembles with the ribosome. Unprocessed MrpL32 on the other hand is unable to assemble into the mitochondrial ribosome and therefore cannot perform its function – resulting in a protein translation incompetent ribosomal complex (Fig. 9.5) [56]. Thus, yeast *m*-AAA protease plays a critical role in activity control, not only of MrpL32, but also of the ribosome. The crystal structure of a bacterial large ribosomal subunit provides an explanation for the inability of unprocessed yeast MrpL32 to assemble into the mitochondrial ribosome large complex. The N-terminal region (~27 residues) of bacterial ribosome subunit L32 is buried, extending into the interior of the ribosome with the Cys metal binding motif residing on the ribosome surface [139]. The 71 amino acid presequence of MrpL32 cannot be accommodated in such a structural arrangement and thus it fails to assemble in the mitochondrial large ribosomal subunit. An incompletely assembled mitochondrial ribosome missing the MrpL32 subunit is inactive and yeast cannot synthesise mitochondrial-encoded polypeptides. Not surprisingly, *mrpl32* null mutants are respiratory deficient. As *m*-AAA protease is responsible for the generation of assembly competent MrpL32, yeast lacking active *m*-AAA protease are also respiratory deficient, unable to synthesise subunits of the respiratory complexes encoded by mitochondrial DNA [56, 140].

Another potential post-translation regulatory mechanism involving MrpL32 and the *m*-AAA protease is oxidative stress sensing [54]. It appears that the cysteine fold of MrpL32 is sensitive to oxidative stress whereby folding is prevented. Unfolded MrpL32 is still recognised by the *m*-AAA protease but instead of undergoing productive maturation controlled by the folded domain, it is completely degraded into peptides. Whether this serves only as a PQC activity for the removal of a non-functional unfolded protein from mitochondria or additionally as a mechanism to regulate mitochondrial protein translation in response to the oxidative state, remains to be determined. The critical role of the *m*-AAA protease in mitochondria is exemplified by human diseases such as spinocerebellar ataxia (SCA28) and a form of hereditary spastic paraplegia. These diseases are caused by loss of function mutations in AFG3L2 (homolog of Yta10/Afg3) and paraplegin (homolog of yeast Yta12/Rca1) respectively, subunits of the human *m*-AAA protease [141, 142].

The Role of i-AAA Protease in Lipid Homeostasis

Recently, the role of *i*-AAA protease in quality control of lipid homeostasis proteins in the mitochondrial IMS was revealed [143]. Three membrane-associated IMS proteins Ups1, Ups2, and Ups3 collectively play a role in the metabolism of phosphatidylethanolamine and cardiolipin although they seem to contribute unique functions [144–146]. A loss of these proteins alters mitochondrial lipid composition and effects the stability of inner membrane protein complexes such as the TIM23 complex, as well as mitochondrial morphology [144–146]. Each of these three proteins forms a stable complex with a small IMS protein known as Mdm35 [143, 147]. This protein also acts to drive import of Ups proteins into the IMS upon translocation through the TOM complex. In the absence of Mdm35 (i.e. in *mdm35* deletion strain), Ups1 and Ups2 are unstable and degraded more rapidly than in its presence [143]. This change in stability is attributed to the action of *i*-AAA protease and Atp23 [143]. In the case of uncomplexed Ups2, its degradation is mediated by *i*-AAA protease. Degradation of free Ups1 on the other hand is mediated by *i*-AAA protease and Atp23. It seems that the degradation of Ups proteins at least serves a PQC function to avoid the accumulation of non-functional protein. However, it is possible that the *i*-AAA protease plays a regulatory role in the activity control of Ups proteins and hence mitochondrial lipid homeostasis, for example by competing with Mdm35 for binding of Ups proteins. In such a situation adaptor proteins could play a leading role in the targeted delivery of the substrate, however it seems that putative *i*-AAA protease adaptor proteins Mgr1 and Mgr3 [148, 149] are not involved in *i*-AAA protease-mediated degradation of Ups1 and Ups2 [143]. A potential role of *i*-AAA protease and Atp23 in regulatory control of phosphatidylethanolamine and cardiolipin metabolism awaits further analysis. Regulation of lipid biogenesis is also required for normal function in *E. coli* and the *i*-AAA protease homolog FtsH plays a critical role. In this case, FtsH-mediated degradation modulates the levels of two enzymes, KdtA and LpxC in the lipopolysaccharide biosynthetic pathway (for a recent review see [126]).

Regulation of Metazoan Biosynthetic Pathways by Lon

Until very recently it seemed that the mitochondrial matrix protease Lon only contributed to PQC and not protein activity control. However, recent studies in human cells (and other animals) have revealed important contributions of LONM to the regulation of mitochondrial biogenesis and metabolic pathways. One of these pathways is the heme biosynthetic pathway. Mitochondria are the site of several steps of the cellular heme biosynthetic pathway. Heme is generated from substrates glycine and succinyl-CoA. The enzyme 5-aminolevulinic acid synthase (known as ALAS-1 in humans), catalyses the first of eight enzymatic steps. ALAS-1 activity, the rate-limiting step of the pathway, is regulated at many levels through expression and import control by a heme-mediated negative feedback mechanism. Heme inhibits transcription and import of the ALAS-1 precursor into mitochondria while enhancing mRNA degradation and turnover of the endogenous protein in mitochondria [150–153]. All of these effects lead to a decrease in the steady state levels of ALAS-1, allowing the levels of heme to be adjusted according to need. The ability to sense and regulate heme levels is important as both an excess or a deficiency is toxic. The proteolytic element of regulation allows a rapid post-translational adjustment of protein levels. Recently, it was revealed that LONM is responsible for the turnover of ALAS-1 in hepatic cells however the mechanism by which heme targets ALAS-1 for degradation is currently unknown. At this point, partial functional redundancy with other proteases, i.e. the mitochondrial ClpXP protease, have not been ruled out [152]. A regulatory role of Lon has also been identified in *Drosophila* Schneider cells where it degrades mitochondrial transcription factor A (TFAM) [154]. A major component of mitochondrial nucleoids, TFAM contributes to mtDNA maintenance and transcription [155]. Thus, proteolysis may also contribute to the transcriptional expression of the mitochondrial genome. Steroidogenic acute regulatory protein (StAR) is a protein found in mitochondria of mammalian cells in the adrenal cortex, gonads and placenta. It is a key enzyme in steroid hormone biogenesis from cholesterol. Due to its high turnover rates after import into the matrix compartment of mitochondria it has been suspected to be a substrate of LONM, a hypothesis which has been supported by experiments *in vitro* and *in vivo* examining murine StAR with mammalian LONM [156]. However, so far it is unclear if its degradation represents a PQC process to prevent the accumulation of large quantities of unwanted protein in mitochondria or has a more important role in regulation of steroid biogenesis. In mammals, LONM has also been implicated in the degradation of cytochrome oxidase subunit 4-1 (Cox4-1) under conditions of hypoxia permitting exchange with the low oxygen efficient isoform Cox4-2 [157]. However, direct degradation of Cox4-1 by LONM was not demonstrated in these studies and the addition of MG132, which is known to inhibit LONM as well as the proteasome [156], didn't change the steady state levels of the protein [157]. Thus, further studies are required to understand the full contribution of mitochondrial matrix proteases to the elimination of hypoxia sensitive Cox4-1 and thus regulation of the cellular response to changes in oxygen concentrations.

Mitochondrial Protein Degradation via an N-End Rule Pathway?

Some mitochondrial precursor proteins are cleaved by Icp55 or Oct1, following MPP processing to generate the mature protein (*see section* "Maturation of Mitochondrial Proteins"). In the absence of Icp55 or Oct1, intermediate forms of the preprotein are unstable relative to mature counterparts and possess either Tyr, Phe or Leu at their N-terminus [41]. This is suggestive of an N-end rule pathway of protein degradation being operational in mitochondria. In bacteria, amino acid residues Phe, Leu, Tyr and Trp can act as destabilising signals when exposed at the N-terminus of a protein [158]. They mediate recognition of the protein substrate directly by the pathways degradation machinery and as such are referred to as primary destabilising residues [51, 52, 158]. Basic residues Arg and Lys and in a specific case Met are classified as secondary destabilising residues as they act as acceptors for the non-ribosomal attachment of primary destabilising residues Leu and Phe by the leucyl/phenylalanyl-tRNA protein transferase thereby tagging the protein for degradation [51, 52, 158, 159]. In *E. coli*, primary destabilising residues bind the adaptor protein ClpS and are directly delivered to the AAA+ protease ClpAP for degradation [159–162]. In the absence of mitochondrial peptidases Icp55 or Oct1, substrates retain N-terminal residues Tyr, Leu and Phe and are unstable. It is suspected that these proteins are removed by a proteolytic pathway but this has not been determined experimentally. Also, it is yet to be determined if this is a PQC mechanism to remove immature proteins from mitochondria or it serves as a regulatory mechanism whereby the activity of Icp55 and Oct1 or molecular components of the hypothetical degradation pathway are regulated to enable post-translational expression control of a subpopulation of mitochondrial proteins. In general the N-end rule pathway of degradation contributes to regulatory networks rather than PQC tasks [52, 163, 164]. Thus, it will be interesting to determine if such a pathway is present in mitochondria and what role it plays.

The Mitochondrial Unfolded Protein Response (UPR^{mt})

In addition to the contribution that AAA+ proteases make directly to protein quality and activity control, some members participate in stress response pathways. In particular, in metazoan mitochondria it has been established that the matrix peptidase ClpP is a core component of the mitochondrial unfolded protein response (*see later*). This signal transduction pathway allows mitochondria-to-nuclear communication ensuring that the capacity of the mitochondrial protein quality control machinery matches demand.

The constitutive expression of numerous members of the PQC network ensures that cells are under constant molecular surveillance for unfolded, damaged and misassembled proteins. Whilst this may contribute to protein homeostasis under normal cellular conditions, imbalances in the protein-folding environment can place additional demands on the chaperone and protease machinery. For example, when protein damage is amplified through stress or other physiological insults, the resulting substrate load may exceed the capacity of these systems. This leads to the accumulation of unfolded and aggregated proteins. If the unfolded protein stress is excessive or prolonged, cellular function may be compromised [107, 165, 166]. Thus, to alleviate the additional substrate burden, cells have evolved signalling mechanisms to upregulate members of their PQC networks [107, 166–172]. By enabling the cell to manage unfolded or damaged proteins more effectively, protein homeostasis is expected to be restored.

Elevated levels of misfolded proteins can arise through a variety of cellular stresses including errors in protein biogenesis, exposure to reactive oxygen species, metabolic deficiencies and thermal stress. Although perturbations in protein folding can occur broadly throughout the cell, in some situations, protein damage may be restricted to specific organelles. Importantly, cells have dedicated pathways for managing unfolded protein stress within their different subcellular compartments. While the mechanisms that the cell employs for sensing and responding to unfolded protein stress within the cytosol and endoplasmic reticulum are well characterised, the discovery of an equivalent regulatory pathway in metazoan mitochondria has been more recent [168, 169, 173].

Discovery of the Mitochondrial Unfolded Protein Response

Early indications of a mitochondrial specific stress response came from studies examining the regulation of chaperones in cells depleted of mtDNA. Using cultured mammalian cells, Hoogenraad and co-workers demonstrated that the loss of mtDNA resulted in the transcriptional activation and increased expression of mitochondrial chaperonins HSP60 and HSP10 [174]. Interestingly, as there were no detectable changes in the expression of selected cytosolic chaperones, this response was suggested to be compartment-specific. While the primary stress signal was not identified in this study, the authors proposed that the stress response could be linked to the accumulation of unfolded protein [174].

To establish whether the specific upregulation of mitochondrial chaperones was connected to protein misfolding within mitochondria, a mammalian cell-based model was developed for the overexpression of a folding deficient mutant of the mitochondrial protein, ornithine transcarbamylase (OTC) [169]. In comparison to cells expressing wild type OTC, the expression of the OTC mutant resulted in the selective induction of a number of mitochondrial stress proteins including HSP60, HSP10, mtDnaJ/Tid-1 (HSP40), and the CLPP peptidase. Consistent with a role in PQC, both HSP60 and CLPP co-immunoprecipitated with the OTC mutant, implicating them in the clearance of the substrate. Importantly, like other stress response pathways, the removal of non-native protein corresponded with the transcriptional attenuation of chaperone genes [169]. However, the protease responsible for degradation of mutant OTC is yet to be established.

Since the initial discovery of this unfolded protein response in mammals, an equivalent pathway has been identified in *C. elegans*. Like the observations made in mammalian cells, the loss of mtDNA resulted in the upregulation of genes encoding HSP60 and mtHSP70, but did change the expression profile of chaperones in other subcellular compartments [168]. A similar response was also observed upon inactivation of genes suggested to function in the biogenesis or turnover of mitochondrial proteins [168]. Taken together, these findings demonstrate that the selective upregulation of mitochondrial chaperones in both *C. elegans* and mammalian cells is triggered by disturbances in the mitochondrial protein folding environment. This pathway has been coined, the mitochondrial unfolded protein response (UPR^{mt}). As yeast appear to lack a mitochondrial specific unfolded protein response, this section describes the pathway in metazoa. A key component of this signal transduction pathway in the nematode worm is the mitochondrial matrix protease ClpXP.

Components of the UPR^{mt} Signalling Pathway in C. elegans

The mitochondrial unfolded protein response (UPR^{mt}) is defined by the ability of cells to selectively alter the level of PQC machinery in response to protein misfolding within mitochondria [168, 169]. Importantly, for target genes in the nucleus to be up-regulated in response to an extrinsic mitochondrial stress, a mechanism for communication between the two organelles must exist. To dissect this pathway and identify factors involved in relaying the stress signal from the mitochondrion to the nucleus, Ron and colleagues performed a series of gene silencing and deletion studies in *C. elegans* [165]. For these experiments, transgenic *C. elegans* reporter strains harbouring a temperature sensitive mutation (zc32 *II*) were used. At non-permissive temperatures, these animals experience unfolded protein stress within the mitochondrion, which leads to the transcriptional upregulation of a chaperone-GFP reporter (consisting of a DNA fragment containing the promoter region and the coding sequence for the first few amino acid residues of a mitochondrial chaperone) [23, 165]. Genes whose inactivation caused a reduction in reporter activity were predicted to encode proteins that contribute to UPR^{mt} signalling [165]. Using this model, a number of cellular components that are critical for UPR^{mt} signalling were identified [23, 24, 165]. These proteins can be broadly classified into two groups: those that are involved in the sensing and transmission of stress signals from the mitochondria (ClpX1, ClpX2, ClpP, HAF-1), and those that adjust their nuclear distribution or expression profiles to promote chaperone gene activation (ZC376.7, DVE-1, UBL-5).

The mitochondrial matrix ClpP protease together with its anticipated cognate partners, ClpX1 (encoded by *KO7A3.3*) and ClpX2 (encoded by *D2030.2*), [175] are thought to be responsible for the initial activation of the UPR^{mt} signalling pathway [23, 24]. The *C. elegans* ClpXP machinery has been implicated in the ATP-dependent proteolysis of unfolded substrates [24]. While evidence suggests that the ClpP homologue forms homo-oligomeric complexes composed of 14 subunits, little is known about the composition of the ClpX oligomer [23, 24]. *In vivo*, a hexameric form of ClpX is likely to be involved in both substrate recognition and docking to the ClpP protease. Thus, different compositions or oligomeric arrangements of ClpX1 and ClpX2 could potentially serve as a mechanism for regulating substrate detection and delivery. In this regard, it is interesting to note that the loss of either ClpX1 or ClpX2 alone, did not inhibit UPR^{mt} signalling. This functional redundancy suggests that the substrate specificity of ClpX1 and ClpX2 is overlapping [24]. Although the substrates of ClpX are currently unknown, their delivery to the ClpP protease appears to be critical for transmission of the UPR^{mt} stress signal.

The requirement for ClpP in an upstream regulatory role was established through a series of knock-down experiments that examined the position of ClpP relative to other components within the signalling network. Consistent with a position near the top of the signalling hierarchy, depletion of ClpP in stressed worms prevented the activation, recruitment and nuclear redistribution of all known UPR^{mt} transcription factors [23, 24, 165]. To elucidate the mechanism by which ClpP controls the signalling pathway, a chemical inhibitor of ClpP was employed. These studies revealed that the intrinsic proteolytic activity of ClpP was an essential element in UPR^{mt} signal progression [23]. Not surprisingly, this raised questions regarding the identity of the signalling molecules generated by ClpP degradation, and the mechanisms by which these molecules transduce the signal. Although a number of scenarios are possible, there is evidence to suggest that peptides generated from the proteolysis of matrix proteins may be involved in at least one arm of the stress response pathway [24].

The discovery of HAF-1 as a component of the UPR^{mt} signalling pathway suggested that peptides generated from ClpXP proteolysis may be involved in signal transmission. HAF-1 is a member of the ABC transporter family, which appears to be located in the mitochondrial inner membrane [24]. In a process that is thought to be mechanistically equivalent to that reported for the yeast Mdl1p homologue, HAF-1 actively transports peptides from the matrix into the IMS. Peptides are then

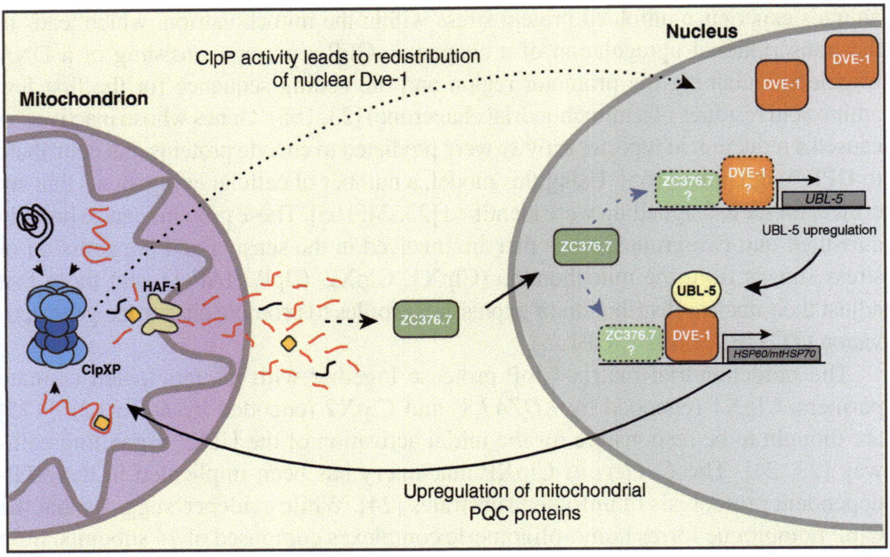

Fig. 9.6 The mitochondrial unfolded protein response in *C. elegans*. Depicted are the known components and model of mitochondria-to-nuclear signalling pathway of the unfolded protein response. A disruption to protein homeostasis within the mitochondrial matrix leads to the proteolytic processing of an unknown substrate(s) by ClpXP. This might be the promiscuous degradation of any unfolded protein or the regulated degradation of specific proteins that are conditional substrates under stress conditions. The products of ClpXP-mediated protein degradation are peptides (*red and black lines*) but for cofactor-bound substrates, prosthetic groups would also be released (*orange square*). These peptides and/or prosthetic groups could act as signals to initiate the pathway outside of mitochondria following transport through HAF-1 and pores in the outer membrane. Other mechanisms may exist for a signalling path independent of HAF-1. The redistribution of pathway transcription factors initiated by the released signalling molecule(s) leads to the transcriptional up-regulation of specific genes, such as those encoding mitochondrial chaperones Hsp60 and Hsp70, that act to restore mitochondrial protein homeostasis

thought to diffuse across the outer mitochondrial membrane into the cytosol [176, 177]. Importantly, in contrast to wild type worms experiencing stress, deletion of *haf-1* resulted in reduced mitochondrial peptide efflux and attenuated mitochondrial chaperone gene induction [24]. Based on these findings, a model that integrates ClpP-mediated proteolysis with transmission of the stress signal across the mitochondrial membranes has been proposed (Fig. 9.6). Although HAF-1 has been shown to transport a broad spectrum of peptides derived from a range of mitochondrial matrix proteins, the precise nature of the stress signal is currently unknown [24]. It is not known whether signal transmission is dependent upon a unique type of peptide, or whether the rate of peptide flux is a contributing factor. Further, a role for HAF-1 in the transport of free prosthetic groups cannot be excluded. Although additional work is required to elucidate the sensory mechanisms within the cytosol, it appears that the introduction of these peptides or prosthetic groups into the signalling pathway results in the activation of downstream transcription factors. Specifically, the ZC376.7 (bZIP) transcription factor is recruited into the UPR^{mt}

pathway in a process that is dependent upon both ClpP and HAF-1. Under normal cellular conditions, ZC376.7 is located in the cytosol. Upon induction of UPR^{mt} however, the protein translocates to the nucleus, where it accumulates. Although the gene targets of ZC376.7 are currently unknown, its translocation to the nucleus is closely linked to the upregulation of mitochondrial chaperone genes [24].

Whilst the incorporation of HAF-1 into the UPR^{mt} signalling pathway provides a feasible model for transmission of the stress signal across the mitochondrial membranes, current evidence suggests that an alternative mechanism for UPR^{mt} activation may also be at play. In what appears to be a separate arm of the response, ClpP has been shown to transmit signals independently of HAF-1 (Fig. 9.6). While the loss of ClpP prevented the nuclear redistribution of the transcription factor DVE-1 under conditions of mitochondrial stress, the loss of HAF-1 had no effect [23, 24]. Although the signalling mechanisms that govern this response are currently unknown, the redistribution of DVE-1 within the nucleus correlates with its binding to chaperone gene promoters [23]. Importantly, to enhance the transcription of chaperone genes, DVE-1 is proposed to function with a small ubiquitin-like protein called UBL-5 [23]. Upon induction of UPR^{mt}, UBL-5 is upregulated and accumulates within the nucleus [165]. As demonstrated through pull-down experiments, this enrichment enhances the formation of stable UBL-5/DVE-1 complexes [23]. Interestingly, as DVE-1 redistribution and promoter binding can occur independently of UBL-5 [23], it is tempting to speculate that UBL-5 may be recruited into transcription complexes as a downstream event.

The above findings support a model whereby the stress-induced nuclear reorganisation of DVE-1 and UBL-5 serves as a mechanism for upregulating mitochondrial chaperone genes. Importantly however, within this model, the transcriptional induction of UBL-5 also requires regulation. In what appears to be an amplification circuit, RNAi knock-down studies revealed that the transcriptional induction of UBL-5 is also dependent upon DVE-1. It is currently unclear whether DVE-1 is directly involved in *ubl-5* gene activation, or whether it exerts this control via an indirect mechanism [23]. Moreover, in worms experiencing mitochondrial stress, the loss of HAF-1 also circumvented *ubl-5* gene activation [24]. Thus, the upregulation of UBL-5 also requires a mediator that is transmitted via the HAF-1 signalling pathway. While the factors that contribute to this event are currently unknown, it is possible to speculate that ZC376.7 may fulfil this role. Further analysis of this transcription factor and the identification of additional UPR^{mt} signalling components may assist in elucidating this mechanism.

What Is the UPR^{mt} Signalling Factor Generated by ClpXP Action?

The ability of cells to selectively respond to protein conformational stress within specific subcellular compartments is a conserved paradigm that exists in both prokaryotic and eukaryotic organisms [166, 170, 175, 176, 178]. These response pathways are underpinned by signalling cascades that are tailored towards sensing unfolded protein stress and transmitting stress signals to designated transcription factors.

Interestingly, a common theme shared by many of these response pathways is the participation of cellular proteases in regulating stress signal transmission [166, 178]. The discovery that ClpXP plays a pivotal role in controlling the mitochondrial unfolded protein response in *C. elegans* provides yet another example of how proteolytic processing contributes to signal transduction [23, 24]. Although significant progress has been made in identifying components of the *C. elegans* UPR^{mt} signalling pathway, a number of key elements have yet to be defined. In particular, the substrates that undergo ClpXP-mediated proteolysis and the type of the signalling molecules that are generated through this process require elucidation. Importantly, given that HAF-1 homologues have been implicated in the transport of both peptides and prosthetic groups [176, 177, 179], a range of potential signalling mechanisms can be conceived. Based on current knowledge of the UPR^{mt} signalling pathway (Fig. 9.6), a number of hypothetical models describing the contribution that ClpXP could make to stress signal transmission have been proposed [23, 175, 176, 178]. It has been suggested that ClpXP could function in the promiscuous degradation of unfolded and/or misfolded proteins that arise from perturbations in mitochondrial protein folding. Transduction of the stress signal could involve a peptide/cofactor receptor in the cytosol, or may rely on a sensor molecule that monitors the rate of peptide efflux [23, 175, 176]. It is also possible that some matrix proteins may be conditional substrates of ClpXP – displaying specific recognition motifs under conditions of stress. In this scenario, a specific peptide or prosthetic group could be liberated from the protein following its targeted degradation. Again, signal transduction may rely on a specific cytosolic receptor that recognises mitochondrially-derived ligands [23, 175, 176].

The above models not only provide insight into the mechanisms that ClpXP may employ, but also pave the way for future studies. The identification of ClpXP substrates and the molecules involved in downstream signalling may assist in further defining the molecular mechanisms that control the *C. elegans* UPR^{mt} signalling pathway. In addition, understanding the sensory components within the cytosol may also shed light on the mechanism by which ClpP is able to transmit signals independently of HAF-1. Whilst it is possible that ABC transporters that are similar to HAF-1 may have escaped detection in original screens, the existence of an alternative signalling route that also negotiates the mitochondrial membrane barrier cannot be discounted.

Despite *C. elegans* and mammalian UPR^{mt} pathways being functionally equivalent, the transcription factors that control these stress response pathways appear to be quite different [169, 180]. Whilst studies have addressed the mechanisms by which UPR^{mt} gene targets are selectively upregulated in response to mitochondrial stress [169, 181], the identity of the stress sensor component has remained elusive in mammalian mitochondria.

Future Perspectives

Most of our understanding of mitochondrial AAA+ proteases has been achieved via genetic, biochemical and proteomic studies, particularly in relation to PQC functions. A detailed understanding of the recognition motifs in substrates and substrate docking sites on mitochondrial AAA+ proteases will help to refine their substrate repertoire. It is anticipated that much will be revealed about the substrates of these proteases in higher eukaryotes where cell type dependent processes take place. Research on the role of AAA+ proteases in the mitochondrial unfolded protein response, cell death and aging are just a few areas that are expected to expand in the near future. An understanding of the mitochondrial AAA+ proteases in the context of human mitochondrial diseases will be important in the development of successful therapies for disease intervention.

References

1. Todde V, Veenhuis M, van der Klei IJ (2009) Autophagy: principles and significance in health and disease. Biochim Biophys Acta 1792(1):3–13
2. Wong E, Cuervo AM (2010) Integration of clearance mechanisms: the proteasome and autophagy. Cold Spring Harb Perspect Biol 2(12):a006734
3. Yang Z, Klionsky DJ (2009) An overview of the molecular mechanism of autophagy. Curr Top Microbiol Immunol 335:1–32
4. Buchberger A (2013) Roles of Cdc48 in regulated protein degradation in yeast. In: Dougan DA (ed) Regulated proteolysis in microorganisms. Springer, Subcell Biochem 66:195–222
5. Kwasniak M, Pogorzelec L, Migdal I, Smakowska E et al (2012) Proteolytic system of plant mitochondria. Physiol Plant 145(1):187–195
6. Ezaki J, Kominami E, Ueno T (2011) Peroxisome degradation in mammals. IUBMB Life 63(11):1001–1008
7. Druyan R, DeBernard B, Rabinowitz M (1969) Turnover of cytochromes labeled with delta-aminolevulinic acid-3H in rat liver. J Biol Chem 244(21):5874–5878
8. Swick RW, Rexroth AK, Stange JL (1968) The metabolism of mitochondrial proteins. 3. The dynamic state of rat liver mitochondria. J Biol Chem 243(13):3581–3587
9. Wheeldon LW, Dianoux AC, Bof M, Vignais PV (1974) Stable and labile products of mitochondrial protein synthesis in vitro. Eur J Biochem 46(1):189–199
10. Desautels M, Goldberg AL (1982) Demonstration of an ATP-dependent, vanadate-sensitive endoprotease in the matrix of rat liver mitochondria. J Biol Chem 257(19):11673–11679
11. Desautels M, Goldberg AL (1982) Liver mitochondria contain an ATP-dependent, vanadate-sensitive pathway for the degradation of proteins. Proc Natl Acad Sci U S A 79(6):1869–1873
12. Kalnov SL, Novikova LA, Zubatov AS, Luzikov VN (1979) Participation of a mitochondrial proteinase in the breakdown of mitochondrial translation products in yeast. FEBS Lett 101(2):355–358
13. Yasuhara T, Mera Y, Nakai T, Ohashi A (1994) ATP-dependent proteolysis in yeast mitochondria. J Biochem 115(6):1166–1171
14. Arlt H, Tauer R, Feldmann H, Neupert W et al (1996) The YTA10-12 complex, an AAA protease with chaperone-like activity in the inner membrane of mitochondria. Cell 85(6):875–885
15. Kang SG, Ortega J, Singh SK, Wang N et al (2002) Functional proteolytic complexes of the human mitochondrial ATP-dependent protease, hClpXP. J Biol Chem 277(23):21095–21102
16. Kutejova E, Durcova G, Surovkova E, Kuzela S (1993) Yeast mitochondrial ATP-dependent protease: purification and comparison with the homologous rat enzyme and the bacterial ATP-dependent protease La. FEBS Lett 329(1–2):47–50
17. Kuzela S, Goldberg AL (1994) Mitochondrial ATP-dependent protease from rat liver and yeast. Methods Enzymol 244:376–383
18. Leonhard K, Herrmann JM, Stuart RA, Mannhaupt G et al (1996) AAA proteases with catalytic sites on opposite membrane surfaces comprise a proteolytic system for the

ATP-dependent degradation of inner membrane proteins in mitochondria. EMBO J 15(16):4218–4229
19. Nakai T, Yasuhara T, Fujiki Y, Ohashi A (1995) Multiple genes, including a member of the AAA family, are essential for degradation of unassembled subunit 2 of cytochrome c oxidase in yeast mitochondria. Mol Cell Biol 15(8):4441–4452
20. Pajic A, Tauer R, Feldmann H, Neupert W et al (1994) Yta10p is required for the ATP-dependent degradation of polypeptides in the inner membrane of mitochondria. FEBS Lett 353(2):201–206
21. Pearce DA, Sherman F (1995) Degradation of cytochrome oxidase subunits in mutants of yeast lacking cytochrome c and suppression of the degradation by mutation of yme1. J Biol Chem 270(36):20879–20882
22. Watabe S, Kimura T (1985) ATP-dependent protease in bovine adrenal cortex. Tissue specificity, subcellular localization, and partial characterization. J Biol Chem 260(9):5511–5517
23. Haynes CM, Petrova K, Benedetti C, Yang Y et al (2007) ClpP mediates activation of a mitochondrial unfolded protein response in C. elegans. Dev Cell 13(4):467–480
24. Haynes CM, Yang Y, Blais SP, Neubert TA et al (2010) The matrix peptide exporter HAF-1 signals a mitochondrial UPR by activating the transcription factor ZC376.7 in C. elegans. Mol Cell 37(4):529–540
25. Heo JM, Rutter J (2011) Ubiquitin-dependent mitochondrial protein degradation. Int J Biochem Cell Biol 43(10):1422–1426
26. Karbowski M, Youle RJ (2011) Regulating mitochondrial outer membrane proteins by ubiquitination and proteasomal degradation. Curr Opin Cell Biol 23(4):476–482
27. Scott I, Youle RJ (2010) Mitochondrial fission and fusion. Essays Biochem 47:85–98
28. Drago I, Pizzo P, Pozzan T (2011) After half a century mitochondrial calcium in- and efflux machineries reveal themselves. EMBO J 30(20):4119–4125
29. Smith DJ, Ng H, Kluck RM, Nagley P (2008) The mitochondrial gateway to cell death. IUBMB Life 60(6):383–389
30. West AP, Shadel GS, Ghosh S (2011) Mitochondria in innate immune responses. Nat Rev Immunol 11(6):389–402
31. Gray MW, Burger G, Lang BF (2001) The origin and early evolution of mitochondria. Genome Biol 2(6):reviews 1018.1–reviews 1018.5
32. Friedman JR, Lackner LL, West M, DiBenedetto JR et al (2011) ER tubules mark sites of mitochondrial division. Science 334(6054):358–362
33. Harner M, Korner C, Walther D, Mokranjac D et al (2011) The mitochondrial contact site complex, a determinant of mitochondrial architecture. EMBO J 30(21):4356–4370
34. Hoppins S, Collins SR, Cassidy-Stone A, Hummel E et al (2011) A mitochondrial-focused genetic interaction map reveals a scaffold-like complex required for inner membrane organization in mitochondria. J Cell Biol 195(2):323–340
35. Kornmann B, Currie E, Collins SR, Schuldiner M et al (2009) An ER-mitochondria tethering complex revealed by a synthetic biology screen. Science 325(5939):477–481
36. Kornmann B, Osman C, Walter P (2011) The conserved GTPase Gem1 regulates endoplasmic reticulum-mitochondria connections. Proc Natl Acad Sci U S A 108(34):14151–14156
37. Stroud DA, Oeljeklaus S, Wiese S, Bohnert M et al (2011) Composition and topology of the endoplasmic reticulum-mitochondria encounter structure. J Mol Biol 413(4):743–750
38. von der Malsburg K, Muller JM, Bohnert M, Oeljeklaus S et al (2011) Dual role of mitofilin in mitochondrial membrane organization and protein biogenesis. Dev Cell 21(4):694–707
39. Sickmann A, Reinders J, Wagner Y, Joppich C et al (2003) The proteome of Saccharomyces cerevisiae mitochondria. Proc Natl Acad Sci U S A 100(23):13207–13212
40. Young JC, Hoogenraad NJ, Hartl FU (2003) Molecular chaperones Hsp90 and Hsp70 deliver preproteins to the mitochondrial import receptor Tom70. Cell 112(1):41–50
41. Vogtle FN, Wortelkamp S, Zahedi RP, Becker D et al (2009) Global analysis of the mitochondrial N-proteome identifies a processing peptidase critical for protein stability. Cell 139(2):428–439
42. Gakh O, Cavadini P, Isaya G (2002) Mitochondrial processing peptidases. Biochim Biophys Acta 1592(1):63–77

43. Chacinska A, Koehler CM, Milenkovic D, Lithgow T et al (2009) Importing mitochondrial proteins: machineries and mechanisms. Cell 138(4):628–644
44. Mokranjac D, Neupert W (2010) The many faces of the mitochondrial TIM23 complex. Biochim Biophys Acta 1797(6–7):1045–1054
45. Ott M, Herrmann JM (2010) Co-translational membrane insertion of mitochondrially encoded proteins. Biochim Biophys Acta 1803(6):767–775
46. Mossmann D, Meisinger C, Vogtle FN (2012) Processing of mitochondrial presequences. Biochim Biophys Acta 1819(9–10):1098–1106
47. Taylor AB, Smith BS, Kitada S, Kojima K et al (2001) Crystal structures of mitochondrial processing peptidase reveal the mode for specific cleavage of import signal sequences. Structure 9(7):615–625
48. Rawlings ND, Morton FR, Kok CY, Kong J et al (2008) MEROPS: the peptidase database. Nucleic Acids Res 36(Database issue):320–325
49. Vogtle FN, Prinz C, Kellermann J, Lottspeich F et al (2011) Mitochondrial protein turnover: role of the precursor intermediate peptidase Oct1 in protein stabilization. Mol Biol Cell 22(13):2135–2143
50. Naamati A, Regev-Rudzki N, Galperin S, Lill R et al (2009) Dual targeting of Nfs1 and discovery of its novel processing enzyme, Icp55. J Biol Chem 284(44):30200–30208
51. Dougan DA, Truscott KN, Zeth K (2010) The bacterial N-end rule pathway: expect the unexpected. Mol Microbiol 76(3):545–558
52. Varshavsky A (2011) The N-end rule pathway and regulation by proteolysis. Protein Sci 20(8):1298–1345
53. Gur E, Ottofuelling R, Dougan DA (2013) Machines of destruction – AAA+ proteases and the adaptors that control them. In: Dougan DA (ed) Regulated proteolysis in microorganisms. Springer, Subcell Biochem 66:3–33
54. Bonn F, Tatsuta T, Petrungaro C, Riemer J et al (2011) Presequence-dependent folding ensures MrpL32 processing by the m-AAA protease in mitochondria. EMBO J 30(13):2545–2556
55. Koppen M, Bonn F, Ehses S, Langer T (2009) Autocatalytic processing of m-AAA protease subunits in mitochondria. Mol Biol Cell 20(19):4216–4224
56. Nolden M, Ehses S, Koppen M, Bernacchia A et al (2005) The m-AAA protease defective in hereditary spastic paraplegia controls ribosome assembly in mitochondria. Cell 123(2):277–289
57. Osman C, Wilmes C, Tatsuta T, Langer T (2007) Prohibitins interact genetically with Atp23, a novel processing peptidase and chaperone for the F1Fo-ATP synthase. Mol Biol Cell 18(2):627–635
58. Zeng X, Neupert W, Tzagoloff A (2007) The metalloprotease encoded by ATP23 has a dual function in processing and assembly of subunit 6 of mitochondrial ATPase. Mol Biol Cell 18(2):617–626
59. Esser K, Tursun B, Ingenhoven M, Michaelis G et al (2002) A novel two-step mechanism for removal of a mitochondrial signal sequence involves the mAAA complex and the putative rhomboid protease Pcp1. J Mol Biol 323(5):835–843
60. Herlan M, Vogel F, Bornhovd C, Neupert W et al (2003) Processing of Mgm1 by the rhomboid-type protease Pcp1 is required for maintenance of mitochondrial morphology and of mitochondrial DNA. J Biol Chem 278(30):27781–27788
61. McQuibban GA, Saurya S, Freeman M (2003) Mitochondrial membrane remodelling regulated by a conserved rhomboid protease. Nature 423(6939):537–541
62. Sesaki H, Southard SM, Yaffe MP, Jensen RE (2003) Mgm1p, a dynamin-related GTPase, is essential for fusion of the mitochondrial outer membrane. Mol Biol Cell 14(6):2342–2356
63. Tatsuta T, Augustin S, Nolden M, Friedrichs B et al (2007) m-AAA protease-driven membrane dislocation allows intramembrane cleavage by rhomboid in mitochondria. EMBO J 26(2):325–335
64. Hartl FU, Bracher A, Hayer-Hartl M (2011) Molecular chaperones in protein folding and proteostasis. Nature 475(7356):324–332
65. Voos W, Rottgers K (2002) Molecular chaperones as essential mediators of mitochondrial biogenesis. Biochim Biophys Acta 1592(1):51–62

66. Marom M, Azem A, Mokranjac D (2011) Understanding the molecular mechanism of protein translocation across the mitochondrial inner membrane: still a long way to go. Biochim Biophys Acta 1808(3):990–1001
67. Voos W (2009) Mitochondrial protein homeostasis: the cooperative roles of chaperones and proteases. Res Microbiol 160(9):718–725
68. Truscott KN, Lowth BR, Strack PR, Dougan DA (2010) Diverse functions of mitochondrial AAA+ proteins: protein activation, disaggregation, and degradation. Biochem Cell Biol 88(1):97–108
69. van Dyck L, Dembowski M, Neupert W, Langer T (1998) Mcx1p, a ClpX homologue in mitochondria of Saccharomyces cerevisiae. FEBS Lett 438(3):250–254
70. Craig EA, Marszalek J (2002) A specialized mitochondrial molecular chaperone system: a role in formation of Fe/S centers. Cell Mol Life Sci 59(10):1658–1665
71. Reinders J, Sickmann A (2007) Proteomics of yeast mitochondria. Methods Mol Biol 372:543–557
72. Sliwa D, Dairou J, Camadro JM, Santos R (2012) Inactivation of mitochondrial aspartate aminotransferase contributes to the respiratory deficit of yeast frataxin-deficient cells. Biochem J 441(3):945–953
73. Erzberger JP, Berger JM (2006) Evolutionary relationships and structural mechanisms of AAA+ proteins. Annu Rev Biophys Biomol Struct 35:93–114
74. Ogura T, Wilkinson AJ (2001) AAA+ superfamily ATPases: common structure–diverse function. Genes Cells 6(7):575–597
75. Van Melderen L, Aertsen A (2009) Regulation and quality control by Lon-dependent proteolysis. Res Microbiol 160(9):645–651
76. Goldberg AL (1992) The mechanism and functions of ATP-dependent proteases in bacterial and animal cells. Eur J Biochem 203(1–2):9–23
77. Suzuki CK, Suda K, Wang N, Schatz G (1994) Requirement for the yeast gene LON in intramitochondrial proteolysis and maintenance of respiration. Science 264(5161):891
78. Van Dyck L, Pearce DA, Sherman F (1994) PIM1 encodes a mitochondrial ATP-dependent protease that is required for mitochondrial function in the yeast Saccharomyces cerevisiae. J Biol Chem 269(1):238–242
79. Wagner I, Arlt H, van Dyck L, Langer T et al (1994) Molecular chaperones cooperate with PIM1 protease in the degradation of misfolded proteins in mitochondria. EMBO J 13(21):5135–5145
80. Savel'ev AS, Novikova LA, Kovaleva IE, Luzikov VN et al (1998) ATP-dependent proteolysis in mitochondria. m-AAA protease and PIM1 protease exert overlapping substrate specificities and cooperate with the mtHsp70 system. J Biol Chem 273(32):20596–20602
81. Hanson PI, Whiteheart SW (2005) AAA+ proteins: have engine, will work. Nat Rev Mol Cell Biol 6(7):519–529
82. Striebel F, Kress W, Weber-Ban E (2009) Controlled destruction: AAA+ ATPases in protein degradation from bacteria to eukaryotes. Curr Opin Struct Biol 19(2):209–217
83. Sauer RT, Baker TA (2011) AAA+ proteases: ATP-fueled machines of protein destruction. Annu Rev Biochem 80:587–612
84. Botos I, Melnikov EE, Cherry S, Tropea JE et al (2004) The catalytic domain of Escherichia coli Lon protease has a unique fold and a Ser-Lys dyad in the active site. J Biol Chem 279(9):8140–8148
85. Wagner I, van Dyck L, Savel'ev AS, Neupert W et al (1997) Autocatalytic processing of the ATP-dependent PIM1 protease: crucial function of a pro-region for sorting to mitochondria. EMBO J 16(24):7317–7325
86. Stahlberg H, Kutejova E, Suda K, Wolpensinger B et al (1999) Mitochondrial Lon of Saccharomyces cerevisiae is a ring-shaped protease with seven flexible subunits. Proc Natl Acad Sci U S A 96(12):6787–6790
87. Cha SS, An YJ, Lee CR, Lee HS et al (2010) Crystal structure of Lon protease: molecular architecture of gated entry to a sequestered degradation chamber. EMBO J 29(20):3520–3530
88. Ondrovicova G, Liu T, Singh K, Tian B et al (2005) Cleavage site selection within a folded substrate by the ATP-dependent lon protease. J Biol Chem 280(26):25103–25110

89. Rep M, van Dijl JM, Suda K, Schatz G et al (1996) Promotion of mitochondrial membrane complex assembly by a proteolytically inactive yeast Lon. Science 274(5284):103–106
90. van Dijl JM, Kutejova E, Suda K, Perecko D et al (1998) The ATPase and protease domains of yeast mitochondrial Lon: roles in proteolysis and respiration-dependent growth. Proc Natl Acad Sci U S A 95(18):10584–10589
91. Major T, von Janowsky B, Ruppert T, Mogk A et al (2006) Proteomic analysis of mitochondrial protein turnover: identification of novel substrate proteins of the matrix protease pim1. Mol Cell Biol 26(3):762–776
92. Moradas-Ferreira P, Costa V, Piper P, Mager W (1996) The molecular defences against reactive oxygen species in yeast. Mol Microbiol 19(4):651–658
93. Raha S, Robinson BH (2000) Mitochondria, oxygen free radicals, disease and ageing. Trends Biochem Sci 25(10):502–508
94. Halliwell B (2006) Oxidative stress and neurodegeneration: where are we now? J Neurochem 97(6):1634–1658
95. Bender T, Leidhold C, Ruppert T, Franken S et al (2010) The role of protein quality control in mitochondrial protein homeostasis under oxidative stress. Proteomics 10(7):1426–1443
96. Bota DA, Davies KJ (2002) Lon protease preferentially degrades oxidized mitochondrial aconitase by an ATP-stimulated mechanism. Nat Cell Biol 4(9):674–680
97. Bota DA, Davies KJ (2001) Protein degradation in mitochondria: implications for oxidative stress, aging and disease: a novel etiological classification of mitochondrial proteolytic disorders. Mitochondrion 1(1):33–49
98. Nystrom T (2005) Role of oxidative carbonylation in protein quality control and senescence. EMBO J 24(7):1311–1317
99. Bukau B, Weissman J, Horwich A (2006) Molecular chaperones and protein quality control. Cell 125(3):443–451
100. Leonhardt SA, Fearson K, Danese PN, Mason TL (1993) HSP78 encodes a yeast mitochondrial heat shock protein in the Clp family of ATP-dependent proteases. Mol Cell Biol 13(10):6304–6313
101. Liberek K, Lewandowska A, Zietkiewicz S (2008) Chaperones in control of protein disaggregation. EMBO J 27(2):328–335
102. Weibezahn J, Tessarz P, Schlieker C, Zahn R et al (2004) Thermotolerance requires refolding of aggregated proteins by substrate translocation through the central pore of ClpB. Cell 119(5):653–665
103. von Janowsky B, Major T, Knapp K, Voos W (2006) The disaggregation activity of the mitochondrial ClpB homolog Hsp78 maintains Hsp70 function during heat stress. J Mol Biol 357(3):793–807
104. Germaniuk A, Liberek K, Marszalek J (2002) A bichaperone (Hsp70-Hsp78) system restores mitochondrial DNA synthesis following thermal inactivation of Mip1p polymerase. J Biol Chem 277(31):27801–27808
105. Rottgers K, Zufall N, Guiard B, Voos W (2002) The ClpB homolog Hsp78 is required for the efficient degradation of proteins in the mitochondrial matrix. J Biol Chem 277(48):45829–45837
106. Bateman JM, Iacovino M, Perlman PS, Butow RA (2002) Mitochondrial DNA instability mutants of the bifunctional protein Ilv5p have altered organization in mitochondria and are targeted for degradation by Hsp78 and the Pim1p protease. J Biol Chem 277(49):47946–47953
107. Morimoto RI (2008) Proteotoxic stress and inducible chaperone networks in neurodegenerative disease and aging. Genes Dev 22(11):1427–1438
108. Bender T, Lewrenz I, Franken S, Baitzel C et al (2011) Mitochondrial enzymes are protected from stress-induced aggregation by mitochondrial chaperones and the Pim1/LON protease. Mol Biol Cell 22(5):541–554
109. Elsasser S, Finley D (2005) Delivery of ubiquitinated substrates to protein-unfolding machines. Nat Cell Biol 7(8):742–749
110. Samanovic M, Li H, Darwin KH (2013) The Pup-Proteasome system of Mycobacterium tuberculosis. In: Dougan DA (ed) Regulated proteolysis in microorganisms. Springer, Subcell Biochem 66:267–295

111. Gottesman S, Roche E, Zhou Y, Sauer RT (1998) The ClpXP and ClpAP proteases degrade proteins with carboxy-terminal peptide tails added by the SsrA-tagging system. Genes Dev 12(9):1338–1347
112. Moore SD, Sauer RT (2007) The tmRNA system for translational surveillance and ribosome rescue. Annu Rev Biochem 76:101–124
113. von Janowsky B, Knapp K, Major T, Krayl M et al (2005) Structural properties of substrate proteins determine their proteolysis by the mitochondrial AAA+ protease Pim1. Biol Chem 386(12):1307–1317
114. Herman C, Prakash S, Lu CZ, Matouschek A et al (2003) Lack of a robust unfoldase activity confers a unique level of substrate specificity to the universal AAA protease FtsH. Mol Cell 11(3):659–669
115. Prakash S, Matouschek A (2004) Protein unfolding in the cell. Trends Biochem Sci 29(11):593–600
116. van Dyck L, Neupert W, Langer T (1998) The ATP-dependent PIM1 protease is required for the expression of intron-containing genes in mitochondria. Genes Dev 12(10):1515–1524
117. Liu T, Lu B, Lee I, Ondrovicova G et al (2004) DNA and RNA binding by the mitochondrial lon protease is regulated by nucleotide and protein substrate. J Biol Chem 279(14):13902–13910
118. Lu B, Yadav S, Shah PG, Liu T et al (2007) Roles for the human ATP-dependent Lon protease in mitochondrial DNA maintenance. J Biol Chem 282(24):17363–17374
119. Chen XJ, Wang X, Butow RA (2007) Yeast aconitase binds and provides metabolically coupled protection to mitochondrial DNA. Proc Natl Acad Sci U S A 104(34):13738–13743
120. Teichmann U, van Dyck L, Guiard B, Fischer H et al (1996) Substitution of PIM1 protease in mitochondria by Escherichia coli Lon protease. J Biol Chem 271(17):10137–10142
121. Balaban RS, Nemoto S, Finkel T (2005) Mitochondria, oxidants, and aging. Cell 120(4):483–495
122. Luce K, Weil AC, Osiewacz HD (2010) Mitochondrial protein quality control systems in aging and disease. Adv Exp Med Biol 694:108–125
123. Luce K, Osiewacz HD (2009) Increasing organismal healthspan by enhancing mitochondrial protein quality control. Nat Cell Biol 11(7):852–858
124. Guelin E, Rep M, Grivell LA (1996) Afg3p, a mitochondrial ATP-dependent metalloprotease, is involved in degradation of mitochondrially-encoded Cox1, Cox3, Cob, Su6, Su8 and Su9 subunits of the inner membrane complexes III, IV and V. FEBS Lett 381(1–2):42–46
125. Weber ER, Hanekamp T, Thorsness PE (1996) Biochemical and functional analysis of the YME1 gene product, an ATP and zinc-dependent mitochondrial protease from S. cerevisiae. Mol Biol Cell 7(2):307–317
126. Okuno T, Ogura T (2013) FtsH protease-mediated regulation of various cellular functions. In: Dougan DA (ed) Regulated proteolysis in microorganisms. Springer, Subcell Biochem 66:53–69
127. Koppen M, Metodiev MD, Casari G, Rugarli EI et al (2007) Variable and tissue-specific subunit composition of mitochondrial m-AAA protease complexes linked to hereditary spastic paraplegia. Mol Cell Biol 27(2):758–767
128. Guelin E, Rep M, Grivell LA (1994) Sequence of the AFG3 gene encoding a new member of the FtsH/Yme1/Tma subfamily of the AAA-protein family. Yeast 10(10):1389–1394
129. Schnall R, Mannhaupt G, Stucka R, Tauer R et al (1994) Identification of a set of yeast genes coding for a novel family of putative ATPases with high similarity to constituents of the 26S protease complex. Yeast 10(9):1141–1155
130. Thorsness PE, White KH, Fox TD (1993) Inactivation of YME1, a member of the ftsH-SEC18-PAS1-CDC48 family of putative ATPase-encoding genes, causes increased escape of DNA from mitochondria in Saccharomyces cerevisiae. Mol Cell Biol 13(9):5418–5426
131. Tzagoloff A, Yue J, Jang J, Paul MF (1994) A new member of a family of ATPases is essential for assembly of mitochondrial respiratory chain and ATP synthetase complexes in Saccharomyces cerevisiae. J Biol Chem 269(42):26144–26151
132. Thorsness PE, Fox TD (1993) Nuclear mutations in Saccharomyces cerevisiae that affect the escape of DNA from mitochondria to the nucleus. Genetics 134(1):21–28

133. Lee S, Augustin S, Tatsuta T, Gerdes F et al (2011) Electron cryomicroscopy structure of a membrane-anchored mitochondrial AAA protease. J Biol Chem 286(6):4404–4411
134. Gerdes F, Tatsuta T, Langer T (2012) Mitochondrial AAA proteases – towards a molecular understanding of membrane-bound proteolytic machines. Biochim Biophys Acta 1823(1):49–55
135. Pearce DA, Sherman F (1995) Diminished degradation of yeast cytochrome c by interactions with its physiological partners. Proc Natl Acad Sci U S A 92(9):3735–3739
136. Weil A, Luce K, Drose S, Wittig I et al (2011) Unmasking a temperature-dependent effect of the P. anserina i-AAA protease on aging and development. Cell Cycle 10(24):4280–4290
137. Graack HR, Bryant ML, O'Brien TW (1999) Identification of mammalian mitochondrial ribosomal proteins (MRPs) by N-terminal sequencing of purified bovine MRPs and comparison to data bank sequences: the large subribosomal particle. Biochemistry 38(50): 16569–16577
138. Graack HR, Wittmann-Liebold B (1998) Mitochondrial ribosomal proteins (MRPs) of yeast. Biochem J 329(Pt 3):433–448
139. Harms J, Schluenzen F, Zarivach R, Bashan A et al (2001) High resolution structure of the large ribosomal subunit from a mesophilic eubacterium. Cell 107(5):679–688
140. Arlt H, Steglich G, Perryman R, Guiard B et al (1998) The formation of respiratory chain complexes in mitochondria is under the proteolytic control of the m-AAA protease. EMBO J 17(16):4837–4847
141. Casari G, De Fusco M, Ciarmatori S, Zeviani M et al (1998) Spastic paraplegia and OXPHOS impairment caused by mutations in paraplegin, a nuclear-encoded mitochondrial metalloprotease. Cell 93(6):973–983
142. Di Bella D, Lazzaro F, Brusco A, Plumari M et al (2010) Mutations in the mitochondrial protease gene AFG3L2 cause dominant hereditary ataxia SCA28. Nat Genet 42(4):313–321
143. Potting C, Wilmes C, Engmann T, Osman C et al (2010) Regulation of mitochondrial phospholipids by Ups1/PRELI-like proteins depends on proteolysis and Mdm35. EMBO J 29(17):2888–2898
144. Osman C, Haag M, Potting C, Rodenfels J et al (2009) The genetic interactome of prohibitins: coordinated control of cardiolipin and phosphatidylethanolamine by conserved regulators in mitochondria. J Cell Biol 184(4):583–596
145. Sesaki H, Dunn CD, Iijima M, Shepard KA et al (2006) Ups1p, a conserved intermembrane space protein, regulates mitochondrial shape and alternative topogenesis of Mgm1p. J Cell Biol 173(5):651–658
146. Tamura Y, Endo T, Iijima M, Sesaki H (2009) Ups1p and Ups2p antagonistically regulate cardiolipin metabolism in mitochondria. J Cell Biol 185(6):1029–1045
147. Tamura Y, Iijima M, Sesaki H (2010) Mdm35p imports Ups proteins into the mitochondrial intermembrane space by functional complex formation. EMBO J 29(17):2875–2887
148. Dunn CD, Lee MS, Spencer FA, Jensen RE (2006) A genomewide screen for petite-negative yeast strains yields a new subunit of the i-AAA protease complex. Mol Biol Cell 17(1): 213–226
149. Dunn CD, Tamura Y, Sesaki H, Jensen RE (2008) Mgr3p and Mgr1p are adaptors for the mitochondrial i-AAA protease complex. Mol Biol Cell 19(12):5387–5397
150. Dailey TA, Woodruff JH, Dailey HA (2005) Examination of mitochondrial protein targeting of haem synthetic enzymes: in vivo identification of three functional haem-responsive motifs in 5-aminolaevulinate synthase. Biochem J 386(Pt 2):381–386
151. Handschin C, Lin J, Rhee J, Peyer AK, Chin S, Wu PH, Meyer UA, Spiegelman BM (2005) Nutritional regulation of hepatic heme biosynthesis and porphyria through PGC-1alpha. Cell 122(4):505–515
152. Tian Q, Li T, Hou W, Zheng J et al (2011) Lon peptidase 1 (LONP1)-dependent breakdown of mitochondrial 5-aminolevulinic acid synthase protein by heme in human liver cells. J Biol Chem 286(30):26424–26430
153. Yoshino K, Munakata H, Kuge O, Ito A et al (2007) Haeme-regulated degradation of delta-aminolevulinate synthase 1 in rat liver mitochondria. J Biochem 142(4):453–458

154. Matsushima Y, Goto Y, Kaguni LS (2010) Mitochondrial Lon protease regulates mitochondrial DNA copy number and transcription by selective degradation of mitochondrial transcription factor A (TFAM). Proc Natl Acad Sci U S A 107(43):18410–18415
155. Hallberg BM, Larsson NG (2011) TFAM forces mtDNA to make a U-turn. Nat Struct Mol Biol 18(11):1179–1181
156. Granot Z, Kobiler O, Melamed-Book N, Eimerl S et al (2007) Turnover of mitochondrial steroidogenic acute regulatory (StAR) protein by Lon protease: the unexpected effect of proteasome inhibitors. Mol Endocrinol 21(9):2164–2177
157. Fukuda R, Zhang H, Kim JW, Shimoda L et al (2007) HIF-1 regulates cytochrome oxidase subunits to optimize efficiency of respiration in hypoxic cells. Cell 129(1):111–122
158. Tobias JW, Shrader TE, Rocap G, Varshavsky A (1991) The N-end rule in bacteria. Science 254(5036):1374–1377
159. Ninnis RL, Spall SK, Talbo GH, Truscott KN et al (2009) Modification of PATase by L/F-transferase generates a ClpS-dependent N-end rule substrate in Escherichia coli. EMBO J 28(12):1732–1744
160. Erbse A, Schmidt R, Bornemann T, Schneider-Mergener J et al (2006) ClpS is an essential component of the N-end rule pathway in Escherichia coli. Nature 439(7077):753–756
161. Schmidt R, Zahn R, Bukau B, Mogk A (2009) ClpS is the recognition component for Escherichia coli substrates of the N-end rule degradation pathway. Mol Microbiol 72(2):506–517
162. Schuenemann VJ, Kralik SM, Albrecht R, Spall SK et al (2009) Structural basis of N-end rule substrate recognition in Escherichia coli by the ClpAP adaptor protein ClpS. EMBO Rep 10(5):508–514
163. Dougan DA, Micevski D, Truscott KN (2012) The N-end rule pathway: From recognition by N-recognins, to destruction by AAA+ proteases. Biochim Biophys Acta 1823(1):83–91
164. Sriram SM, Kim BY, Kwon YT (2011) The N-end rule pathway: emerging functions and molecular principles of substrate recognition. Nat Rev Mol Cell Biol 12(11):735–747
165. Benedetti C, Haynes CM, Yang Y, Harding HP et al (2006) Ubiquitin-like protein 5 positively regulates chaperone gene expression in the mitochondrial unfolded protein response. Genetics 174(1):229–239
166. Ron D, Walter P (2007) Signal integration in the endoplasmic reticulum unfolded protein response. Nat Rev Mol Cell Biol 8(7):519–529
167. Vabulas RM, Raychaudhuri S, Hayer-Hartl M, Hartl FU (2010) Protein folding in the cytoplasm and the heat shock response. Cold Spring Harb Perspect Biol 2(12):a004390
168. Yoneda T, Benedetti C, Urano F, Clark SG et al (2004) Compartment-specific perturbation of protein handling activates genes encoding mitochondrial chaperones. J Cell Sci 117(Pt 18):4055–4066
169. Zhao Q, Wang J, Levichkin IV, Stasinopoulos S et al (2002) A mitochondrial specific stress response in mammalian cells. EMBO J 21(17):4411–4419
170. Barchinger SE, Ades SE (2013) Regulated proteolysis: control of the Escherichia coli σE-dependent cell envelope stress response. In: Dougan DA (ed) Regulated proteolysis in microorganisms. Springer, Dordrecht, pp xxx–xxx
171. Micevski D, Dougan DA (2013) Proteolytic regulation of stress response pathways in Escherichia coli. In: Dougan DA (ed) Regulated proteolysis in microorganisms. Springer, Subcell Biochem 66:105–128
172. Molière N, Turgay K (2013) General and regulatory proteolysis in Bacillus subtilis. In: Dougan DA (ed) Regulated proteolysis in microorganisms. Springer, Subcell Biochem 66:73–103
173. Buchberger A, Bukau B, Sommer T (2010) Protein quality control in the cytosol and the endoplasmic reticulum: brothers in arms. Mol Cell 40(2):238–252
174. Martinus RD, Garth GP, Webster TL, Cartwright P et al (1996) Selective induction of mitochondrial chaperones in response to loss of the mitochondrial genome. Eur J Biochem 240(1):98–103

175. Kirstein-Miles J, Morimoto RI (2010) Peptides signal mitochondrial stress. Cell Metab 11(3):177–178
176. Haynes CM, Ron D (2010) The mitochondrial UPR – protecting organelle protein homeostasis. J Cell Sci 123(Pt 22):3849–3855
177. Young L, Leonhard K, Tatsuta T, Trowsdale J et al (2001) Role of the ABC transporter Mdl1 in peptide export from mitochondria. Science 291(5511):2135–2138
178. Truscott KN, Bezawork-Geleta A, Dougan DA (2011) Unfolded protein responses in bacteria and mitochondria: a central role for the ClpXP machine. IUBMB Life 63(11):955–963
179. Shirihai OS, Gregory T, Yu C, Orkin SH et al (2000) ABC-me: a novel mitochondrial transporter induced by GATA-1 during erythroid differentiation. EMBO J 19(11):2492–2502
180. Horibe T, Hoogenraad NJ (2007) The chop gene contains an element for the positive regulation of the mitochondrial unfolded protein response. PLoS One 2(9):e835
181. Aldridge JE, Horibe T, Hoogenraad NJ (2007) Discovery of genes activated by the mitochondrial unfolded protein response (mtUPR) and cognate promoter elements. PLoS One 2(9):e874

Part IV
Ubiquitin-Like Protein Modification and Protein Degradation in Microorganisms

Chapter 10
The Pup-Proteasome System of *Mycobacterium tuberculosis*

Marie I. Samanovic, Huilin Li, and K. Heran Darwin

Abstract Proteasomes are ATP-dependent protein degradation machines present in all archaea and eukaryotes, and found in several bacterial species of the order Actinomycetales. *Mycobacterium tuberculosis* (*Mtb*), an Actinomycete pathogenic to humans, requires proteasome function to cause disease. In this chapter, we describe what is currently understood about the biochemistry of the *Mtb* proteasome and its role in virulence. The characterization of the *Mtb* proteasome has led to the discovery that proteins can be targeted for degradation by a small protein modifier in bacteria as they are in eukaryotes. Furthermore, the understanding of proteasome function in *Mtb* has helped reveal new insight into how the host battles infections.

Introduction

Mycobacterium tuberculosis (*Mtb*) is the causative agent of tuberculosis (TB) and kills nearly two million people every year (http://www.who.int/). The infectious process starts with the inhalation of air-borne droplets containing *Mtb* bacilli. Bacteria replicate in professional phagocytes in the lungs where they must combat numerous anti-microbial molecules. If the host cannot control the infection, *Mtb* growth will result in the destruction of lung tissues and, ultimately, the death of the host.

M.I. Samanovic • K.H. Darwin (✉)
Department of Microbiology, New York University School of Medicine,
550 First Avenue, MSB 236, New York, NY 10016, USA
e-mail: heran.darwin@med.nyu.edu

H. Li
Department of Biochemistry and Cell Biology, Stony Brook University,
Stony Brook, NY 11794, USA

Brookhaven National Laboratory Biology Department, Brookhaven National Laboratory,
50 Bell Ave, Upton, Brookhaven, NY 11973-5000, USA

D.A. Dougan (ed.), *Regulated Proteolysis in Microorganisms*, Subcellular Biochemistry 66,
DOI 10.1007/978-94-007-5940-4_10,

Despite the astounding mortality caused by TB, most individuals infected with *Mtb* can control mycobacterial growth for much of their lives. Among the host's arsenal of antimicrobial effectors is nitric oxide (NO), which is produced by activated macrophages and is toxic to numerous microbes [1]. Evidence that supports the notion that NO is critical to controlling *Mtb* has come from mouse studies. Inactivation of the macrophage associated inducible NO synthase, (iNOS) also known as NOS2, dramatically sensitizes mice to *Mtb* infections [2]. The cytotoxic effects of NO are likely to be dependent on the formation of highly reactive nitrogen intermediates (RNIs). It is thought that in host cells NO is oxidized to nitrite, which can be protonated to nitrous acid in the phagosomes of activated macrophages. Nitrous acid dismutates to reform NO, which can penetrate bacterial membranes and cell walls to combine with reactive oxygen intermediates (ROIs) such as superoxide to generate peroxynitrite. RNIs and ROIs can induce lethal injuries including DNA and protein damage as well as lipid peroxidation [3, 4].

Regardless of the apparent protective effects of host-produced NO during *Mtb* infections, humans, as well as experimentally infected animals, are rarely sterilized of *Mtb* [1, 2]. This observation was the basis for the hypothesis that *Mtb* encodes proteins required for resistance to NO toxicity. Due to the emergence of multi-drug resistant (MDR) and extensively-drug resistant (XDR) *Mtb* strains, researchers around the world are looking for novel ways to target TB. Drugs that inhibit bacterial defenses against mammalian antimicrobial effectors like NO could help the host win the war against this disease.

It has long been a technical challenge to identify and characterize pathways important for the pathogenesis of *Mtb*, a highly infectious and slow growing (doubling time ~20 h) Actinobacterium with a high GC-content. Over the last 20 years, improved molecular genetic tools and bio-safety provisions have greatly facilitated studies into understanding the pathogenesis of this challenging organism. With the advent of efficient transposon mutagenesis, it became feasible to perform a screen to identify genes required for NO resistance *in vitro*. After screening over 10,000 transposon mutants for NO sensitivity, Nathan and colleagues identified five *Mtb* mutants of the virulent laboratory strain H37Rv with independent insertions in Rv2115c and Rv2097c, two genes that were predicted to be associated with proteasome function [5]. Rv2115c was named *mpa* (*Mycobacterium* proteasomal ATPase) due to its high similarity with eukaryotic and archaeal proteasomal ATPases [6]. Mpa is 81 % identical to ARC (AAA ATPase forming ring-shaped complexes) of *Rhodococcus erythropolis*, the first biochemically characterized bacterial proteasomal ATPase [7]. In contrast to Rv2115c/Mpa, Rv2097c did not exhibit similarity to any known proteins at the time, however, was proposed to participate in proteasomal function, and as such was named *paf* for proteasome accessory factor (later termed *pafA*). Importantly, mutations in *mpa* and *pafA* severely attenuate *Mtb* virulence in mice [5].

Proteasomes are multi-subunit barrel-shaped protease complexes that were first discovered in eukaryotes over 20 years ago [8]. In eukaryotes the 26S proteasome is composed of two functionally distinct sub-complexes: the 20S core particle (CP), required for degradation of the substrate, and a 19S regulatory particle (RP) located at either or both ends of the CP, responsible for substrate unfolding and translocation into the CP [9, 10]. The RP is composed of numerous proteins, the composition of which varies depending on its function. The 19S RP contains 19 subunits, including a ring of six distinct AAA (ATPases associated with different cellular activities) proteins that contact the CP, and non-ATPase subunits, which function in various aspects of substrate recognition and processing [11]. The CP is composed of four stacked rings with catalytic activity located within the central rings. The two inner rings are composed of seven distinct catalytic β-subunits sandwiched between two outer rings composed of seven distinct α-subunits [10]. The β-subunits have several proteolytic activities that allow the proteasome to cleave most types of peptide bonds. Protein fragments are estimated to range in size from 8 to 10 residues [12]. The α-subunit rings form a gated channel that controls the passage of substrates and cleaved peptides, and also serves as a docking surface for protein complexes such as the RP [10, 13].

Proteasomes are enzymatically and structurally distinct from the ATP-dependent, chambered bacterial proteases ClpP, Lon, FtsH and HslV [14, 15]. The first clue that *bona fide* proteasomes were present in prokaryotes came from electron microscopy studies on the thermoacidophilic archaeon, *Thermoplasma acidophilum*, in which CP-like particles were obtained from *T. acidophilum* lysates [16]. Ultimately, *T. acidophilum* CPs were purified and crystallized, and shown to be highly similar in structure to eukaryotic CPs [17]. The first bacterial proteasome to be characterized was the *R. erythropolis* proteasome [18]. Later, 20S proteasomes were characterized from *Mycobacterium smegmatis* [19], *Streptomyces coelicolor* [20] and *Frankia* [21]. Additionally, genomic sequencing revealed the presence of proteasomal genes in the pathogens *Mtb* [22] and *Mycobacterium leprae* [23]. Bacterial proteasomes were thought to be confined to Actinobacteria until studies from the Banfield group (reviewed in [24]), discovered two actinomycete-like proteasome genes clusters in a non-culturable Gram-negative bacterium called *Leptospirillum*. To date, Actinomycetes and *Leptospirillum* are the only known bacterial lineages with a proteasome system, and may have acquired this protease complex via lateral gene transfer events [24, 25]. In contrast to the eukaryotic CPs, most prokaryotic CPs are composed of homo-heptameric rings; two β-subunits (PrcB) rings, flanked by two α-subunit (PrcA) rings (reviewed in [26, 27]). For the most part, the presence of only one type of β-subunit limits the proteolytic activity of the prokaryotic 20S CP to chymotryptic activity.

Since the initial identification by Darwin *et al.* of genes required for NO resistance in *Mtb* [5], it was later shown (using two separate *prcBA* mutant *Mtb* strains) that the CP was also needed for resistance to NO [28, 29]. This provided evidence of a functional link between Mpa, PafA and the CP. However, when compared to the wild type, *mpa* or *pafA Mtb* strains, the *prcBA* mutants grow much more slowly in rich broth (~20–30 % lower optical density at stationary phase) and take longer to form colonies on solid media [28, 29]. This growth defect in these genetically manipulated strains is similar to that observed for wild type *Mtb* strains treated with a mammalian proteasome inhibitor, *N*-(4-morpholine)carbonyl-b-(1-naphthyl)-L-alanine-L-leucine boronic acid (MLN-273) or epoxomicin [5]. Collectively, these data suggest that the *Mtb* CP can also degrade proteins in an Mpa/PafA independent manner; or that the CP has other, possibly protease-independent functions important for growth (to be discussed later). The notion that the *Mtb* CP is needed for normal growth is supported by a study that attempted to delineate genes essential for *Mtb* growth *in vitro.* In this study, Rubin and colleagues found that *prcBA*, but not *mpa* or *pafA*, were required for normal growth *in vitro* [30], therefore it is, not surprising that *prcBA*-defective *Mtb* strains are also highly attenuated in a mouse model of infection [28, 29].

The identification of a bacterial proteasome associated with a virulence phenotype piqued the interest of numerous laboratories to better characterize this protease. Here, we summarize what is currently understood about the structure and function of the *Mtb* proteasome and discuss its potential roles in pathogenesis.

Mycobacterium tuberculosis Proteasome Structure and Function

Structure of the Mtb 20S Core Particle

The overall architecture of the *Mtb* 20S CP is similar to the CP of archaeal [31] and eukaryotic [32] proteasomes. All CPs form a barrel-shaped structure consisting of four stacked, seven-subunit rings that are arranged into two central β-rings, flanked by an α-ring at either end. Like other prokaryotic proteasomes, *Mtb* CPs are arranged into a four-ringed α7β7β7α7 cylinder composed of 14 identical α-subunits and 14 identical β-subunits, ~150 Å in height with a diameter of ~115 Å (Fig. 10.1a, b). The *Mtb* proteasome shares modest sequence identity with the archaeal CP from *Thermoplasma* (~32 % identity for both α- and β-subunits) and high identity with the bacterial CP from *Rhodococcus* (~65 % identity). Despite this, with the exception of helix 2 in both the α- and β-subunits, the three-dimensional structures of all three prokaryotic CPs are virtually super-imposable [33, 34]. Although, only the relative position of helix 2 is altered in the bacterial CPs, this small upward tilt of ~10° creates a wider axial substrate channel in the bacterial CPs when compared to the archaeal CPs [33].

The proteolytic active sites of the CP are housed in the β-subunits (in eukaryotes only three of the seven β-subunits contain functional sites: β-1, -2, and -5 [35]). These active subunits are synthesized with N-terminal pro-peptides, the autocleavage of which exposes the catalytic nucleophile, an N-terminal threonine (Thr-1) [36]. Similarly, in *Rhodococcus,* there are two β-subunits (β1 and β2) both of which are translated with long propeptides (65 and 59 residues respectively). In this case, recombinant *Rhodococcus* α- and β-subunits only assemble into an active proteasomes when all subunits are combined, while separately expressed components remain monomeric [34, 37–39]. Collectively, these findings suggest that the β-subunit propeptide not only facilitates the formation of the first assembly intermediate (the α/β hetero-dimer) but also shields the catalytic Thr residue during proteasomal assembly, preventing undesired protein degradation [37, 40]. In *Thermoplasma*, the

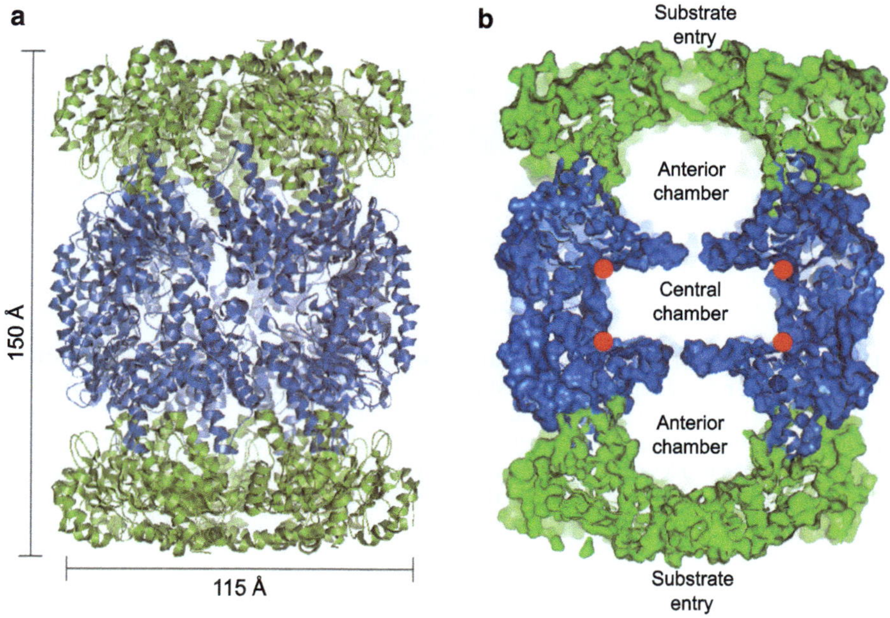

Fig. 10.1 Structure of the *Mtb* 20S core particle (CP). (**a**) Side view of the *Mtb* 20S CP structure rendered in cartoon view (PDB ID 2FHG). The two α-rings at the *top* and the *bottom* are displayed in *green*, and the two catalytic middle β-rings in *blue*. (**b**) The same side view as in (**a**), but the structure is rendered in surface display. Only a central slab of 30 Å is shown such that the two anterior chambers and one central chamber of the CP are visible. The *red circles* mark the proteolytic sites

eight amino acid propeptide of the β-subunit seems to be dispensable for assembly of the CP, as the α-subunits can assemble spontaneously into seven subunit rings when produced in *Escherichia coli* [41]. In contrast the α-ring appears to serve as a template for assembly of the β-ring in the formation of active CP in *Rhodococcus*. In *Mtb* the 56-residue propeptide of PrcB appears to inhibit rather than promote CP assembly [42], as a cryo-EM study of the *Mtb* half proteasome revealed that the propeptide is located outside of the β-ring rather than between the α- and β-rings [43]. Because assembly of the mature CP is a result of the apposition of two half-proteasomes, it seems that the *Mtb* β-propeptides, which are auto-catalytically removed, could be a barrier to the assembly process. However, it may be that simply more time is required to overcome this barrier.

Chamber of Doom: Core Protease Activity

Although several prokaryotic CPs characterized so far have significant peptidase activity [17, 40, 44, 45], the *in vitro* peptidase activity of *Mtb* CP is relatively low [42]. This low activity suggests that the substrate gate is closed in a manner similar

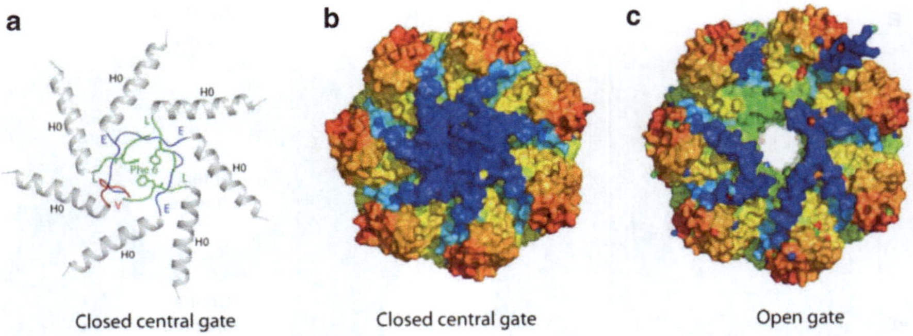

Fig. 10.2 Top view of the *Mtb* 20S CP. (**a**) and (**b**) Closed gate conformation (PDB ID 3MKA). The gate formed by the α-subunits is ordered and closed with a mechanism different to the eukaryotic CP. The seven subunits are chemically identical but adopt a total of three different conformations at their N-termini as indicated by the *different colors* in (**a**). A phenylalanine side chain from each of the green octapeptides contributes to the gate closure. (**c**) Deletion of the N-terminal residues 2–9 results in the open gate conformation (PDB ID 3MFE)

to the eukaryotic CP. In the eukaryotic CP, the substrate entrance to the α-ring is closed to prevent uncontrolled proteolytic activity [46]. In the eukaryotic CP, the gate to the catalytic chamber is blocked by the N-terminal sequences of the seven different α-subunits, which adopt different conformations and seal the entry portal [46]. In *Mtb*, a high-resolution crystal structure of the CP containing a mutant β-subunit (PrcB T1A, which prevents propeptide cleavage) revealed that the seven identical N-terminal peptides that form the gate were ordered, but exhibit three different conformations to tightly seal its substrate entrance at the seven-fold symmetry axis [43] (Fig. 10.2a, b). Deletion of the N-terminal residues 2-9 of PrcA results in an "open gate" conformation that increases peptidolytic activity to small peptides *in vitro* [42] (Fig. 10.2c).

All CPs are N-terminal Thr hydrolases [14, 35, 36]. While most prokaryotic CPs seem to exclusively hydrolyze hydrophobic small synthetic peptide substrates (so-called "chymotryptic" activity), the *Mtb* CP has broad substrate specificity, targeting not only hydrophobic targets but also basic ("tryptic" activity) and acidic peptides ("peptidyl-glutamyl-peptide-hydrolyzing", "caspase-like" or "post-acidic" activity) [42]. Unlike eukaryotic CPs in which substrate preference is determined by several different β-subunits [35, 47, 48], the *Mtb* 20S CP contains a single type of β-subunit, raising the question as to how it displays such broad substrate specificity. Structural analysis revealed that the substrate-binding pocket in the *Mtb* proteasome combines features found in different eukaryotic β-subunits: a hydrophobic upper surface similar to that of formed by a eukaryotic β5 subunit, and a hydrophilic lower surface similar to eukaryotic β1 and β2 subunits [33]. This composite feature of the substrate-binding pocket in the *Mtb* CP likely accounts for its broad substrate specificity [42].

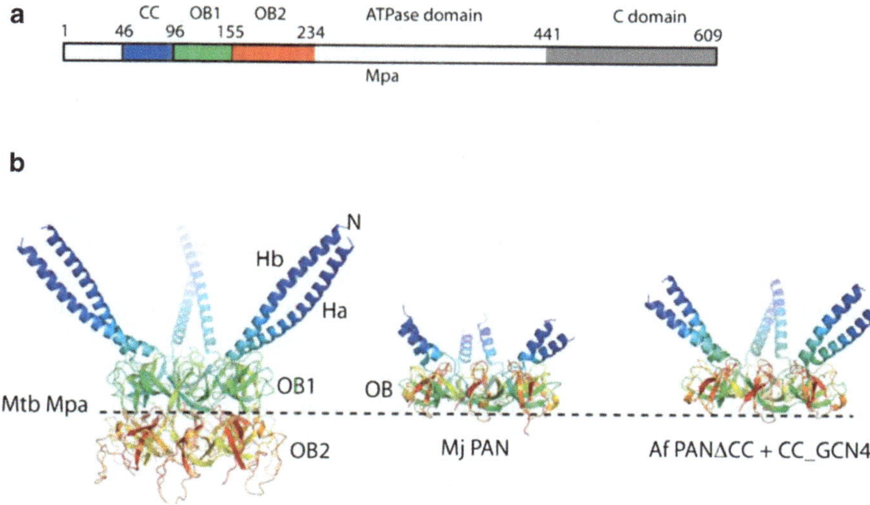

Fig. 10.3 Comparison of Mpa with PAN. (**a**) Linear map of the domain structure of Mpa. *CC*=coiled coil; *OB*=oligosaccharide/oligonucleotide binding domain. (**b**) The structures of *Mtb* Mpa1-234 (PDB ID 3M9B), *Archaeoglobus fulgidus* PAN-ΔCC_CC_{GCN4} hybrid (PDB ID 2WG5), and *M. jannaschii* PAN (PDB ID 3H43) are aligned and displayed individually (Figure adapted from [54])

Mpa: Gateway to Doom

Like other chambered proteases, proteasomal proteolysis requires an ATP-dependent chaperone to unfold structured proteins for delivery into the proteasome core where they are degraded. In archaea the best-characterized proteasomal ATPase is *Methanococcus jannaschii* PAN (proteasome activating nucleotidase) [49]. PAN can facilitate the degradation of artificial substrates by CPs [6, 49] but it remains to be determined how PAN recognizes archaeal proteins. Mpa forms homo-hexamers and is homologous to PAN. Like other AAA ATPases, Mpa has characteristic Walker A and B motifs for ATP binding and hydrolysis, respectively [50]. Mpa has relatively low ATPase activity: ATP hydrolysis is about four times slower than that of ARC, the Mpa orthologue in *Rhodococcus* (V_{max} of 62 versus 268 pmol min^{-1} μg^{-1}) [7, 51], or PAN [49]. As predicted (and will be discussed in detail below), a major function of Mpa is to deliver proteins into the CP for destruction.

The full-length structure of Mpa is currently unknown, however, crystal structure analysis of partial Mpa polypeptides has yielded highly informative insight into its activity. Mpa and ARC each contain two domains of the oligosaccharide/oligonucleotide-binding (OB) fold in tandem, with the second OB fold appearing to play a major role in Mpa oligomerization [52, 53] (Fig. 10.3a). In contrast, the archaeal PAN has only one OB domain [55], and it is not clear why Mpa and ARC have two. Immediately preceding the OB folds in Mpa is a 75 Å long α-helix [54] (Fig. 10.3b).

Remarkably, helices from neighbouring subunit pairs form a coiled coil, thereby reducing the six-fold symmetry at the intermediate OB domain region to three-fold. This structural feature may be important for the specific protein recognition and unfolding activity of this class of ATPases.

By analogy to the eukaryotic 26S proteasome, the prokaryotic proteasomal ATPases are also expected to physically interact with the CP in order to couple protein unfolding with delivery into the CP. A major distinction between the prokaryotic ATPases and eukaryotic 19S RP is that the 19S binds the CP with an affinity strong enough for the entire 26S complex (19S RP+20S CP) to be co-purified. In stark contrast, prokaryotic ATPases, despite strong phenotypic associations with CP activity *in vivo*, only bind CPs either weakly or transiently *in vitro* [52, 56]. *Methanococcus* PAN weakly interacts with *Thermoplasma* CPs [56], and Mpa can directly associate, although flexibly and weakly, with open-gate mutant *Mtb* CPs [52] and can degrade proteins [57]. Because Mpa only weakly interacts with an altered proteasome *in vitro*, the precise nature of how any bacterial proteasome interacts with its cognate ATPase *in vivo* is a mystery. It is notable that proteasomal ATPases from bacteria to mammals contain a "HbYX (hydrophobic amino acid-tyrosine-X) motif" at their C-termini, and this motif is crucial for degradation but not ATPase activity [51, 56, 58, 59]. This motif is needed for PAN-mediated activation of proteolysis [60] and is implicated in proteasome assembly in yeast [59]. Mpa also has a HbYX-like motif, which is critical for the degradation of proteins in *Mtb* [51, 58]. Like PAN, Mpa lacking this motif is impaired in its interaction with the *Mtb* 20S CP *in vitro* [57].

A Pup-y Tale

In eukaryotes, proteins that are destined for proteasomal degradation are usually post-translationally modified with the small protein ubiquitin (Ub) (reviewed in [61]). Ubiquitin is synthesized as part of a precursor protein, processed to form a 76 amino acid protein containing a C-terminal diglycine motif (Gly-Gly), with a compact β-grasp fold [62–64]. The C-terminal Gly is subject to a series of reactions that result in the conjugation of Ub to a Lys residue on the target protein [65–67]. In the first step, the C-terminal Gly of Ub is adenylated by an Ub activating enzyme (E1) using ATP. Ub is then transferred to the active site Cys residue on the E1 enzyme. Next, Ub is transferred to an Ub conjugating enzyme (E2) and delivered to an Ub ligase (E3), which catalyzes the formation of an isopeptide bond with a Lys residue on the target protein (Fig. 10.4). In eukaryotes there are numerous E2 activating enzymes and E3 ligases (which contain a variety of substrate binding activities) to provide substrate specificity to the Ub proteasome system (UPS) (reviewed in [68, 69]). In general, proteins that are targeted to the proteasome have several Ub molecules added to the substrate, usually resulting in the formation of polyubiquitin (polyUb) chains [66, 70, 71]. The polyUb chains are recognized by the RP of the proteasome, and removed by proteasome-associated deubiquitinases

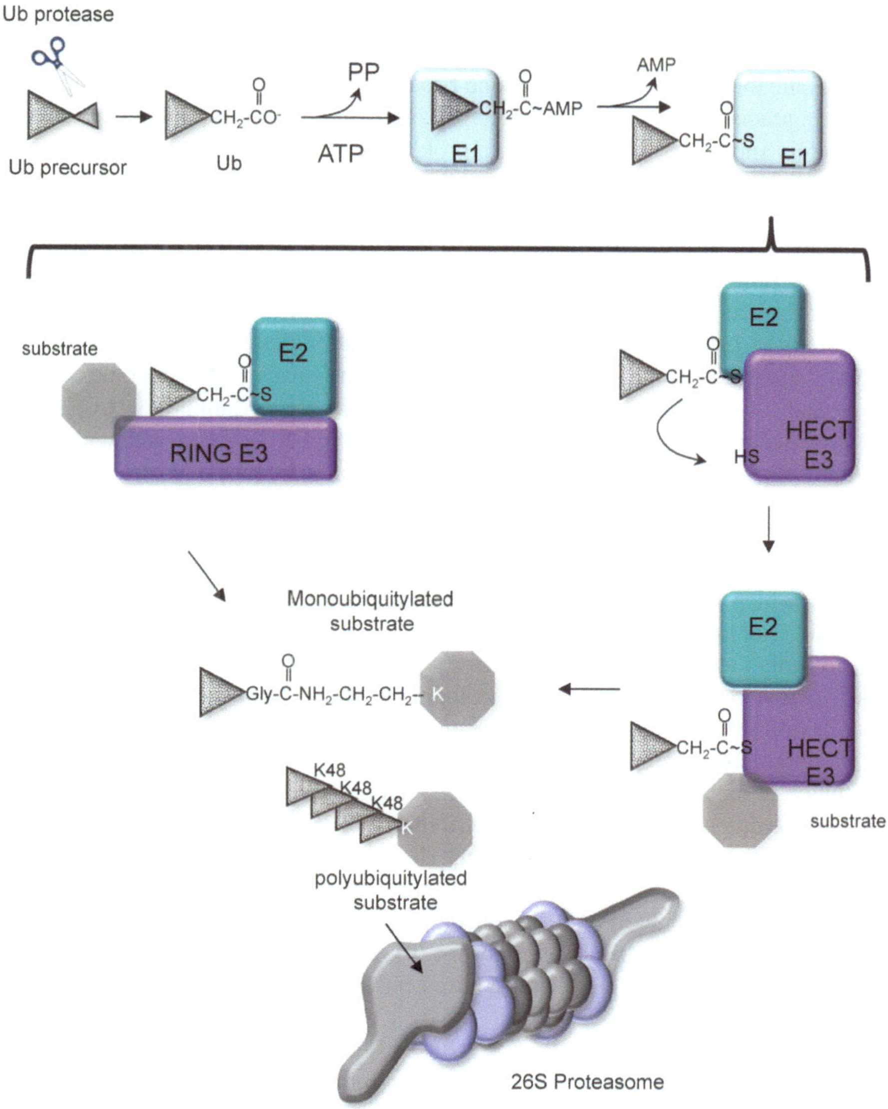

Fig. 10.4 Eukaryotic ubiquitin-proteasome system. Ubiquitin (Ub) is encoded as part of a larger polypeptide. Proteases expose a C-terminal Gly-Gly motif that is activated by adenylation with an E1 enzyme. The E1 enzyme transfers Ub to an E2 enzyme, where a thioester bond is formed. The E2 then transfers Ub to any number of E3 ligases. The E3 ligase family can be sub-divided into HECT and RING domain ligases: RING ligases interact with both the E2 and substrate, and facilitate the direct transfer of Ub from the E2 to the substrate. In contrast, HECT ligases form a thioester bond with Ub prior to transfer to a substrate Lys. E3 ligases dictate the type of Ub linkages that are formed. Ubiquitylated protein with Lys48 (K48) linked Ub chains are targeted for degradation by the 26S proteasome. Other types of Ub linkages (mono- and poly-K63 and others) generally do not result in degradation but serve other functions

(DUBs) for recycling of Ub. The deubiquitylated substrate is then delivered to the CP for destruction (reviewed in [11]).

Despite the presence of archaeal and bacterial CPs that are almost biochemically indistinguishable from the eukaryotic CP, an Ub-like system for the degradation of protein substrates was not found in prokaryotes for many years. A major hurdle at this time was an inability to reconstitute the proteolytic activity of bacterial proteasomes. This strongly suggested that other co-factors were required for proteolysis or that specific model substrates were required. It was speculated that bacterial proteasomal substrates only required intrinsic signals for degradation, as there was no evidence for the existence of any post-translational small protein modifiers in bacteria. To examine this, Darwin and colleagues set out to identify natural *Mtb* proteasome substrates by comparing the steady-state proteomes of wild type and *mpa Mtb* strains using two-dimensional polyacrylamide electrophoresis [58]. Although the proteomic profiles of untreated and NO-treated stationary phase cultures showed limited differences between the two *Mtb* strains, two proteins, FabD (malonyl CoA-acyl carrier protein transacylase) and PanB (ketopantoate hydroxymethyltransferase), accumulate significantly in the *mpa* strain. Importantly, the transcript levels *fabD* and *panB* are nearly identical in both wild type and *mpa Mtb* strains, which supported the notion that Mpa is required for FabD and PanB turnover, and not *fabD* or *panB* expression.

To further test if FabD and PanB were potential proteasome substrates, *Mtb fabD* and *panB* were expressed from a heterologous *M. bovis hsp60* promoter in *Mtb* [72]. In addition to ruling out potential differences in gene expression, this system would also address the possibility that *fabD* and *panB* mRNA had differences in translation initiation in the *mpa* or *pafA* strains. Each recombinant gene also encoded a FLAG and His_6 epitope tag at the N- and C-termini, respectively. Similar to that observed in 2D-PAGE analysis, both FLAG-FabD-His_6 and FLAG-PanB-His_6 accumulate in the *mpa* and *pafA* mutants [58]. Consistently, treatment of wild type *Mtb* with a eukaryotic proteasome inhibitor stabilized FabD and PanB [58] and deletion of the 20S CP permitted PanB accumulation [29]. Collectively, these data supported the idea that Mpa and PafA were required for the degradation of FabD and PanB by the CP.

In addition to the identification of FabD and PanB as proteasomal substrates, an unexpected observation was made during these studies: Mpa itself was also identified as a putative proteasomal substrate [58]. Chemical inhibition of the CP results in the accumulation of Mpa in *Mtb*. This finding was consistent with an earlier observation that mutations in *mpa*, which disrupt the ATPase activity or the HbYX motif, increase the steady-state levels of Mpa [51]. Similarly, Mpa accumulates in a *pafA* mutant strain, supporting a role for PafA in substrate degradation [58]. Thus it appears that the proteasome may “cannibalize” its ATPase to regulate its levels and hence activity.

Despite the identification of these putative substrates, their degradation, using purified Mpa and CP, could not be reconstituted *in vitro*. Therefore it was proposed that other factors were needed to facilitate proteasomal degradation. To identify these factors a bacterial two-hybrid screen [73] was used to search for proteins that bind to Mpa, with the reasoning that Mpa interacts not only with substrates but also with other proteins that promote proteolysis. From a library of ~100,000 *Mtb* genomic DNA fragments, Darwin and colleagues identified a protein encoded upstream of the CP genes *prcBA*, termed Rv2111c [74]. Importantly, recombinant Mpa and Rv2111c interacted *in vitro*, however, the addition of purified Rv2111c to CP and Mpa was unable to stimulate degradation of FabD.

The inability to reconstitute proteasomal degradation *in vitro* suggested that additional co-factors were still needed for proteolysis. Because *E. coli* does not encode a proteasome system it was reasoned that these co-factors were likely missing and possibly *Mycobacterium*-specific. A mycobacterial two-hybrid system [75] was thus used to interrogate interactions between proteasome subunits, substrates and the newly identified Rv2111c. In this system, *M. smegmatis* (*Msm*), a non-pathogenic relative of *Mtb*, was used as the host organism to identify protein-protein interactions. An unexpected interaction was detected between the substrate FabD and Rv2111c. To validate the genetic result, recombinant *fabD* and Rv2111c were co-expressed in *Msm* and found to co-purify as a heat stable complex. Mass spectrometry revealed that the C-terminal residue on Rv2111c formed an isopeptide bond with the side chain of Lys173 in FabD [74]. However, in contrast to Ub and related modifiers, Rv2111c did not have a C-terminal Gly-Gly motif, but instead has a Gly-Gly-Gln motif. Moreover, the C-terminal Gln was deamidated to Glu, before conjugation to FabD [74]. In a subsequent study, an orthologue of Rv2111c in *Msm* (MSMEG_3896) was identified as the protein modifier in that species. Thus, these two studies showed that this post-translational protein modification is conserved in both pathogenic and non-pathogenic mycobacteria [74, 76].

To determine if Rv2111c targeted proteins for degradation, Darwin and co-workers mutated the modified lysine residue (Lys173) in FabD [74]. Consistent with the idea that attachment with Rv2111c was the signal for degradation by the *Mtb* proteasome, mutagenesis of Lys173 to alanine in *Mtb* FabD stabilized the protein substrate. Based on its functional similarity to Ub, Rv2111c was named "Pup" for prokaryotic ubiquitin-like protein. Polyclonal antibodies to Pup recognize numerous proteins in *Mtb* H37Rv demonstrating that "pupylation" is widespread. Immunoblot analysis of the *pafA* mutant shows no anti-Pup reactive proteins, suggesting that PafA was the only Pup ligase in *Mtb* [74]. This result was somewhat unexpected as there are several hundred different Ub ligases in eukaryotes. In a subsequent study, Weber-Ban and colleagues demonstrated that PafA and ATP were sufficient to conjugate deamidated Pup to proteasomal substrates *in vitro* [77]. Interestingly, in contrast to Ub and other Ub-related modifiers, which all form a compact β-grasp fold, Pup is an intrinsically disordered protein with a propensity for helicity [78–80].

Lack of a Pupylation Motif

Although sequence recognition motifs for Ub ligases and other accessory factors have been identified, there is currently no known sequence motif surrounding the Lys on which Ub is attached (reviewed in [81]). In contrast, SUMO (small ubiquitin-like modifier) often attaches to a Lys residue that is part of a tetrapeptide motif,

ΨKxD/E, where Ψ is a large hydrophobic residue and x is any amino acid (reviewed in [67]). The identification of a sequence that could predict pupylation could be useful for understanding how Pup regulates proteins. Therefore, to identify a possible pupylation motif, the "pupylome" was determined by several independent groups by purifying an epitope tagged Pup from *Mtb* [82] or *Msm* [83, 84]. In the *Mtb* study 604 proteins, representing ~15 % of the total predicted proteome, were identified, but only 55 proteins, including Mpa, were confirmed to harbor a site of Pup attachment [82]. In *Msm*, two independent studies identified 103 and 243 proteins, with 52 and 41 proteins, respectively, having confirmed Pup attachment sites [83, 84]. In all cases, Pup was attached to Lys, and there was little to no evidence of Pup chains, although Pup contains three Lys residues. In one study, Song and colleagues observed pupylation of Lys31 and Lys61 on Pup [84]; however, the authors of this study speculated this might not be physiologically relevant as Pup was overproduced. The authors also noted that pupylation was dynamic and changed depending on the growth condition examined. In all of the "pupylome" studies most, if not all, of the proteins identified are involved in housekeeping functions or stress responses.

Despite the successful identification of numerous pupylation targets, a motif is yet to be identified. It has however, been speculated that an intrinsic sequence is required to signal pupylation because PafA has a much higher affinity for at least one proteasome substrate, PanB, than for free Lys [85]. Nevertheless, it is hard to imagine how PafA specifically recognizes its targets because so many different proteins can be pupylated. In *E. coli*, a bacterial species that does not encode a Pup-proteasome system, pupylation can be reconstituted by expressing Pup with a C-terminal Glu (PupGlu) and PafA. Over 50 *E. coli* proteins can be pupylated using only PafA and a Pup mutant containing a C-terminal Glu (Pup_{Glu}), suggesting the notion that an intrinsic *Mycobacterium* specific sequence is required for pupylation is unlikely [86]. Consistently, PtsI (an *E. coli* pupylated substrate) was pupylated by a native mycobacterial system when produced in *Msm* [86]. Thus, signals for PafA target recognition are not expected to be *Mycobacterium* specific.

Based on these *E. coli* studies, it appears that pupylation is partly stochastic. However, it seems unlikely that mere over-expression of *pup* and *pafA* could determine the fate of so many proteins. Firstly, not all Lys containing proteins are pupylated, e.g. pupylation of the *Mtb* protein DlaT (dihydrolipoamide acyltransferase), which contains 27 Lys residues, has not been observed in either *Mtb* [74] or *E. coli* [86]. Secondly, not all Lys residues within the target protein are modified. Strikingly, *Mtb* FabD is preferentially modified on a single Lys residue, Lys173 [74, 82], despite the fact that *Mtb* FabD contains eight surface exposed Lys residues [87]. Although, two additional Lys residues can, to a lesser extent, be pupylated in *Mtb* FabD when produced in *E. coli* [86]. A simple explanation may be that over-production of *Mtb* FabD in the *E. coli* system merely gives PafA access to other Lys residues. Alternatively, these data may suggest that other factors regulate how and when FabD is pupylated in mycobacteria. Indeed, Darwin and colleagues speculated that the Lys residues in FabD may be involved in interactions with other enzymes in the fatty acid synthesis II (FASII) pathway [86] protecting them from modification. Interestingly, most enzymes in the FASII pathway, several of which are encoded in an operon with *fabD*, are pupylation targets [82, 83]. However, under normal culture conditions, only some of the proteins in this pathway appear to be proteasome substrates in *Mtb*. Recombinant FabG (3-ketoacyl-Acp-reductase), KasA and KasB (3-oxoacyl-Acp-synthases 1 and 2, respectively) do not accumulate in *mpa* or *pafA* mutants under routine culture conditions [82]. It is possible that FabG, KasA, and KasB are degraded by the proteasome under different conditions or are degraded more slowly than FabD. Taken together, there may still be additional *Mycobacterium* specificity factors that regulate pupylation or the delivery of certain pupylated proteins to the proteasome.

Pupylation: An Enzymatic Process That Resembles Glutamine Synthesis

At the time of its identification, PafA did not resemble any protein of known function [5]. Shortly after the discovery that PafA was involved in pupylation, Aravind and colleagues performed a detailed bioinformatic analysis that predicted PafA to have structural similarity to glutamine synthetase (GS) and glutamine cysteine synthetase (GCS) [88]. GS catalyzes the formation of Gln from Glu and ammonia, while GCS catalyzes the formation of γ-glutamyl-cysteine from Glu and Cys; both processes occur via a phosphorylated Glu intermediate. It was therefore proposed that the side chain carboxylate group of the C-terminal Glu in Pup would be phosphorylated or "activated" by PafA. This phosphorylated intermediate would then be primed for nucleophilic attack by the ε-amino group of a side chain Lys on a substrate, resulting in an isopeptide bond between C-terminal Glu of Pup and an internal Lys on the substrate. Indeed, as predicted by Iyer et al. [88], Weber-Ban and colleagues could show that Pup ~ substrate conjugates are generated via activation of the carboxylate group on the C-terminal Glu of Pup [77]. Consistently, site directed mutagenesis of residues in PafA predicted to coordinate ATP or Mg^{2+} disrupted PafA function both *in vivo* [89] and *in vitro* [77].

But how is Pup deamidated prior to activation by PafA? The first evidence came from the Weber-Ban group, who identified a homologue of PafA in H37Rv, (Rv2112c) near the proteasome core genes *prcBA* [77]. They demonstrated, using purified recombinant proteins, that Rv2112c was responsible for Pup deamidation, rendering Pup competent for ligation to FabD or PanB by PafA. Rv2112c was therefore named Dop for deamidase of Pup [77] (Fig. 10.5). In contrast to PafA, which needs ATP to phosphorylate Pup, Dop can deamidate Pup in the presence of ATP, ADP or non-hydrolyzable ATP analogues, suggesting ATP/ADP are allosteric activators of deamidation. Pupylation could also be achieved, in the absence of Dop, when Pup was replaced with a mutant form of Pup (Pup_{Glu}, that does not require deamidation for activation), obviating the need for Dop *in vitro* [77]. This demonstrated that Dop and PafA catalyze independent reactions: deamidation of Pup and conjugation of Pup, respectively. Curiously some Dop-containing bacteria encode Pup_{Glu}, presumably eliminating the need for Dop, which suggests that Dop may play

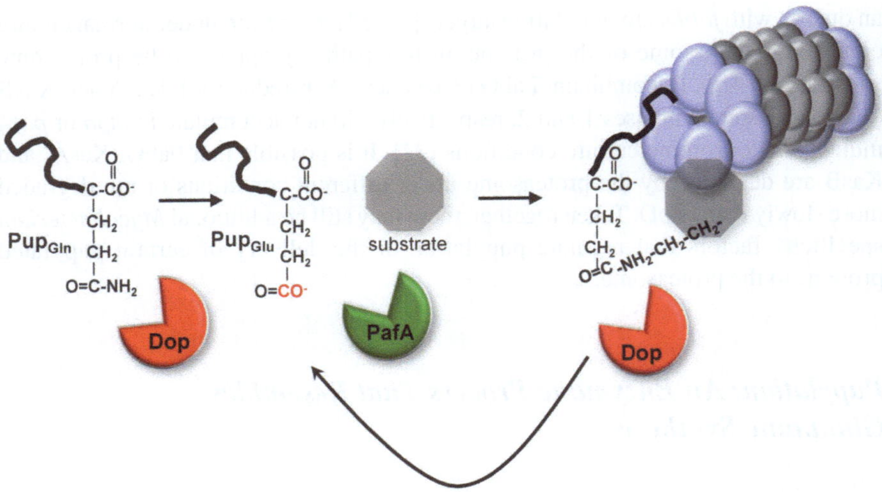

Fig. 10.5 Overview of the Pup-proteasome system in mycobacteria. Pup is deamidated at the C-terminal Gln by Dop. PafA phosphorylates the α-carboxylate of the C-terminal Glu of Pup, priming it for attack by the ε-amino group of a substrate Lys. Pup can be removed by Dop prior to degradation to potentially rescue a substrate from destruction or possibly facilitate its degradation

a different role in these organisms. Nevertheless, for mycobacteria, deamidation is required for pupylation to take place as disruption of *dop* impairs pupylation and proteasomal substrate degradation [89, 90].

Degradation by the Mtb Proteasome: End of the Road…or Is It?

How are pupylated proteins recognized by Mpa prior to degradation? In eukaryotes, Ub receptors are present in the RP of the 26S proteasome, ready to receive ubiquitylated proteins (reviewed in [11]). To date, an equivalent Pup receptor has not been found, however, Pup has a strong affinity for Mpa both *in vitro* and *in vivo* [74] and pull-down experiments using Pup-decorated beads showed that the N-terminal coiled-coil domain of Mpa interacts with Pup [79]. Unlike Ub, Pup is a mostly unstructured, intrinsically disordered protein [78–80], which raised the question: how does Mpa specifically recognize Pup?

A series of *in vitro* and *in vivo* experiments determined that Pup is a two-part degron where the N-terminal ~30 residues are required for Mpa to start the unfolding process and the C-terminal ~30 residues, which have the propensity for helicity as determined by NMR, are needed to interact with Mpa [57, 91]. Darwin and colleagues showed that the N-terminal half of Pup is essential for degradation, but dispensable for pupylation *in vivo* [91], suggesting the C-terminal half of Pup is necessary and sufficient for interaction with PafA and Dop. Weber-Ban and colleagues showed that Mpa could unfold green fluorescent protein if fused to Pup, and this required the N-terminal half of Pup [57]. Importantly, this study also showed that a pupylated substrate can be degraded, albeit somewhat slowly, by Mpa and the proteasome. Interestingly, Pup itself is degraded with the substrate in this system, showing removal of Pup is not essential for degradation.

The molecular details of the interactions required for substrate degradation were ultimately revealed when the three-dimensional structure of the Pup-Mpa complex was solved [54]. Analysis of this complex revealed that the central part of Pup (residues 21–51) becomes ordered upon binding to Mpa [54]. Indeed, the central part of Pup forms an α-helix, using the coiled-coil region of Mpa as a template (Fig. 10.6a). This interaction positions the disordered N-terminus of Pup towards the central channel of the hexameric ATPase, apparently priming the initial threading of Pup into the narrow unfolding pore [54] (Fig. 10.6b). This is consistent with previous studies that suggested the N-terminal half of Pup is needed to facilitate substrate unfolding and degradation [57, 91]. Both hydrophobic and electrostatic interactions are formed between Pup and Mpa, and disruption of either type of interaction abolishes Pup-mediated degradation by the proteasome in *Msm* [54]. Furthermore, consistent with other AAA proteases, mutagenesis of the conserved hydrophobic "pore loop" (Val342Ala) in Mpa abolishes degradation [52].

Importantly, although there are three coiled-coils in the Mpa hexamer (each potentially capable of interacting with Pup or a pupylated substrate) it appears that only one Pup associates per Mpa hexamer [54, 78, 79]. This arrangement would prevent multiple substrates from being recruited to the same Mpa hexamer at any one time, and hence would eliminate potential substrate aggregation or jamming at the proteasome.

Depupylation: What Goes On, Must Come Off

In the eukaryotic UPS, DUBs play an important role in protein degradation. Some DUBs, by removing Ub, are responsible for reversing the fate of a protein destined for degradation, while other DUBs located at the 19S RP, remove Ub to facilitate degradation (reviewed in [92]). DUBs are also responsible for the recycling of Ub to permit new ubiquitylation reactions. It was therefore predicted that pupylation would also be reversible. In a series of elegant experiments the Pup deamidase (Dop) was ultimately identified as a "depupylase" (DPUP) in both *Mtb* [91] and *Msm* [93]. DPUP activity was first demonstrated by Darwin and colleagues, using purified pupylated substrates (Pup ~ FabD and Pup ~ Ino1) with lysates from wild type *Mtb*. Although wild type *Mtb* lysates demonstrated DPUP activity, the *dop* mutant strain did not [91]. These data suggested that Dop was responsible not only for deamidation of Pup, but also for depupylation (Fig. 10.5). Indeed, Dop-mediated DPUP activity was then confirmed *in vitro* using a variety of pupylated substrates [91, 93]. It was then shown by Weber-Ban and colleagues that Dop cleaves specifically the isopeptide bond between Pup and the proteasomal substrate Lys and that Pup_{Glu} is released from the depupylation reaction, suggesting Pup could be recycled [93].

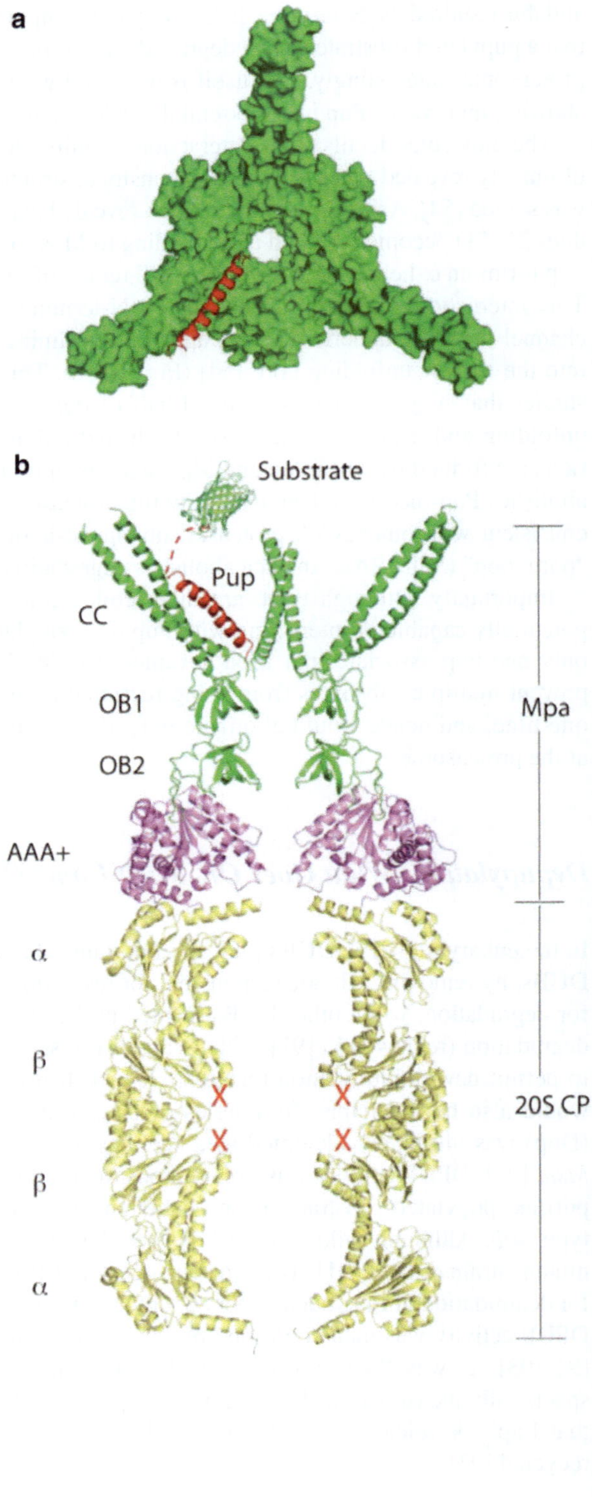

Fig. 10.6 Binding of Pup to Mpa induces a helical structure at the C-terminus of Pup. (**a**) Top view of the crystal structure of the hexameric Mpa1-234 complex with Pup (PDB ID 3M9D). The binding-induced Pup helix is shown in the cartoon view in *red*, and the Mpa coiled-coil and OB domains are shown in surface view in *green*. (**b**) Model for the targeting of pupylated proteins for degradation by Mpa and CP. The Pup:Mpa1-234 complex structure (*red and green*) was placed over the homologous PAN AAA+domain structure (PDB ID 3H4M, *magenta*), which was overlaid on the *Mtb* CP (PDB ID 2FHH, *yellow*). A vertical central slice of the complex structure is shown for clarity. Pup is in *red*, and a model substrate in *green*

Dop is a strict isopeptidase because it cannot remove Pup from a longer polypeptide or linear fusion protein [91, 93]. This is in contrast to some Ub processing enzymes (DUBs) that remove Ub from larger polypeptides [64]. In contrast to DUBs, which tend to be either cysteine or zinc metalloproteases (reviewed in [92]), the catalytic motif of Dop is currently unknown; it does not have a nucleophilic Cys in its partially modeled active site [89], and is not known to require zinc for function. Like PafA, Dop is predicted to adopt a GS/GCS fold [88]. Dop binds Pup well *in vitro* unless Pup's penultimate Gly-Gly motif is mutated, or additional amino acids are added to the C-terminus. Interestingly, however the Gly-Gly motif is not required for substrate attachment [89]. Taken together, it seems that the Gly-Gly motif is important for access to the active site of Dop but not for conjugation to substrates.

It is challenging to assess the role of Dop's DPUP activity *in vivo* because Dop is required for Pup deamidation prior to pupylation in *Msm* and *Mtb* [89, 93]. Weber-Ban and colleagues fully restored pupylation in an *Msm dop* mutant by ectopic expression of pup_{Glu} [93]. In striking contrast, however, expression of pup_{Glu} does not restore the pupylome in *Mtb* [89]. However, the *Mtb* pupylome can be restored in the *dop* strain expressing pup_{Glu} if treated with a proteasome inhibitor. This result suggests that Pup, along with its conjugated proteins, is directly degraded by the proteasome in *Mtb* lacking Dop. Importantly, this observation strongly suggests depupylation is needed to maintain Pup levels for normal pupylation. Thus, it seems likely that a critical function of Dop in *Mtb* is to act as a DPUP, an activity that is essential for Pup recycling. Furthermore, DPUP activity could also be used to regulate protein stability by altering the fate of a once doomed pupylated substrate. As with PafA and pupylation, it is not known how Dop selects substrates for depupylation. Finally, in bacteria that encode Pup_{Glu}, the primary function of Dop must be as a DPUP. It remains to be determined if certain bacteria have evolved to use Pup deamidation as a regulatory step in pupylation. Additionally, we do not yet understand why *dop* mutants in *Msm* and *Mtb* have such different phenotypes.

A curious observation was made in a study that examined the stability of Ino1 in *Msm* [94]. Over-expression of *pup* in wild type *Msm* results in virtually undetectable levels of endogenous Ino1, presumably due to its accelerated turnover. Interestingly, over-expression of *pup* in a *prcBA* mutant results in the accumulation of unpupylated Ino1. This finding suggests that in the *prcBA* mutant either pupylation of Ino1 is inhibited or depupylation prevents detection of Pup~Ino1. In contrast, a follow up study showed that Pup~Ino1 accumulates dramatically in an *Msm mpa* mutant overproducing Pup [91], which led to the hypothesis that Mpa helps to unfold a pupylated substrate in order for it to be depupylated. Consistent with this idea, Weber-Ban and co-workers showed that Mpa increases depupylation of a substrate *in vitro* [93]. Interestingly, it has been noted that corynebacteria, a distant relative of mycobacteria, encode *pup*, *mpa*, *pafA,* and *dop* orthologues but do not have proteasomes [90]. It is tempting to speculate that pupylated proteins are degraded by a different protease, or that pupylation-depupylation is involved in regulating protein activity in this bacterial genus.

The notion that protein unfolding by Mpa is coupled to depupylation poses some challenges to the current model of proteasome-mediated degradation in mycobacteria.

It is well established that unfolding of proteasome substrates starts with the engagement of Pup with Mpa [54, 57]. Presumably, Pup is threaded through the channel in Mpa and, as previously shown, can itself be destroyed by the proteasome along with the substrate *in vitro* [57, 89]. However it also appears that substrates can be depupylated prior to degradation, presumably as they exit from the proximal end of the Mpa hexamer. This scenario implies that Dop interacts with Mpa or substrates at the interface between Mpa and the CP where the unfolded protein is being funnelled into the degradation chamber. One wonders if the conserved, but poorly understood, symmetry mismatch between the six-fold ATPase and the seven-fold CP evolved to prevent tight binding, and allow a gap for the removal of Pup by Dop. Clearly, much needs to be done in order to understand how, when and where Dop coordinates depupylation with degradation.

In eukaryotes, DUBs, ATPases, ligases and other proteins are associated with the eukaryotic 19S RP to remodel or recycle Ub chains on substrates. The *Mycobacterium* proteasome system appears to have been streamlined in such a way that Mpa plays multiple roles in the Pup-proteasome pathway by acting as a substrate receptor, unfoldase and a facilitator of depupylation. It remains to be determined if the *Mtb* proteasome requires additional co-factors to catalyze proteolysis. Because the *in vitro* degradation rate of a pupylated protein seems unusually slow [57, 91], it seems likely that other undetermined factors may be needed to facilitate degradation in *Mtb*.

Proteasomes and Pathogenesis

As discussed earlier in this chapter, *mpa* and *pafA* were identified as genes required for NO resistance and virulence in an animal model of infection [5]. Later studies (to be discussed below) identified additional components of the Pup-proteasome system that are also needed for *Mtb* pathogenesis. How does the *Mtb* proteasome protect against NO toxicity and promote TB pathogenesis? Perhaps the proteasome provides bacteria with a critical pool of amino acids through protein degradation during the chronic phase of infection. Alternatively, the bacterial proteasome may modify the host machinery to its advantage by degrading proteins that alter the recognition of the pathogen by the host's immune system. In the next section we will attempt to address these complex questions by discussing the *in vitro* and *in vivo* phenotypes of proteasome associated mutants in more detail.

Characterization of Proteasome Pathway Mutants

The mouse model of TB is characterized by two phases, an acute phase, during which *Mtb* replicates exponentially within the lungs for about 3 weeks and a chronic or latent phase that is brought about by the emergence of acquired cell-mediated immunity. During the chronic phase bacterial numbers are stabilized in the lungs. Eventually, all mice that are experimentally inoculated with wild type *Mtb* die of TB, in contrast to humans that are naturally infected with *Mtb*. In a low dose aerosol model of *Mtb* infection, mice can survive for more than a year before succumbing to TB; in contrast, mice infected with either an *mpa* or *pafA* mutant show no symptoms of TB [51, 58]. Similar to the *mpa* and *pafA* mutants, an *Mtb dop* mutant is sensitive to RNI *in vitro* and severely attenuated in mice [89]. The degree of attenuation in mice (bacterial load and histopathology) is similar among the *dop*, *mpa* and *pafA* mutants, consistent with the notion that pupylation and Mpa-dependent proteolysis are functionally linked [5, 89]. The attenuation of symptoms is likely due to the presence of 100-1,000 times fewer recoverable mutant bacilli in the lungs, spleens and livers during the persistent phase of infection [5]. However, it is also possible that the *Mtb* proteasome regulates one or more factors that affect the host's response to infection.

Targeted gene disruptions in the *Mtb* CP genes dramatically slow *Mtb* growth on solid media [28, 29]. In C57BL/6 mice infected with either Δ*prcBA::hyg* or P_{tetO}-*prcBA Mtb* strains, the number of bacilli recovered from the lungs is approximately 100-fold lower (compared to wild type or non-silenced *Mtb*) after 3 weeks of infection and continues to decline after this time. This is not completely surprising based on the severe *in vitro* growth defects associated with *Mtb* CP mutations. In contrast to the *prcBA* strains, *mpa, pafA* and *dop Mtb* mutants grow more similarly to wild type *Mtb* in rich broth [5, 89]. These observations suggest that the CP may have critical functions independent of pupylation-dependent proteolysis in *Mtb*. Interestingly, the *mpa*, *pafA* and *prcBA*-defective *Mtb* strains are more resistant to hydrogen peroxide than wild type bacteria, suggesting there is an increase in activity or expression of one or more anti-oxidant pathways in the absence of proteasome function [5, 28]. However, it is currently not understood how loss of proteasome activity could lead to increased resistance to ROIs.

Because proteasome pathway mutants are sensitive to NO *in vitro*, Nathan and colleagues questioned if mice defective in NO production would be more susceptible to infection with *mpa* or *pafA* mutants. In mice and humans NO is produced by three different isoforms of nitric oxide synthase (NOS): endothelial NOS (eNOS), neuronal NOS (nNOS) and inducible or immune NOS (iNOS). iNOS is expressed in activated macrophages and is critical for the control of numerous microbial infections (reviewed in [1]). In comparison to wild type mice, mice genetically inactivated for iNOS ($iNOS^{-/-}$) or treated with chemical inhibitors of NO production are extremely susceptible to *Mtb* [2]. Low dose aerosol infection of $iNOS^{-/-}$ mice with wild type *Mtb* (~200 bacteria/mouse) results in death within 3 months [5, 51, 58]. In contrast, $iNOS^{-/-}$ mice live significantly longer when infected with an *mpa* or *pafA Mtb* strain (~200–500 days post-infection) compared to infection with wild type *Mtb* (~60–80 days post-infection) [51, 58]. Because disruption of iNOS does not fully restore the virulence of the *mpa* and *pafA* mutants, it appears that the role of the *Mtb* proteasome extends beyond protection against RNIs *in vivo*.

In another study Bishai and colleagues identified three clones from a collection of random transposon insertion mutants in CDC1551 (a clinical isolate of *Mtb*) that were consistently smaller than the wild type strain [95]. All three independent mutants contained insertions in MT2175, which is identical to *mpa* in *Mtb* H37Rv. The CDC1551 mutants grow slower (doubling time of ~22 h) than the wild type strain (doubling time of 18 h) in standard 7H9 medium and fail to reach the same final culture density as wild type *Mtb*. Complementation of this CDC1551 mutant strain with *mpa* restores wild type colony morphology. Infection of BALB/c mice with a CDC1551 *mpa* mutant results in similar infection profiles as previously observed with the H37Rv *mpa* mutant. During the chronic phase of infection, bacterial numbers gradually decrease. Mice infected with CDC1551 *mpa* mutants survive without any signs of disease until 180 days (the latest time point assessed), while in contrast, mice infected with wild type *Mtb* succumb within 70 days. Lungs of mice infected with CDC1551 *mpa* mutants have attenuated pathological symptoms, such as less inflammation and fewer granuloma-like foci, and no weight loss compared to mice infected with wild type *Mtb*. Interferon gamma (IFN-γ) production fails to rise after 3 weeks of infection with the mutant compared to the wild type *Mtb* strain, hence *mpa* mutants seem to elicit a milder Th1 immune response in mice [95].

Is Mtb Proteasome Protease Activity Necessary for All Phenotypes?

Ehrt and colleagues made the puzzling observation that proteasomes containing a mutation in the active site can complement a *prcBA* null mutation in *Mtb* for RNI sensitivity and slow growth, but were unable to rescue impaired bacterial persistence in mice [29]. It is unclear how CPs that are proteolytically inactive could restore certain defects but not others. However, it may be possible that the CP has activities that have not yet been identified by routine biochemical assays. For example, the CP may act as a dock or scaffold for other proteins in order to function in specific stress responses. Taken together, these results suggest that the CP, proteolytically active or not, has a broader role for normal cell growth in virulent mycobacteria compared to its non-pathogenic relative *Msm* in which the CP appears to be dispensable under all conditions tested so far [19] (K.H.D., unpublished observations). The genome of *Msm* (7 Mb for *Msm* mc^{2}155) is considerably larger than that of *Mtb* (4.4 Mb for *Mtb* H37Rv) and, unlike *Mtb*, *Msm* encodes another ATP-dependent compartmentalized protease (Lon protease) that may be able to compensate for deletion of *prcBA* in *Msm* (reviewed in [14]).

Proteasome Function and NO Resistance: An Unsolved Mystery

Although it is clear that the lack of proteasome function is a disadvantage for *Mtb* fitness during an infection, it remains to be established how proteasomal proteolysis is linked to pathogenesis. It seems likely that the inability to regulate proteins through degradation compromises bacterial survival when adapting to a new environment, i.e. within activated macrophages. There are several hypotheses that could explain why proteasome function is protective against RNI stress and important for survival in an animal host. Perhaps the simplest explanation is that the proteasome degrades damaged proteins. Oxidative and nitrosative damage of proteins can result in misfolding and aggregation, which is potentially lethal to cells. This damage could possibly be a signal for pupylation. It is also possible that specific accumulated proteasome substrates are particularly dangerous in the presence of NO. For example, iron-sulfur (Fe-S) clusters or copper (Cu) in metal binding proteins can be displaced by NO [96]. The liberation of Fe^{2+} or Cu^{+} is highly toxic to the cell as it can catalyze Fenton chemistry, resulting in the production of ROI. The observation that *mpa*, *pafA* and *prcBA* mutants are hyper-resistant to hydrogen peroxide, suggests that other anti-oxidant pathways may already be induced in an attempt to compensate for loss of the Pup-proteasome system. Currently, however, there is no evidence for the presence of increased amounts of metal-binding or damaged proteins in proteasomal degradation-defective mutants treated with NO.

Regulation of Transcription: Meddling with Metals

Another potential function of the proteasome is in transcriptional regulation. Almost all (if not all) compartmentalized proteases have been shown to regulate gene expression (reviewed in [97]). A microarray study comparing wild type *Mtb* with *mpa* and *pafA* mutants grown under standard culture conditions revealed that a common set of genes was differentially regulated (Table 10.1) [98]. Notably, none of the identified genes appears to be associated with the NO sensitive phenotype of the *mpa* and *pafA* mutants. Among the up-regulated genes in the *mpa* and *pafA* mutants were members of the zinc uptake regulator (Zur) regulon. In the presence of Zn, Zur is released from operators in at least three promoters in *Mtb*, and gene expression is induced [99]. One of the Zur-regulated promoters identified in the microarray drives the expression of the *esx-3* (ESAT-6, region 3) operon. The *esx*-3 locus is, for the most part, essential for the growth of *Mtb* under normal culture conditions and is proposed to encode a type VII secretion system that is involved in zinc and iron acquisition [100, 101]. In addition, Zur regulates the expression of genes that encode homologues of Zn-binding ribosomal proteins. Ribosomes are comprised of numerous small proteins, several of which bind Zn. Under Zn-limiting conditions, these Zn-binding proteins are thought to be replaced with non-metal binding components [102, 103], allowing the bacteria to gain access to a large pool of zinc. If mutations in *mpa* or *pafA* result in deregulation of the Zur regulon *in vivo* as they do *in vitro*, these data would suggest that metal homeostasis is critical during infection and an inappropriate increase in expression of the Zur regulon during infection may also have deleterious affects on bacterial survival for other reasons.

Transcriptome analysis also identified a set of genes repressed in the *mpa* and *pafA* mutants that are regulated by copper [98]. Several of these genes form a copper-inducible regulon, which is under the control of RicR (regulated in copper repressor). During copper depleted conditions, RicR represses five promoters

Table 10.1 Zur and RicR regulons are differentially regulated in *mpa* and *pafA* mutants when compared to wild type *Mtb* (Adapted from [98])

	CDC1551[a]	H37Rv[a]	Gene	Name/function[b]	*mpa*[c]	*pafA*[c]
Zur regulon[d]	MT0115	Rv0106		Conserved hypothetical protein. Unknown function	3.25	4.21
	MT0292	Rv0280	*ppe3*	PPE family protein. Unknown function. Unknown function	3.85	4.1
	MT0293	Rv0281		Conserved hypothetical protein. Possible methlytransferase	1.94	2.46
	MT0295	Rv0282		Conserved hypothetical protein. Unknown function	2.15	2.49
	MT0296	Rv0283		Possible conserved membrane protein. Unknown function	2.06	2.34
	MT0297	Rv0284		Possible conserved membrane protein. Unknown function	2.33	2.51
	MT0298	Rv0285	*pe5*	PE family protein. Unknown function	1.91	2.19
	MT0299	Rv0286	*ppe4*	PPE family protein. Unknown function	1.85	2.16
	MT0300	Rv0287	*esxG*	ESAT-6 like protein. Unknown function	1.61	2.1
	MT0302	Rv0289		Conserved hypothetical protein. Unknown function	2	2.2
	MT0303	Rv0290		Probable conserved transmembrane protein. Unknown function	1.94	1.96
	MT035	Rv0292		Probable conserved transmembrane protein. Unknown function	1.89	2.13
	MT2115	Rv2055c	*rpsR2*	Probable ribosomal protein S18. involved in translation, amino-acyl tRNA binding	2.65	3.81
	MT2117	Rv2056c	*rpsN2*	Probable ribosomal protein S14. Involved in translation	2.43	4.21
	MT2117.1	Rv2057c	*rpmG1*	Probable ribosomal L33. Involved in translation	2.91	4.2
	MT2118	Rv2058c	*rpmB2*	Probable 50S ribosomal protein L28. Involved I ribosome activity	3.47	6.12
	MT2119	Rv2059		Conserved hypothetical protein. Unknown function	n.i	2.17
	MT2428	Rv2359	*zur*	Zinc uptake regulator (formally *furB*)	0.54	0.94

(continued)

Table 10.1 (continued)

	CDC1551[a]	H37Rv[a]	Gene	Name/function[b]	*mpa*[c]	*pafA*[c]
RicR regulon[e]	MT0196	Rv0186a	*mymT*	Metallothionein	0.71	0.47
	MT0200	Rv0190	*ricR*	Regulated in copper repressor	n.i.	n.i.
	MT0870	Rv0847	*lpqS*	Probable lipoprotein. Unknown function	0.52	0.33
			socA	small ORF. Unknown function	n.i.	n.i.
	MT1746.1		*socB*	small ORF. Unknown function	n.i.	n.i.
	MT3039	Rv2963		Probable integral membrane protein. Unknown function	0.6	0.38

[a]Locus number in *Mtb* strains CDC1551 and H37Rv
[b]Functional annotations are taken from http://genolist.pasteur.fr/TubercuList/
[c]Numbers represent expression in *pafA* and *mpa* strains relative to wild type *Mtb* H37Rv grown to early stationary phase in 7H9 media, as determined by microarray analysis. *n.i*: not-identified in the microarray [98]
[d]Zur regulon is described in detail elsewhere [99]
[e]RicR regulon is described in detail elsewhere [98]

that drive the expression of *ricR* itself, *mymT* (a copper methallothionein) and several genes of unknown function (Table 10.1). Disruption of *ricR* results in hyper-resistance of *Mtb* to normally toxic levels of copper, presumably due to the constitutive expression of one or more copper resistance genes like *mymT* [96]. It is worth noting that several of the RicR-regulated genes (*mymT, lpqS, socAB*) are unique to pathogenic mycobacteria, suggesting that copper regulation is important for virulence. Thus, the attenuated phenotype of Pup-proteasome pathway mutants may in part be explained by the incomplete derepression of the RicR regulon during infection. These data support an emerging notion that copper has an important antimicrobial role during TB infection and possibly other infections [96, 98, 104–107].

It is interesting that two metal-dependent regulons are deregulated in *mpa* and *pafA* mutants. In both cases, the mutant bacteria appear to be responding to low metal concentrations. These data also suggest *Mtb* (and possibly other bacteria) need to adapt to changes in metal homeostasis in the host. As with Zur, it is currently not understood how the proteasome affects the expression of the RicR regulon.

Remaining Questions

As with other organisms, regulated proteolysis is critical for numerous aspects of TB biology. *Mtb* possesses a proteasome highly similar to those found in other domains of life, and uses it to resist host derived stresses like NO and regulate pathways that may be needed for pathogenesis. Proteasomal proteolysis is controlled, in part, by pupylation, which is functionally, if not biochemically, similar to ubiquitylation. The characterization of proteasome biochemistry and biology will undoubtedly allow researchers to gain valuable insight into the lifestyle of one of the most successful human pathogens. Among the numerous questions that remain to be asked of the young field of bacterial proteasome biology include:

1. How are proteins selected for pupylation and depupylation?
2. How does Mpa interact with the 20S CP? Why is Mpa itself a proteasome substrate?
3. Does pupylation have functions independent of targeting proteins to the proteasome?
4. How is proteasome function linked to NO resistance?
5. Are misfolded or damaged proteins degraded in a proteasome dependent manner?
6. Why and how are the Zur and RicR regulons affected by proteasome activity?

Acknowledgements We are grateful to Nadine Bode and Andrew Darwin for critical review of this manuscript. K.H.D is supported by NIH grants AI065437 and HL92774, the Irma T. Hirschl Trust, and holds an Investigators in the Pathogenesis of Infectious Disease Award from the Burroughs Wellcome Fund. H.L. is supported by NIH grant AI070285 and Brookhaven National Laboratory LDRD grant 10-016.

References

1. Nathan C, Shiloh MU (2000) Reactive oxygen and nitrogen intermediates in the relationship between mammalian hosts and microbial pathogens. Proc Natl Acad Sci U S A 97(16): 8841–8848
2. MacMicking J, Xie QW, Nathan C (1997) Nitric oxide and macrophage function. Annu Rev Immunol 15:323–350
3. Alvarez B, Radi R (2003) Peroxynitrite reactivity with amino acids and proteins. Amino Acids 25(3–4):295–311
4. Szabo C (2003) Multiple pathways of peroxynitrite cytotoxicity. Toxicol Lett 140–141: 105–112
5. Darwin KH, Ehrt S, Gutierrez-Ramos JC, Weich N et al (2003) The proteasome of Mycobacterium tuberculosis is required for resistance to nitric oxide. Science 302(5652): 1963–1966
6. Zwickl P, Ng D, Woo KM, Klenk HP et al (1999) An archaebacterial ATPase, homologous to ATPases in the eukaryotic 26S proteasome, activates protein breakdown by 20S proteasomes. J Biol Chem 274(37):26008–26014
7. Wolf S, Nagy I, Lupas A, Pfeifer G et al (1998) Characterization of ARC, a divergent member of the AAA ATPase family from Rhodococcus erythropolis. J Mol Biol 277(1):13–25
8. Arrigo AP, Tanaka K, Goldberg AL, Welch WJ (1988) Identity of the 19S 'prosome' particle with the large multifunctional protease complex of mammalian cells (the proteasome). Nature 331(6152):192–194
9. Pickart CM, Cohen RE (2004) Proteasomes and their kin: proteases in the machine age. Nat Rev Mol Cell Biol 5(3):177–187

10. Coux O, Tanaka K, Goldberg AL (1996) Structure and functions of the 20S and 26S proteasomes. Annu Rev Biochem 65:801–847
11. Finley D (2009) Recognition and processing of ubiquitin-protein conjugates by the proteasome. Annu Rev Biochem 78:477–513
12. Rock KL, Goldberg AL (1999) Degradation of cell proteins and the generation of MHC class I-presented peptides. Annu Rev Immunol 17:739–779
13. Schmidt M, Hanna J, Elsasser S, Finley D (2005) Proteasome-associated proteins: regulation of a proteolytic machine. Biol Chem 386(8):725–737
14. Butler SM, Festa RA, Pearce MJ, Darwin KH (2006) Self-compartmentalized bacterial proteases and pathogenesis. Mol Microbiol 60(3):553–562
15. Gur E, Ottofuelling R, Dougan DA (2013) Machines of destruction – AAA+proteases and the adaptors that control them. In: Dougan DA (ed) Regulated proteolysis in microorganisms. Springer, Subcell Biochem 66:3–33
16. Dahlmann B, Kopp F, Kuehn L, Niedel B et al (1989) The multicatalytic proteinase (prosome) is ubiquitous from eukaryotes to archaebacteria. FEBS Lett 251(1–2):125–131
17. Lowe J, Stock D, Jap B, Zwickl P et al (1995) Crystal structure of the 20S proteasome from the archaeon T. acidophilum at 3.4 A resolution. Science 268(5210):533–539
18. Tamura T, Nagy I, Lupas A, Lottspeich F et al (1995) The first characterization of a eubacterial proteasome: the 20S complex of Rhodococcus. Curr Biol 5(7):766–774
19. Knipfer N, Shrader TE (1997) Inactivation of the 20S proteasome in Mycobacterium smegmatis. Mol Microbiol 25(2):375–383
20. Nagy I, Tamura T, Vanderleyden J, Baumeister W et al (1998) The 20S proteasome of Streptomyces coelicolor. J Bacteriol 180(20):5448–5453
21. Pouch MN, Cournoyer B, Baumeister W (2000) Characterization of the 20S proteasome from the actinomycete Frankia. Mol Microbiol 35(2):368–377
22. Lupas A, Zuhl F, Tamura T, Wolf S et al (1997) Eubacterial proteasomes. Mol Biol Rep 24(1–2):125–131
23. Cole ST, Brosch R, Parkhill J, Garnier T et al (1998) Deciphering the biology of Mycobacterium tuberculosis from the complete genome sequence. Nature 393(6685):537–544
24. De Mot R (2007) Actinomycete-like proteasomes in a Gram-negative bacterium. Trends Microbiol 15(8):335–338
25. Gille C, Goede A, Schloetelburg C, Preissner R et al (2003) A comprehensive view on proteasomal sequences: implications for the evolution of the proteasome. J Mol Biol 326(5):1437–1448
26. Groll M, Bochtler M, Brandstetter H, Clausen T et al (2005) Molecular machines for protein degradation. Chembiochem 6(2):222–256
27. Cerda-Maira F, Darwin KH (2009) The Mycobacterium tuberculosis proteasome: more than just a barrel-shaped protease. Microbes Infect 11(14–15):1150–1155
28. Gandotra S, Schnappinger D, Monteleone M, Hillen W et al (2007) In vivo gene silencing identifies the Mycobacterium tuberculosis proteasome as essential for the bacteria to persist in mice. Nat Med 13(12):1515–1520
29. Gandotra S, Lebron MB, Ehrt S (2010) The Mycobacterium tuberculosis proteasome active site threonine is essential for persistence yet dispensable for replication and resistance to nitric oxide. PLoS Pathog 6(8):e1001040
30. Sassetti CM, Boyd DH, Rubin EJ (2003) Genes required for mycobacterial growth defined by high density mutagenesis. Mol Microbiol 48(1):77–84
31. Maupin-Furlow JA (2013) Archaeal proteasomes and sampylation. In: Dougan DA (ed) Regulated proteolysis in microorganisms. Springer, Subcell Biochem 66:297–327
32. Buchberger A (2013) Roles of Cdc48 in regulated protein degradation in yeast. In: Dougan DA (ed) Regulated proteolysis in microorganisms. Springer, Subcell Biochem 66:195–222
33. Hu G, Lin G, Wang M, Dick L et al (2006) Structure of the Mycobacterium tuberculosis proteasome and mechanism of inhibition by a peptidyl boronate. Mol Microbiol 59(5):1417–1428

34. Witt S, Kwon YD, Sharon M, Felderer K et al (2006) Proteasome assembly triggers a switch required for active-site maturation. Structure 14(7):1179–1188
35. Groll M, Ditzel L, Lowe J, Stock D et al (1997) Structure of 20S proteasome from yeast at 2.4 A resolution. Nature 386(6624):463–471
36. Seemuller E, Lupas A, Baumeister W (1996) Autocatalytic processing of the 20S proteasome. Nature 382(6590):468–471
37. Zuhl F, Seemuller E, Golbik R, Baumeister W (1997) Dissecting the assembly pathway of the 20S proteasome. FEBS Lett 418(1–2):189–194
38. Zuhl F, Tamura T, Dolenc I, Cejka Z et al (1997) Subunit topology of the Rhodococcus proteasome. FEBS Lett 400(1):83–90
39. Mayr J, Seemuller E, Muller SA, Engel A et al (1998) Late events in the assembly of 20S proteasomes. J Struct Biol 124(2–3):179–188
40. Kwon YD, Nagy I, Adams PD, Baumeister W et al (2004) Crystal structures of the Rhodococcus proteasome with and without its pro-peptides: implications for the role of the pro-peptide in proteasome assembly. J Mol Biol 335(1):233–245
41. Zwickl P, Kleinz J, Baumeister W (1994) Critical elements in proteasome assembly. Nat Struct Biol 1(11):765–770
42. Lin G, Hu G, Tsu C, Kunes YZ et al (2006) Mycobacterium tuberculosis prcBA genes encode a gated proteasome with broad oligopeptide specificity. Mol Microbiol 59(5):1405–1416
43. Li D, Li H, Wang T, Pan H et al (2010) Structural basis for the assembly and gate closure mechanisms of the Mycobacterium tuberculosis 20S proteasome. EMBO J 29(12):2037–2047
44. Groll M, Huber R (2003) Substrate access and processing by the 20S proteasome core particle. Int J Biochem Cell Biol 35(5):606–616
45. Zhang F, Wu Z, Zhang P, Tian G et al (2009) Mechanism of substrate unfolding and translocation by the regulatory particle of the proteasome from Methanocaldococcus jannaschii. Mol Cell 34(4):485–496
46. Groll M, Bajorek M, Kohler A, Moroder L et al (2000) A gated channel into the proteasome core particle. Nat Struct Biol 7(11):1062–1067
47. Groll M, Heinemeyer W, Jager S, Ullrich T et al (1999) The catalytic sites of 20S proteasomes and their role in subunit maturation: a mutational and crystallographic study. Proc Natl Acad Sci U S A 96(20):10976–10983
48. Heinemeyer W, Fischer M, Krimmer T, Stachon U et al (1997) The active sites of the eukaryotic 20 S proteasome and their involvement in subunit precursor processing. J Biol Chem 272(40):25200–25209
49. Benaroudj N, Goldberg AL (2000) PAN, the proteasome-activating nucleotidase from archaebacteria, is a protein-unfolding molecular chaperone. Nat Cell Biol 2(11):833–839
50. Walker JE, Saraste M, Runswick MJ, Gay NJ (1982) Distantly related sequences in the alpha- and beta-subunits of ATP synthase, myosin, kinases and other ATP-requiring enzymes and a common nucleotide binding fold. EMBO J 1(8):945–951
51. Darwin KH, Lin G, Chen Z, Li H et al (2005) Characterization of a Mycobacterium tuberculosis proteasomal ATPase homologue. Mol Microbiol 55(2):561–571
52. Wang T, Li H, Lin G, Tang C et al (2009) Structural insights on the Mycobacterium tuberculosis proteasomal ATPase Mpa. Structure 17(10):1377–1385
53. Djuranovic S, Hartmann MD, Habeck M, Ursinus A et al (2009) Structure and activity of the N-terminal substrate recognition domains in proteasomal ATPases. Mol Cell 34(5):580–590
54. Wang T, Darwin KH, Li H (2010) Binding-induced folding of prokaryotic ubiquitin-like protein on the Mycobacterium proteasomal ATPase targets substrates for degradation. Nat Struct Mol Biol 17(11):1352–1357
55. Zhang F, Hu M, Tian G, Zhang P et al (2009) Structural insights into the regulatory particle of the proteasome from Methanocaldococcus jannaschii. Mol Cell 34(4):473–484
56. Smith DM, Kafri G, Cheng Y, Ng D et al (2005) ATP binding to PAN or the 26S ATPases causes association with the 20S proteasome, gate opening, and translocation of unfolded proteins. Mol Cell 20(5):687–698

57. Striebel F, Hunkeler M, Summer H, Weber-Ban E (2010) The mycobacterial Mpa-proteasome unfolds and degrades pupylated substrates by engaging Pup's N-terminus. EMBO J 29(7): 1262–1271
58. Pearce MJ, Arora P, Festa RA, Butler-Wu SM et al (2006) Identification of substrates of the Mycobacterium tuberculosis proteasome. EMBO J 25(22):5423–5432
59. Kusmierczyk AR, Kunjappu MJ, Kim RY, Hochstrasser M (2011) A conserved 20S proteasome assembly factor requires a C-terminal HbYX motif for proteasomal precursor binding. Nat Struct Mol Biol 18(5):622–629
60. Smith DM, Chang SC, Park S, Finley D et al (2007) Docking of the proteasomal ATPases' carboxyl termini in the 20S proteasome's alpha ring opens the gate for substrate entry. Mol Cell 27(5):731–744
61. Hochstrasser M (1996) Ubiquitin-dependent protein degradation. Annu Rev Genet 30:405–439
62. Vijay-Kumar S, Bugg CE, Wilkinson KD, Cook WJ (1985) Three-dimensional structure of ubiquitin at 2.8 A resolution. Proc Natl Acad Sci U S A 82(11):3582–3585
63. Wilkinson KD (1997) Regulation of ubiquitin-dependent processes by deubiquitinating enzymes. FASEB J 11(14):1245–1256
64. Ozkaynak E, Finley D, Solomon MJ, Varshavsky A (1987) The yeast ubiquitin genes: a family of natural gene fusions. EMBO J 6(5):1429–1439
65. Kerscher O, Felberbaum R, Hochstrasser M (2006) Modification of proteins by ubiquitin and ubiquitin-like proteins. Annu Rev Cell Dev Biol 22:159–180
66. Hershko A, Ciechanover A (1998) The ubiquitin system. Annu Rev Biochem 67:425–479
67. Johnson ES (2004) Protein modification by SUMO. Annu Rev Biochem 73:355–382
68. Rotin D, Kumar S (2009) Physiological functions of the HECT family of ubiquitin ligases. Nat Rev Mol Cell Biol 10(6):398–409
69. Deshaies RJ, Joazeiro CA (2009) RING domain E3 ubiquitin ligases. Annu Rev Biochem 78:399–434
70. Chau V, Tobias JW, Bachmair A, Marriott D et al (1989) A multiubiquitin chain is confined to specific lysine in a targeted short-lived protein. Science 243(4898):1576–1583
71. Hough R, Rechsteiner M (1986) Ubiquitin-lysozyme conjugates. Purification and susceptibility to proteolysis. J Biol Chem 261(5):2391–2399
72. Scholz O, Thiel A, Hillen W, Niederweis M (2000) Quantitative analysis of gene expression with an improved green fluorescent protein. p6. Eur J Biochem 267(6):1565–1570
73. Karimova G, Pidoux J, Ullmann A, Ladant D (1998) A bacterial two-hybrid system based on a reconstituted signal transduction pathway. Proc Natl Acad Sci U S A 95(10):5752–5756
74. Pearce MJ, Mintseris J, Ferreyra J, Gygi SP et al (2008) Ubiquitin-like protein involved in the proteasome pathway of Mycobacterium tuberculosis. Science 322(5904):1104–1107
75. Singh A, Mai D, Kumar A, Steyn AJ (2006) Dissecting virulence pathways of Mycobacterium tuberculosis through protein-protein association. Proc Natl Acad Sci U S A 103(30): 11346–11351
76. Burns KE, Liu WT, Boshoff HI, Dorrestein PC et al (2009) Proteasomal protein degradation in Mycobacteria is dependent upon a prokaryotic ubiquitin-like protein. J Biol Chem 284(5):3069–3075
77. Striebel F, Imkamp F, Sutter M, Steiner M et al (2009) Bacterial ubiquitin-like modifier Pup is deamidated and conjugated to substrates by distinct but homologous enzymes. Nat Struct Mol Biol 16(6):647–651
78. Chen X, Solomon WC, Kang Y, Cerda-Maira F et al (2009) Prokaryotic ubiquitin-like protein pup is intrinsically disordered. J Mol Biol 392(1):208–217
79. Sutter M, Striebel F, Damberger FF, Allain FH et al (2009) A distinct structural region of the prokaryotic ubiquitin-like protein (Pup) is recognized by the N-terminal domain of the proteasomal ATPase Mpa. FEBS Lett 583(19):3151–3157
80. Liao S, Shang Q, Zhang X, Zhang J et al (2009) Pup, a prokaryotic ubiquitin-like protein, is an intrinsically disordered protein. Biochem J 422(2):207–215

81. Jackson PK, Eldridge AG, Freed E, Furstenthal L et al (2000) The lore of the RINGs: substrate recognition and catalysis by ubiquitin ligases. Trends Cell Biol 10(10):429–439
82. Festa RA, McAllister F, Pearce MJ, Mintseris J et al (2010) Prokaryotic ubiquitin-like protein (Pup) proteome of Mycobacterium tuberculosis [corrected]. PLoS One 5(1):e8589
83. Watrous J, Burns K, Liu WT, Patel A et al (2010) Expansion of the mycobacterial "PUPylome". Mol Biosyst 6(2):376–385
84. Poulsen C, Akhter Y, Jeon AH, Schmitt-Ulms G et al (2010) Proteome-wide identification of mycobacterial pupylation targets. Mol Syst Biol 6:386
85. Guth E, Thommen M, Weber-Ban E (2011) Mycobacterial ubiquitin-like protein ligase PafA follows a two-step reaction pathway with a phosphorylated pup intermediate. J Biol Chem 286(6):4412–4419
86. Cerda-Maira FA, McAllister F, Bode NJ, Burns KE et al (2011) Reconstitution of the Mycobacterium tuberculosis pupylation pathway in Escherichia coli. EMBO Rep 12(8): 863–870
87. Ghadbane H, Brown AK, Kremer L, Besra GS et al (2007) Structure of Mycobacterium tuberculosis mtFabD, a malonyl-CoA:acyl carrier protein transacylase (MCAT). Acta Crystallogr Sect F Struct Biol Cryst Commun 63(Pt 10):831–835
88. Iyer LM, Burroughs AM, Aravind L (2008) Unraveling the biochemistry and provenance of pupylation: a prokaryotic analog of ubiquitination. Biol Direct 3:45
89. Cerda-Maira FA, Pearce MJ, Fuortes M, Bishai WR et al (2010) Molecular analysis of the prokaryotic ubiquitin-like protein (Pup) conjugation pathway in Mycobacterium tuberculosis. Mol Microbiol 77(5):1123–1135
90. Imkamp F, Rosenberger T, Striebel F, Keller PM et al (2010) Deletion of dop in Mycobacterium smegmatis abolishes pupylation of protein substrates in vivo. Mol Microbiol 75(3):744–754
91. Burns KE, Cerda-Maira FA, Wang T, Li H et al (2010) "Depupylation" of prokaryotic ubiquitin-like protein from mycobacterial proteasome substrates. Mol Cell 39(5):821–827
92. Komander D, Clague MJ, Urbe S (2009) Breaking the chains: structure and function of the deubiquitinases. Nat Rev Mol Cell Biol 10(8):550–563
93. Imkamp F, Striebel F, Sutter M, Ozcelik D et al (2010) Dop functions as a depupylase in the prokaryotic ubiquitin-like modification pathway. EMBO Rep 11(10):791–797
94. Burns KE, Pearce MJ, Darwin KH (2010) Prokaryotic ubiquitin-like protein provides a two-part degron to Mycobacterium proteasome substrates. J Bacteriol 192(11):2933–2935
95. Lamichhane G, Raghunand TR, Morrison NE, Woolwine SC et al (2006) Deletion of a Mycobacterium tuberculosis proteasomal ATPase homologue gene produces a slow-growing strain that persists in host tissues. J Infect Dis 194(9):1233–1240
96. Gold B, Deng H, Bryk R, Vargas D et al (2008) Identification of a copper-binding metallothionein in pathogenic mycobacteria. Nat Chem Biol 4(10):609–616
97. Gottesman S (2003) Proteolysis in bacterial regulatory circuits. Annu Rev Cell Dev Biol 19:565–587
98. Festa RA, Jones MB, Butler-Wu S, Sinsimer D et al (2011) A novel copper-responsive regulon in Mycobacterium tuberculosis. Mol Microbiol 79(1):133–148
99. Maciag A, Dainese E, Rodriguez GM, Milano A et al (2007) Global analysis of the Mycobacterium tuberculosis Zur (FurB) regulon. J Bacteriol 189(3):730–740
100. Siegrist MS, Unnikrishnan M, McConnell MJ, Borowsky M et al (2009) Mycobacterial Esx-3 is required for mycobactin-mediated iron acquisition. Proc Natl Acad Sci U S A 106(44): 18792–18797
101. Serafini A, Boldrin F, Palu G, Manganelli R (2009) Characterization of a Mycobacterium tuberculosis ESX-3 conditional mutant: essentiality and rescue by iron and zinc. J Bacteriol 191(20):6340–6344
102. Nanamiya H, Akanuma G, Natori Y, Murayama R et al (2004) Zinc is a key factor in controlling alternation of two types of L31 protein in the Bacillus subtilis ribosome. Mol Microbiol 52(1):273–283

103. Natori Y, Nanamiya H, Akanuma G, Kosono S et al (2007) A fail-safe system for the ribosome under zinc-limiting conditions in Bacillus subtilis. Mol Microbiol 63(1):294–307
104. Liu T, Ramesh A, Ma Z, Ward SK et al (2007) CsoR is a novel Mycobacterium tuberculosis copper-sensing transcriptional regulator. Nat Chem Biol 3(1):60–68
105. Ward SK, Abomoelak B, Hoye EA, Steinberg H et al (2010) CtpV: a putative copper exporter required for full virulence of Mycobacterium tuberculosis. Mol Microbiol 77(5):1096–1110
106. Ward SK, Hoye EA, Talaat AM (2008) The global responses of Mycobacterium tuberculosis to physiological levels of copper. J Bacteriol 190(8):2939–2946
107. Wolschendorf F, Ackart D, Shrestha TB, Hascall-Dove L et al (2011) Copper resistance is essential for virulence of Mycobacterium tuberculosis. Proc Natl Acad Sci U S A 108(4): 1621–1626

Chapter 11
Archaeal Proteasomes and Sampylation

Julie A. Maupin-Furlow

Abstract Archaea contain, both a functional proteasome and an ubiquitin-like protein conjugation system (termed sampylation) that is related to the ubiquitin proteasome system (UPS) of eukaryotes. Archaeal proteasomes have served as excellent models for understanding how proteins are degraded by the central energy-dependent proteolytic machine of eukaryotes, the 26S proteasome. While sampylation has only recently been discovered, it is thought to be linked to proteasome-mediated degradation in archaea. Unlike eukaryotes, sampylation only requires an E1 enzyme homolog of the E1-E2-E3 ubiquitylation cascade to mediate protein conjugation. Furthermore, recent evidence suggests that archaeal and eurkaryotic E1 enzyme homologs can serve dual roles in mediating protein conjugation and activating sulfur for incorporation into biomolecules. The focus of this book chapter is the energy-dependent proteasome and sampylation systems of Archaea.

Introduction

Archaea are one of three major lineages of life which share deep evolutionary roots with eukaryotes. Although typically single-celled, many Archaea thrive in conditions once thought to be uninhabitable to life such as temperatures above 100°C, saturating salt and extreme pH. Thus, Archaea provide an exciting opportunity to examine how proteolytic systems can function in maintaining the quality of proteomes when cells are growing at physical conditions considered suboptimal for protein folding and stability.

J.A. Maupin-Furlow (✉)
Department of Microbiology and Cell Science, University of Florida,
Gainesville, FL 32611-0700, USA
e-mail: jmaupin@ufl.edu

D.A. Dougan (ed.), *Regulated Proteolysis in Microorganisms*, Subcellular Biochemistry 66,
DOI 10.1007/978-94-007-5940-4_11,

Although the hydrolysis of peptide bonds is exergonic (releasing energy), many proteolytic machines (e.g. the proteasome) require energy for protein degradation. Of the various types of proteases, it is often the energy-dependent proteases that are important in regulating central processes of the cell such as protein quality, cell division and survival after exposure to stress [1, 2]. The energy-dependence of the proteolytic system adds assurance that the appropriate protein substrate has been selected prior to its destruction, which would otherwise come at a high energy cost.

Energy-dependent proteases (e.g. Lon, Clp, FtsH, HslUV and the proteasome), while not necessarily conserved in primary amino acid sequence, share a common protein architecture of self-compartmentalization [3]. In general, the proteolytic active-sites of the protease are sequestered within a chamber that has narrow openings for substrate entry. These openings are often gated to reduce the non-specific entry of proteins into the chamber [4–7]. Typically, the chamber alone does not degrade folded proteins but requires association with regulatory ATPases (either a protein domain or separate protein complex). The ATPases couple the hydrolysis of ATP to the unfolding and the translocation of the substrate protein, so that the substrate protein can access the proteolytic active sites lining the central chamber of the protease.

The ATPase regulators of energy-dependent proteolysis belong to the AAA+ (ATPases associated with a variety of cellular activities) superfamily [8, 9]. These AAA+proteins form hexameric rings that associate with a compartmentalized cylindrical peptidase. In some cases, the peptidase forms a hexameric ring (e.g. HslV), while in other cases the peptidase exhibits a seven-fold symmetry. Hence, symmetry mismatch can exist between the ATPase and protease components. For example, the hexameric ATPases (ClpA, ClpC and ClpX) associate with ClpP (which is composed of two heptameric rings). Similarly, the proteasomal ATPases (ARC/Mpa of actinobacteria, PAN of Archaea, and Rpt1-6 of eukaryotes) all form hexamers and interact with a 20S core particle (CP), which has seven-fold symmetry. Energy-dependent proteases such as Lon and FtsH differ from most AAA+proteases, in that both the ATPase and proteolytic domains are located on the same polypeptide chain [10] (see also accompanying reviews [11, 12]).

Archaeal Proteasomes

Similarly to eukaryotes, all Archaea are predicted to use proteasomes as one of their major systems for energy-dependent proteolysis [13]. All species of Archaea with sequenced genomes are predicted to encode a 20S CP of α- and β-type subunit composition as well as an AAA+regulator (PAN and/or Cdc48) thought to associate with the CP and mediate protein degradation [13, 14]. Archaea also contain the membrane bound B-type Lon protease [15, 16]. However in contrast to bacteria, Archaea lack homologs of the membrane-spanning FtsH protease and, in most cases also lack HslUV, Clp and the A-type Lon proteases [15, 16]. Like other bacteria, the actinobacteria (including species from *Mycobacterium*, *Rhodococcus* and *Frankia*)

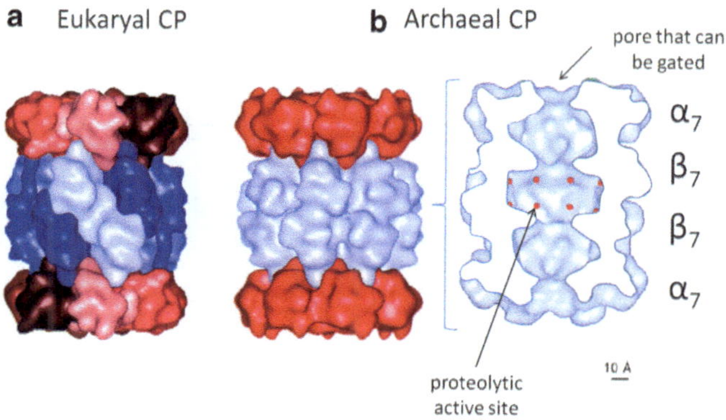

Fig. 11.1 20S proteasome core particles (*CPs*). CPs are composed of four stacked heptameric rings of α- and β-type subunits (indicated by α7 and β7, respectively) that form a cylindrical structure. The central channel of the CP is accessed by gated pores on each end of the cylinder and connects three interior chambers. Proteolytic active sites formed by β-type subunits (*indicated in red*) line the central chamber that is flanked by two antechambers. CPs of (**a**) the eukaryote *Saccharomyces cerevisiae* and (**b**) the archaeon *Thermoplasma acidophilum* are presented as examples. (Figure modified from [22, 23] with permission)

encode many energy-dependent proteases including the proteasomal CPs and its associated AAA+components termed ARC or Mpa, in addition to Lon, Clp and FtsH proteases, but do not encode the HslUV protease [17–20]. Overall, the distribution of energy-dependent proteases in Archaea is more like eukaryotes than either bacteria or bacterial-like organelles of eukaryotes.

Proteasomal Core Particles (CPs)

Proteasomes are composed of a 20S CP that is required for the hydrolysis of unfolded proteins and small peptides [21]. The CP is formed from structurally related α- and β-type subunits that are arranged into four-stacked heptameric rings (Fig. 11.1). The two outer rings are composed of α-subunits, and the two inner rings are of β-subunits. These subunits are arranged to form a cylindrical particle with openings on both ends of the particle, gated by the N-termini of the α-subunits. The gates open into a central channel that connects three interior chambers (two antechambers and one central chamber). The central chamber is lined with the proteolytic active sites that are formed by the N-terminal threonine residues of the β-subunits and are exposed after cleavage of the β propeptide during CP assembly. In contrast to eukaryotic CPs, which are formed from seven different α-subunits (α1 to α7) and seven different β-subunits (β1 to β7) that assemble into a 28-subunit complex of dyad symmetry. Core particles from Archaea and actinobacteria are typically assembled from one to two different α-subunits and one to two different β-subunits.

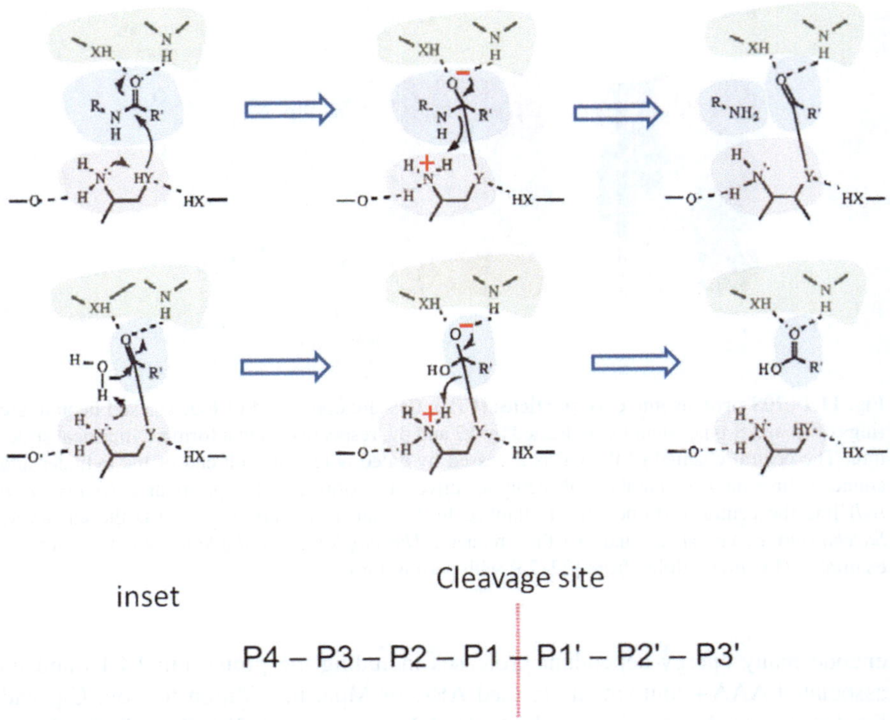

Fig. 11.2 Mechanism of proteasomes and other related Ntn-hydrolyases in the hydrolysis of an amide bond. Substrate, *shaded in blue*; Ntn active site residue, *shaded in red*; oxyanion hole, *shaded in green*; charged state, indicated by + or – in *red font*; Y, oxygen or sulfur; X, nitrogen or oxygen; R-R', substrate bond cleaved (comparable to the P1–P1' peptide bond cleaved by proteases as depicted in the *inset*) (Figure modified from [26] with permission)

Proteolytic Active Sites of Proteasomes

The proteolytic active sites of proteasomes, sequestered within the CP, are of the N-terminal nucleophile (Ntn) hydrolase superfamily [24–26]. This superfamily includes penicillin acylases, aspartylglucosaminidases and other hydrolytic enzymes. While Ntn hydrolases are not necessarily conserved in primary sequence, these enzymes hydrolyze amide bonds and undergo autocatalytic post-translational processing to expose an N-terminal residue (Thr, Ser or Cys) that is used as nucleophile (the side chain OH or SH group) and proton donor (the Nα-amino group) in a structurally related active site configuration. The Ntn hydrolase reaction is initiated when the Ntn residue donates the proton of its side chain to its own α-amino group for attack of the carbonyl carbon of the substrate (Fig. 11.2). A negatively charged tetrahedral intermediate is formed that is stabilized by hydrogen bonding to the oxyanion hole. The Ntn α-amino group completes the acylation step by donating a proton to the nitrogen of the substrate bond undergoing hydrolysis. While part of the cleaved substrate (the portion with the R or P1 group) is released, a covalent bond is formed between the nucleophile of the Ntn and the remaining substrate (with the R' or P1' group) that must be deacylated for product release. Deacylation is initiated when the hydroxyl group of water attacks the carbonyl carbon of the covalent acyl-enzyme intermediate. The Ntn α-amino group accepts a proton from water and stabilizes the negatively charged enzyme intermediate. Reprotonation of the nucleophile by the Ntn α-amino group completes this reaction.

While the mechanism of peptide hydrolysis by the CP active site resembles serine proteases, only one amino acid (Thr1) serves as the catalytic center, rather than the three amino acids (Asp–His–Ser) that constitute the catalytic triad of serine proteases [27]. Both the nucleophile (hydroxyl side chain) and the base (α-amino group) of the CP proteolytic reaction exist in the same amino acid (Thr). Like β-lactamase and some esterases, there is no acid component in the CP active site [28]. Serine can substitute for threonine as the Ntn active site in CPs (T1S variants) for the hydrolysis of peptide amide substrates such as N-succinyl-Leu-Leu-Val-Tyr-7-amido-4-methyl coumarin (Suc-LLVY-amc). However, CP T1S variants are severely impaired in their ability to cleave longer peptides, revealing the importance of the threonine side chain methyl group in active site conformation [29, 30].

Cleavage of N-Terminal Propeptide to Expose Proteolytic Active Site

Proteasomal active sites are formed through intramolecular autolysis of β-type precursor proteins during CP assembly [31]. The CP β-subunits are typically translated as precursor proteins with N-terminal propeptides that are removed during formation of the CP. The β-subunits remain monomeric, unprocessed and inactive until assembled with α-subunits into the CP. In most cases, the propeptide of the β-subunit is autocatalytically removed at a conserved G↓T motif (where ↓ represents the site of peptide bond cleavage) to generate a mature β-subunit that contains an active N-terminal threonine. However, some β-subunits that assemble into CPs are not cleaved at conserved G↓T motifs [32, 33]. For example, the eukaryotic β7 and β6 precursors are cleaved at N↓T and H↓Q sites, respectively, resulting in the removal of N-terminal propeptides and formation of inactive β subunits in CPs. Likewise, the eukaryotic β3 and β4 subunits are not cleaved but are assembled into CPs as inactive subunits.

The position and residues surrounding the cleavage site of the propeptide of proteasomal β-subunits is predicted to be highly diverse among archaea. Like eukaryotes and actinobacteria, Archaea are known to synthesize β-type precursor proteins that are cleaved at G↓T motifs resulting in the removal of N-terminal propeptides of variable length (6–49 residues) and the formation of mature β-subunits with active CPs. This knowledge is based on N-terminal sequencing of CP subunits purified from Archaea of the phylum Euryarchaeota (i.e., *Thermoplasma acidophilum, Methanosarcina thermophila, Haloferax volcanii and Pyrococcus furiosus*) [34–37]. Similarly, most archaea are predicted to encode at least one β-type precursor protein that is cleaved at a G↓T motif to release an N-terminal propeptide and generate an active β-type subunit. Interestingly, select Crenarchaeota (i.e., *Thermoproteus* and

Pyrobaculum species) and Candidatus *Caldiarchaeum subterraneum* of the novel division Aigarchaeota have β-subunit homologs devoid of propeptides (only an initiator methionine residue, which may be removed by a methionine aminopeptidase, is predicted to precede the conserved active site threonine residue). Furthermore, most Crenarchaeota are predicted to encode two different β-type proteins with at least one of these proteins harbouring an N-terminal extension that lacks the classical G↓T cleavage site. Whether or not these latter β-subunits associate in CPs and form active sites remains to be determined.

The propeptides of β-type precursors ('β-propeptides') can modulate the function and assembly of proteasomal CPs. The β-propeptides of yeast protect the N-terminal threonine active sites of the β-subunits from N^{α}-acetylation and promote CP assembly [38]. In actinobacteria, the β-propeptide of *Rhodococcus erythropolis* is important for the formation of CPs, while the β-propeptide of *Mycobacterium tuberculosis* mediates temperature-dependent inhibition of CP assembly [41, 42]. By contrast to the actinobacteria and yeast, the β-propeptides of archaea appear to be dispensable for CP assembly. Archaeal β-subunits can be produced in recombinant *E. coli* without an N-terminal propeptide (βΔpro) and then reconstituted with α-type subunits *in vitro* to form active CPs (*i.e., Methanocaldococcus jannaschii* CPs) [43]. Likewise, archaeal CPs can form active complexes when βΔpro variants are co-expressed with α-type subunits in recombinant *E. coli* (*i.e., T. acidophilum* and *M. thermophila* CPs) [30, 31]. By modulating the ionic strength of the buffer, haloarchaeal CPs (of *H. volcanii*) can also be disassembled into monomers and reconstituted into active CPs (based on gel filtration and activity assays) [37].

While archaeal β-propeptides are not required for the generation of active CPs *in vitro* or when produced in *E. coli*, β-propeptides are likely to be important in the proper formation of proteasomes within archaeal cells. In support of a function for β-propeptides in Archaea, genetic deletion of the 49-residue propeptide of the *H. volcanii* β-subunit results in undetectable levels of this protein (βΔpro) (either by immunoblot or His-tag purification) when expressed in its native host *H. volcanii* (Kaczowka and Maupin-Furlow, unpublished). In contrast, *H. volcanii* βΔpro is readily detected and purified (using a His-tag) when expressed in a heterologous host, *E. coli* [44]. Likewise, the β-subunit is synthesized and purified at high levels when the β-propeptide is added to the *H. volcanii* expression system [44]. Whether or not these findings are unique to *H. volcanii* remains to be determined.

Notably, the N-terminal propeptides (typically over 30 residues) of the β-subunits of haloarchaea are predicted to be the longest amongst all Archaea. Haloarchaea are unusual in that they maintain osmotic balance with hypersaline environments by accumulating molar concentrations of KCl in their cytoplasm [45]. In order to cope with these harsh conditions, haloarchaea synthesize proteins with a high negative surface charge, which makes their proteins more soluble and flexible at high concentrations of salt compared to non-halophilic proteins [46]. Thus, the extended N-terminal propeptide of β-subunits may be needed to facilitate folding and assembly of the CP in the 'harsh' cytosolic environment found in haloarchaeal cells.

Alternatively, given that N^{α}-acetylation is quite common in haloarchaea [47–49], the N-terminal propeptide of β-subunits may prevent N^{α}-acetylation of the active site threonine residue as has been observed in yeast.

CP Active Site Number and Subunit Complexity

While all proteasomes to date have 14 α-type and 14 β-type subunits assembled into each CP, the ratio of β-type subunits with N-terminal threonine active site residues per CP varies among the three domains of life. These differences in active site number do not appear to influence the overall size of peptide products generated nor the rate of peptide bond hydrolysis, based on comparison of archaeal and eukaryotic proteasomes [50, 51].

In general, proteasomal CPs from eukaryotes have less proteolytic active sites per particle than CPs from either archaea or actinobacteria. Eukaryotic CP subtypes have six active sites per particle formed by three of the seven different β-type subunits per heptameric β-ring [52]. For example, within the housekeeping CP of eukaryotes only the β1, β2 and β5 subunits are proteolytically-active (the β3, β4, β6 and β7 subunits are inactive). In response to cytokines and microbial infections, β1, β2 and β5 are replaced by β1i, β2i and β5i, which are also active and alter CPs to generate peptide products optimized for MHC class I loading in hematopoietic cells [53].

In contrast to eukaryotes, all archaeal (and actinobacterial) CPs purified to date have 14 active sites per particle and are typically purified as a cylindrical ~600–700 kDa complex of a single type of β subunit that associates with a single type of α subunit (Fig. 11.1). Thus, most archaeal CPs have identical proteolytic active sites with some determined by N-terminal sequencing of the β-subunits [34, 36, 37, 54]. Examples of CPs purified from Archaea (or recombinant *E. coli* expressing these archaeal genes) composed of only a single type of α- and β-subunits, include CPs from *Thermoplasma acidophilum* [55], *Thermoplasma volcanium* [56], *Methanosarcina thermophila* TM-1 [30, 36], *Methanosarcina mazei* [57], *Methanocaldococcus jannaschii* [43], and *Archaeoglobus fulgidus* [58].

Although Archaea with CPs composed of only a single type of α- and β-subunit have received most attention, as simplified models for biochemistry, the majority of Archaea are predicted to encode multiple α- and/or β-subunit homologs based on genome sequence. Most Crenarchaeota encode single α-type and two different β-type subunit homologs, while Thaumarchaeota, Korarchaeota and select families of Euryarchaeota (*Halobacteriaceae* and *Thermococcaceae*) commonly encode multiple α-type and/or β-type subunit homologs. Interestingly, the recently described Candidatus *Caldiarchaeum subterraneum* is predicted to encode three different β-subunits and a single α-subunit. Many of the Archaea with two or more coding sequences for β-subunits harbour one gene that encodes a β-subunit devoid of the classical G↓T cleavage motif, suggesting that the protein is synthesized either as an inactive β-subunit or has a propeptide cleavage site that differs from the CPs characterized from Euryarchaeota and eukaryotes.

Archaeal CPs composed of a single type of α subunit and two different β-type subunits (both containing a G↓TTT cleavage motif) have been characterized at the biochemical level. Initially, CPs of only a single α/β composition were purified from the archaeon *Pyrococcus furiosus* [34]. However, DNA sequencing of the complete genome of this hyperthermophile predicted that two β-type subunits (β1 and β2) and a single α-type subunit [59] existed in this species. Further analysis by microarray revealed that the level of the β1 transcripts were upregulated after cells were exposed to heat shock, which might explain why only one β-type protein was detected in CPs purified from *P. furiosus* grown under normal (non-stressed) conditions [60]. To examine the function of CPs which contain multiple β-type proteins, the α-type and both β-type proteins were produced in *E. coli* and reconstituted at different temperatures *in vitro* [60]. Using this approach, an increased ratio of β1 to β2 was observed in CPs assembled at higher temperatures, which correlated with more thermostable properties. Surprisingly, the α and β1 proteins could not be reconstituted into CPs or form particles with peptide hydrolyzing activity independent of β2. Based on these results, the *P. furiosus* β1 is thought to stabilize α2/β CPs at high temperature and to influence the biocatalytic properties of the proteasome.

CP subtypes, of different α- and β-type subunit composition, have also been directly purified from an archaeon. The haloarchaeon *H. volcanii* encodes a single β-type and two α-type (α1 and α2) CP subunits [37]. Not surprisingly, the β-type gene is essential for growth and the presence of one of the two α-type genes is required for growth (based on conditional mutation) [61]. All three of the different CPs (α1β, α2β and α1α2β) have been purified from *H. volcanii* [37, 44, 62]. Each of these different CP subtypes are active in the hydrolysis of peptides and unfolded proteins with the α1α2β-proteasome apparently asymmetric based on the detection of homooligomeric rings of α1 and α2 [44, 62]. The population of CP subtypes is likely altered during growth based on the finding that levels of α2 increase several-fold as cells transition to stationary phase, while α1 levels remain relatively constant [63]. Although it remains to be established, differences in α-type subunit composition are thought to influence the type of regulator that associates with the CPs in *H. volcanii*. Interestingly, in *H. volcanii* the CPs (and PAN ATPases) are also modified post-/co-translationally (phosphorylation, N^{α}-acetylation and methylation) suggesting an additional layer of regulation in modulating the populations of proteasomes in archaeal cells [64–66].

Proteasomal Regulators

Eukaryotic proteasomal CPs co-purify with a variety of accessory proteins including ATPase regulators, such as the 19S regulatory particle (RP), and non-ATPase regulators such as Blm10/PA200 and the 11S regulators PA28, PA26 and REG [67]. Of these regulators, the 19S RP can associate with either end of the CP cylinder to form the 26S proteasome that catalyzes the ATP-dependent degradation of proteins (Fig. 11.3) [22]. Subunits of the RP are grouped into two major categories including

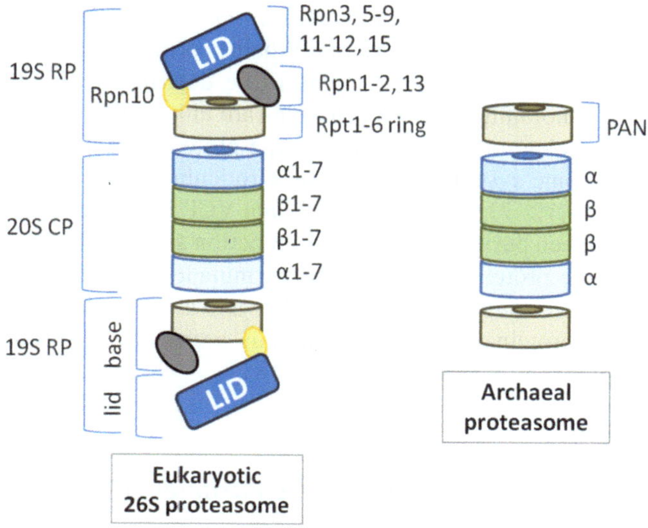

Fig. 11.3 ATP-dependent proteasomes of eukaryotes compared to archaea. Proteasomal CPs can associate on each end of their cylindrical structure with heptameric rings of AAA+proteins including eukaryotic Rpt1-6 and archaeal PAN. Rpt1-6 are AAA+subunits of the 19S regulatory particle (*RP*) that associates with CPs to form 26S proteasomes. In yeast, the 19S RP can be dissociated into lid and base subcomplexes by deletion of the *RPN10* gene. PAN is an AAA ATPase that forms a hexameric ring and associates with CPs *in vitro*. PAN is common to many but not all archaea

the RP non-ATPase subunits (Rpn) and the RP AAA+(triple-A+) subunits (Rpt) of the ATPases associated with various cellular activities (AAA) family within the AAA+superfamily [68]. In yeast, RPs are dissociated into base and lid subcomplexes by deletion of the gene encoding the Rpn10 subunit [69] (Fig. 11.3). The base subcomplex is composed of Rpn1-2, Rpn13 and six different Rpt subunits (Rpt1-6) that form a hexameric ring in direct contact with the outer rings of the CP complex [69].

While archaeal CPs have yet to be purified in complex with regulatory proteins from its native host, archaeal CPs can associate with ATPase regulators *in vitro* (Fig. 11.3). Studies of this association were initiated after release of the first archaeal genome sequence (*M. jannaschii*), which was predicted to encode a close homolog of the Rpt1-6 subunits of eukaryotic 26S proteasomes [70]. The archaeal Rpt homolog (termed PAN for its role as a proteasome-activating nucleotidase) was synthesized and purified from recombinant *E. coli* and shown to be required for the degradation of substrate proteins (e.g., casein) by CPs including those of *T. acidophilum* [71] and *M. jannaschii* [43]. *M. jannaschii* (Mj) PAN forms a relatively simple homohexameric ATPase complex and, thus, has served as an ideal model to understand how Rpt-like proteins function in the degradation of proteins by proteasomes (see later). While all Archaea encode proteasomal CPs based on genome sequence, many do not encode Rpt/PAN homologs including archaea of the Thaumarchaeota, Korarchaeota, *Thermoplasmata* class of Euryarchaeota, Aigarchaeota (Candidatus *Caldiarchaeum subterraneum*), and Thermoproteales order of Crenarchaeota. Whether or not ATPases beyond PAN can regulate these archaeal CPs remains to be determined. Members of the Cdc48/VCP/p97 subfamily of AAA+proteins are universally distributed among archaea, and, in eukaryotes, these Cdc48-type proteins are associated with the ubiquitin-proteasome system in endoplasmic reticulum-associated protein degradation (ERAD) [72, 73]. The *T. acidophilum* VCP is an archaeal Cdc48-type ATPase that has been purified and shown to function as an ATP-dependent unfoldase and stimulate protein degradation using mutant CPs with open gates [74] (see below). More recently, VCP has been shown to associate with CPs and stimulate peptide and protein degradation [75]. Thus, archaeal CPs appear to be regulated by multiple types of AAA+proteins.

Gated Openings of Proteasomal Core Particles (CPs)

The proteolytic active sites of proteasomes are sequestered within the interior of the chambered CP, which has gated openings to control substrate entry. The first 3-D structure of the proteasome, which provided insight into its compartmentalization, came from the CP of the archaeon *T. acidiophilum* (TaCP) that was produced in *E. coli* [75, 76]. The TaCP structure revealed narrow (13 Å) openings on each end of the cylindrical particle (113 Å diameter by 148 Å length) that connected a central channel leading to an interior chamber lined with proteolytic active sites. Subsequent, X-ray structures of yeast and bovine CPs revealed particles with no apparent openings on the ends for substrates to access the proteolytic active sites or for products to exit the chambered particle [77, 78]. The N-terminal tails of α-type subunits within the eukaryotic CP structures appeared to block the channel openings and, thus, were thought to gate substrate entry (Fig. 11.4).

With the yeast CP as a model, genetic modifications to residues associated with the gate and structural analysis of CPs associated with regulators thought to open the gate provided evidence for a gating mechanism in eukaryotes. In particular, genetic deletion of the N-terminal tail of the yeast α3 (α3ΔN) was found to open the channel and derepress the peptide hydrolysis mediated by CPs [79]. The α3ΔN mutation also increased the size of CP-generated peptide products suggesting CP gating controls product release in addition to substrate entry [80]. Binding of regulators to CPs *in vivo,* was thought to relieve the inhibition mediated by the α-subunit tails and open the CP gates [79]. Indeed, deletion of the gate (N-terminal tail of α3) in mutant proteasomes containing an active site mutation in the ATPase subunit (Rpt2) was sufficient to overcome defects in peptide hydrolysis [80]. Likewise, crystal structures of yeast CPs in complex with the non-ATPase activator PA26 revealed, that loops in PA26 could induce conformational changes in the α-subunit tails to open the gate, separating the CP interior from the intracellular environment [81].

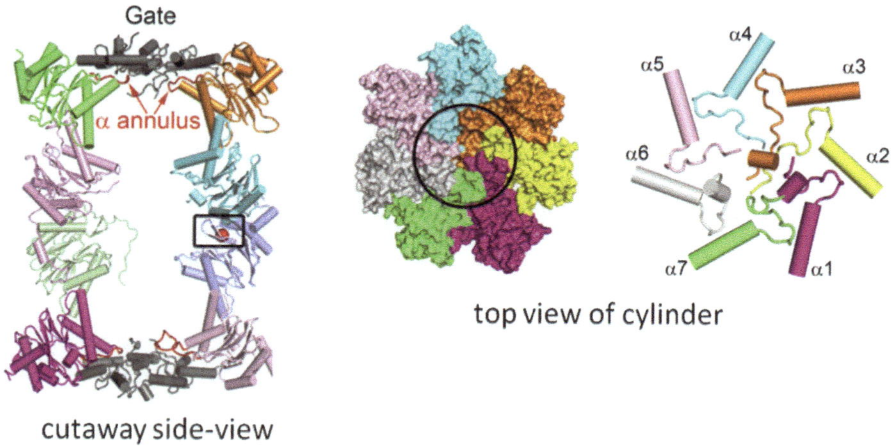

Fig. 11.4 Eukaryal proteasomal CPs are gated. In X-ray crystal structures of eukaryotic (e.g., yeast and bovine) CPs, the central channel used for protein substrate entry is gated by the N-terminal tails of α-type subunits (Figure modified from [67] with permission)

Although gated openings were not evident in the early archaeal CP crystal structures (most likely due to the disordered structure of the α subunit N-terminal tail) [58, 76], several lines of evidence now support a gating mechanism for archaeal CPs. For example, deletion of the N-terminal tail of the *T. acidophilum* α-subunit (αΔN) reduces the central mass of heptameric rings formed by the α-subunit in electron micrographs, supporting the localization of these tails near the central channel of mature CPs [82]. Increased rates of unfolded protein hydrolysis are also observed for TaCPs containing αΔN-subunits compared to wild-type suggesting that deletion of the N-terminal residues of the α-subunit generates a "gateless" CP [82, 83]. The recent crystal structure of a CP assembled from *M. jannaschii* α- and β-subunits produced in *E. coli* also provides evidence that an archaeal CP can be stabilized in closed gate conformation [84]. Furthermore, addition of specific peptides (based on the C-terminal tail of MjPAN) causes a conformational change in TaCPs that is consistent with gate opening (Fig. 11.5) (see later) [85, 86]. Interestingly, the N-terminal tails of the α-subunits that form the gates can be highly dynamic and extend inside and outside the CP cylinder, based on methyl-transverse relaxation optimized nuclear magnetic resonance spectroscopy experiments of TaCPs [87].

Functional Association of Proteasomal ATPases and CPs

While proteasomal CPs have yet to be purified in association with defined regulatory proteins from archaeal cells, MjPAN (an archaeal Rpt homolog) can be reconstituted with CPs *in vitro*. Formation of MjPAN:TaCP complexes has been detected by immunoprecipitation, electron microscopy and surface plasmon

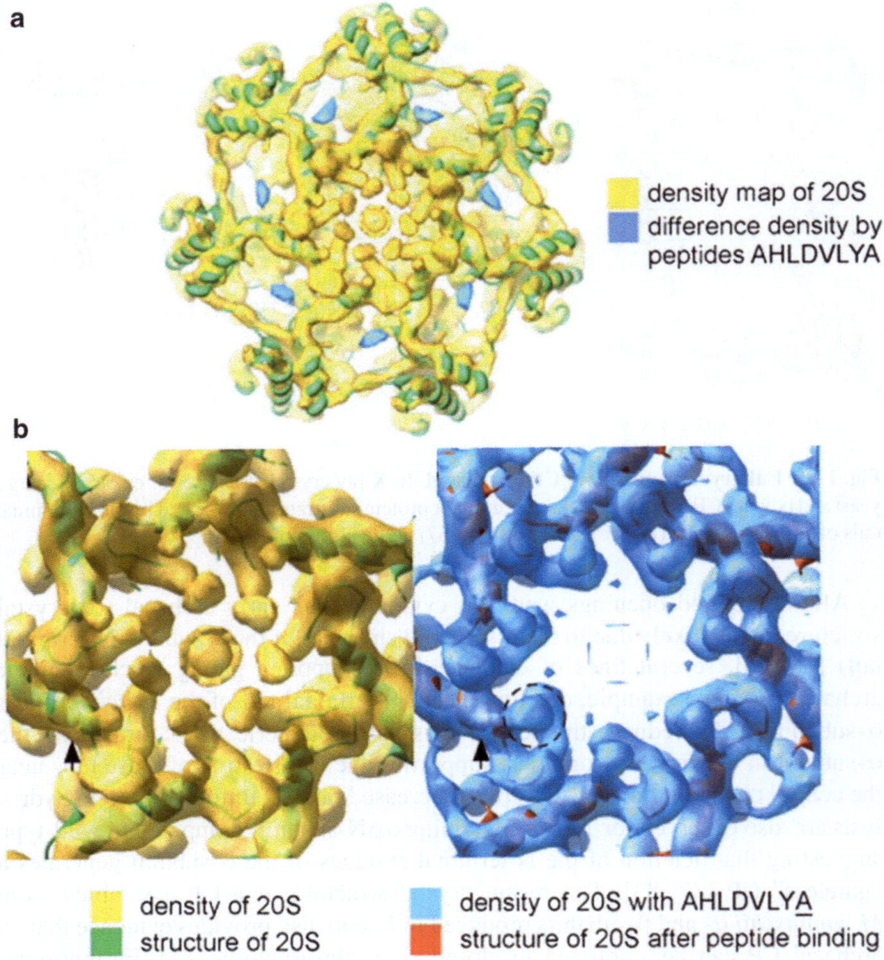

Fig. 11.5 Archaeal proteasomal CPs are gated. Cryo-electron microscopy reveals: (**a**) sites within the intersubunit pockets of α subunits in the TaCP that are bound by peptides mimicking the C-terminal tail of MjPAN and (**b**) structures of the TaCP gate in the closed and open forms induced by these peptides (Figure modified from [85] with permission)

resonance. These complexes form when recombinant MjPAN and TaCP are incubated in the presence of ATP or a non-hydrolyzable ATP analogue (AMPPNP or ATPγS), but not when incubated with ADP or in the absence of nucleotide [83]. Thus, ATP binding is required for MjPAN:TaCP association. In electron micrographs of negatively stained complexes, MjPAN is associated as a two-ringed structure that caps either one or both ends of the TaCP [83]. The outer ring of MjPAN (distal to the TaCP) is thought to correspond to the N-terminal domain of the protein based on analogy to a related AAA+complex from bacteria, HslU [88–90]. The C-terminal domain of MjPAN is important for docking with CPs and opening the CP gates (*see* below).

Docking of the Proteasomal ATPase (PAN) to the CP

Proteasomal ATPases such as MjPAN have a C-terminal tail that is essential for docking with CPs and opening the CP gates. A series of elegant studies focused on the association of the archaeal MjPAN with CPs facilitated this discovery [85, 86, 91]. These studies on MjPAN were guided by similarities to the bacterial HslUV protease and eukaryotic non-ATPase 11S regulators (PA28 and PA26) in association with proteasomal CPs. While somewhat controversial at the time, the C-terminal domain of HslU (the AAA+component of the HslUV protease) was thought to be required for association with HslV [90, 92]. Likewise, the extreme C-terminus of the 11S regulator PA28 was known to be required for association with CP [93], and structural analysis revealed that the C-terminal tail of PA26 (a related 11S regulator) docked into pockets of the CP, formed by adjacent α-subunits [81, 94].

Biochemical studies using MjPAN, demonstrated that a tripeptide motif (HbYX, where Hb=a hydrophobic amino acid and x=any amino acid) located at the extreme C-terminus of the proteasomal ATPase was important for proteasome function. Removal of the MjPAN C-terminus dramatically reduced its ability to stimulate the TaCP-mediated degradation of a nine-residue long peptide substrate [83]. Importantly, this MjPAN variant retained wild type ATPase activity, unfoldase activity and ability to stimulate hydrolysis of small peptides by TaCPs [83]. Systematic mutation of the C-terminal residues of MjPAN demonstrated that both the length of this tail and penultimate tyrosine residue of the HbYX motif was absolutely essential for MjPAN to stimulate the TaCP-mediated degradation of LFP and, thus, CP gate opening (where LFP represents a nine residue peptide substrate) [83]. To further investigate this activation step, a tryptophan residue was introduced into either the Hb or X position of the HbYX motif in MjPAN, and analyzed by tryptophan fluorescence and polarization [83]. This approach provided further evidence that MjPAN could associate with TaCP and suggested that the C-terminal tail of MjPAN moves from an aqueous to a hydrophobic environment upon association. Interestingly, peptides based on the C-terminus of MjPAN could compete with MjPAN for TaCP binding and serve as "gate openers" based on their stimulation of TaCP-mediated hydrolysis of LFP [83]. In contrast, peptides based on the C-terminus of the non-ATPase 11S regulator PA26 were unable to open the TaCP gate but could compete with MjPAN for TaCP binding [83]. These findings were consistent with studies, which revealed docking of the C-terminal tail of PA26 to the CP was associated with binding of a distant activation domain of PA26 to a region of the α-subunit N-terminal tails that form the CP gate [81, 94]. Thus, MjPAN and PA26 appear to use a different mechanism to open the CP gate.

Structural studies have been performed to further understand how proteasomal ATPases with C-terminal HbYX motifs open CP gates. In particular, cryoelectron microscopy (cryoEM) has been used to identify sites in the TaCP that are bound by peptides mimicking the C-terminal tail of MjPAN and to determine the structures of the TaCP gate in the "closed" and "open" forms induced by these peptides (Fig. 11.5) (gated and ungated openings of 9 and 20 Å diameter, respectively) [85]. An artificial hybrid activator has also been constructed using the heptameric PA26 structure as a scaffold for attachment of eight residues corresponding to the C-terminal tail of MjPAN [86]. This hybrid activator is heptameric and forms a stable complex with the TaCP in which the activator caps both ends of the TaCP and opens the gate of the CP through an HbYX-dependent mechanism. Using single particle cryoEM and X-ray crystallography, the structure of the hybrid activator was determined in association with TaCP. In both studies [85, 86], residues corresponding to the MjPAN C-terminal tail were found to bind to the α-rings of TaCP in the 'inter-subunit' pocket (between the adjacent α-subunits) and induce and stabilize the open-gate conformation of TaCP. Although PA26 and MjPAN bind to a similar pocket on the CP, PA26 also requires binding to the CP α-subunit N-terminal tails, by a distant activation domain, to open and stabilize the CP gates [94, 95].

Interestingly, most haloarchaea and many methanogens are predicted to encode two distinct homologs of PAN, often one of which is lacking the conserved C-terminal HbYX motif. Thus, the C-terminal motif needed for CP interaction and gate opening, by PAN, may vary among archaeal species or alternatively archaea may synthesize PAN subtypes of different functions (e.g. PANs that unfold proteins but do not bind the CP or PANs that bind the CP but do not open the gates). Among the C-terminal sequences predicted for haloarchaeal PANs, ~43% include HbYX motifs, ~33% possess TFA motifs and ~23% harbour C-terminal tails that lack tyrosine or phenylalanine in the penultimate position. In *H. volcanii*, two PAN proteins have been identified, PAN-A with a C-terminal sequence, AFA and PAN-B with a C-terminal sequence, YQY [63]. As the penultimate tyrosine or phenylalanine of C-terminal tails appears essential for opening CP gates, the PAN-A and PAN-B proteins may have distinct roles in regulating proteasome-mediated degradation. Consistent with this possibility, a number of biological differences have been detected for the *H. volcanii* PAN-A and PAN-B proteins that support distinct functional roles. First, PAN-A is synthesized at high levels throughout growth, while the levels of PAN-B are low and induced several fold as cells transition to stationary phase [63]. In addition, while a phenotype for *panB* mutant strains has yet to be identified, deletion of *panA* renders cells more resistant to thermal stress and more sensitive to nitrogen limitation, hypo-osmotic stress, and exposure to L-canavanine [62]. In further support of functional differences between the two PAN proteins, the *panA* mutation can be complemented *in trans* by providing a wild type copy of *panA* but not *panB* [62].

Multiple Roles of the ATPase (PAN) in CP-Mediated Proteolysis

MjPAN is not only required for CP gate opening, but also for the unfolding and translocation of protein substrates for hydrolysis by the CP. Indeed "gateless" TaCP (containing αΔN subunits) still require MjPAN and ATP for the degradation of globular substrates (such as SsrA-tagged GFP) [82]. Thus, while the N-terminal tails of CP α-subunits appear to serve as a gate to prevent entry of large peptides as well as folded and unfolded proteins, removal of this gate is not sufficient to stimulate the degradation of structured proteins. This type of proteolysis still requires hydrolysis of ATP by an AAA+regulatory component, such as MjPAN in association with the CP.

Archaeal PAN X-ray Crystal Structures

In the quest to solve the overall structure of an energy-dependent proteasome, a number of recent breakthroughs have been made examining the structure of various archaeal PAN subdomains. In one study, MjPAN was subjected to limited proteolysis resulting in the generation of two subcomplexes (I and II) that were amenable to X-ray crystallography [96]. Subcomplex I, which forms a stable hexamer, was derived from a fragment that spanned residues 74–150 including a portion of the predicted N-terminal coiled-coil (CC) domain (Fig. 11.6). In the subcomplex I structure (resolved to 2.1 Å), the CC domains of the six protein monomers formed pairs that protruded from a donut-shaped particle with an axial pore of 13 Å and an unexpected oligonucleotide/oligosaccharide binding (OB)-fold domain (Fig. 11.6) [96]. In contrast, subcomplex II, which dissociated into a mixture of hexamers to monomers, was composed of a fragment spanning residues 155–430 including the conserved AAA+domain and contained the C-terminal HbYX motif (Fig. 11.6). Given the level of conservation between MjPAN and the bacterial HslU, the atomic coordinates of MjPAN were superimposed onto the six subunits of HslU to generate a hexameric model for MjPAN (Fig. 11.6) [88, 90, 96]. Guided by these structures, a mutagenesis study, examined which amino acids and structural motifs were required for MjPAN function [84]. The structures of *Archaeoglobus fulgidus* PAN (AfPAN) N-terminal subdomain containing a eukaryotic GCN4 leucine zipper (in place of its CC-domain) has also helped to elucidate how folded proteins are degraded by archaeal proteasomes [97].

Proteasome-Mediated Proteolysis

From a combination of structural and biochemical information, a complete view of how the proteasomal ATPase component functions in the unfolding of proteins for translocation into the CP for degradation is beginning to emerge. Based on electron micrographs of PAN:CP complexes and mass spectrometry of purified PAN complexes, archaeal PAN is thought to function as a hexameric complex [83, 98], with its C-terminal domain facing the cylindrical ends of four-stacked heptameric rings that form the CP (Fig. 11.6). The central channel of PAN is aligned with the gated central channel of the CP, to form a long tunnel with narrow restrictions for the passage of substrate proteins to the central proteolytic chamber. The N-domain of PAN is positioned distal to the CP with the N-terminal CC domains (one from each subunit) associating in pairs, to form three tentacles that appear to stretch out and

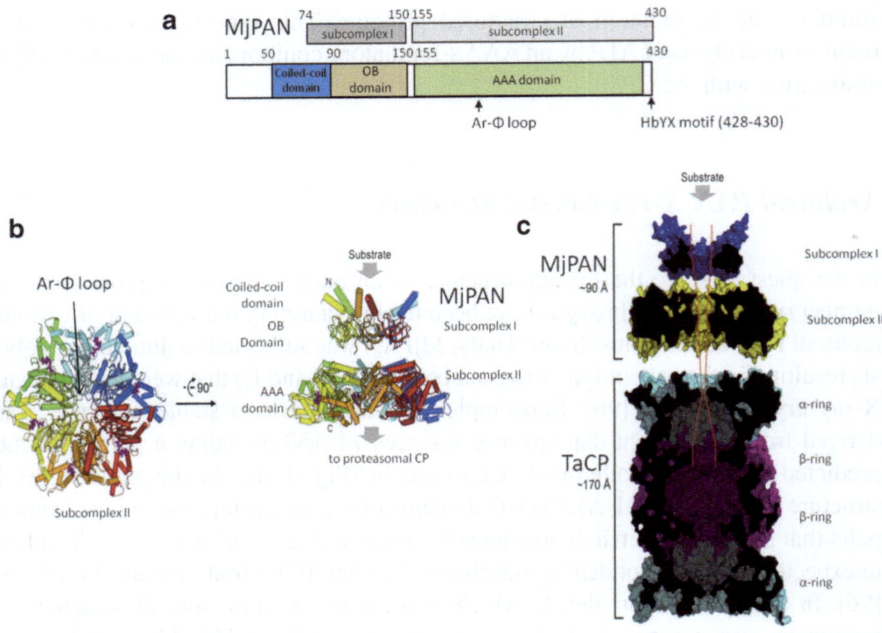

Fig. 11.6 Archaeal PAN subdomain structures provide a foundation for modelling the overall structure of an energy-dependent proteasome. (**a**) and (**b**) Subcomplex I and II generated by limited proteolysis of MjPAN are indicated with the N-terminal coiled-coil (*CC*), OB fold and AAA domains highlighted. (**c**) Evidence supports the docking of MjPAN C-terminal tails with pockets formed at the α-α intersubunit interface of TaCPs. The N-terminal CC-domains are assumed to be distal to the CP and act as tentacles that grab protein substrates, thus, enabling the Ar-Φ loop within subcomplex II to grip regions of the substrate that extend through the pore formed by the OB fold domain. The Ar-Φ loop is also thought to undergo ATP-fuelled conformational changes resulting in pulling and tugging at the protein substrate while the OB fold domain provides a passive force that blocks the movement of folded protein structure through the pore. As the protein is unfolded by this mechanism, it is translocated through the ATPase into the central channel of the CP for degradation (Figure modified from [84] with permission)

search for protein substrates (or possibly protein partners). The OB domain forms a stable ringed-platform for this network of tentacles and a relatively narrow pore (16 Å in diameter) that is thought to function as a molecular sieve restricting access of folded proteins from entering the central channel of the ATPase. While early evidence suggested MjPAN could unfold proteins on its surface in the absence of translocation [99], more recent work supports a model in which proteins are unfolded by energy-dependent translocation through the ATPase ring and that this can be coupled to threading the protein into the CP for destruction [96, 100]. Consistent with a threading model, the Ar-Φ loop (where Ar is any aromatic amino acid and **F symbol** is any hydrophobic amino acid), also known as the pore-1 loop, lines the central passage of the AAA+domain and defines one of the constriction points proximal to the OB domain in the subcomplex II model (Fig. 11.6) [84, 96]. Similar to related AAA+proteins, the Ar-Φ loop of PAN is thought to facilitate protein unfolding by gripping and tugging the substrate protein into the ATPase channel by ATP fuelled motions of the AAA+domain [96, 101–105]. The OB domain is proposed to exert a passive force on protein unfolding by providing a pore with a stable platform that blocks the movement of folded protein structure through the pore as the Ar-Φ loops tug and pull the protein [84]. Ultimately, the unfolded protein is translocated for degradation by the proteolytic active sites sequestered within the central chamber of the CP.

Archaeal PAN – An Ordered Reaction Cycle

Interestingly, subunits of MjPAN appear to bind ATP in pairs, which results in distinct effects on proteasome function that imply an ordered reaction cycle. Subunits of MjPAN are thought to partner with subunits opposite each other, in the hexameric ring, team up and cycle around the ring like a clock between one of three different states including ATP bound, ADP bound and nucleotide free [106]. Thus, MjPAN subunit pairs mediate an ordered reaction cycle of ATP binding, hydrolysis and release that coordinates and drives conformational changes around the MjPAN ring. These ATP-driven conformational changes are proposed to be critical for MjPAN-mediated protein unfolding. In particular, the Ar-Φ loop within the channel of the MjPAN ring is thought to grip proteins, tugging them up and down with repeated cycles of ATP hydrolysis. Tugging the proteins through the pore formed by the OB-fold domain is likely to serve as a rigid platform for resisting the entry of protein structure and, thus, facilitating an unfolding process through the central channel of the PAN ATPase.

Non-ATPases Associated with Archaeal CPs

In addition to AAA+regulatory proteins such as PAN, archaeal CPs are also proposed to associate with non-ATPase components. Early evidence for this possibility, arose when a protein inhibitor was co-purified with *T. acidophilum* CP [107], however the identity of the 20 kDa subunit of the inhibitor remains to be determined. More recently, immature archaeal CPs (composed of β-subunits that retain the N-terminal propeptide due to a Thr to Ala active site mutation) have been shown *in vitro*, to associate with proteins of the DUF75 superfamily [108]. Archaea and eukaryotes typically encode at least two members of the DUF75 superfamily that form heterodimeric complexes [Pba1-Pba2 (yeast), PAC1-PAC2 (mammals) and PbaA-PbaB in the archaeon *Methanococcus maripaludis*] [108–111]. Interestingly, the archaeal PbaA and eukaryotic Pba1/PAC1 proteins all have a conserved C-terminal HbYX motif that is required for binding the immature CPs [108]. Based on site-directed mutagenesis, this C-terminal tail docks to the same surface pocket on the α-subunit of the CP that is used by ATPase activators such as MjPAN [108]. Unlike MjPAN which opens the CP gate, the DUF75 proteins do not appear to alter CP activity and instead are thought to shield proteasomal subunit intermediates from nonproductive associations until assembly is complete or protect the cell from misassembled complexes [109–111]. While DUF75 protein homologs are also found in actinobacteria, their function is thought to be distinct from the DUF75 proteins of Archaea and eukaryotes due to the addition of a large C-terminal domain and absence of a C-terminal HbYX motif [108].

Protein Modifications

Ubiquitylation and Its Role in Marking Proteins for Proteolysis by Proteasomes in Eukaryotes

Ubiquitin is a highly conserved small protein modifier that is covalently attached to substrate proteins by a process termed ubiquitylation (Fig. 11.7a) [112]. Ubiquitylation involves the covalent attachment of the C-terminal carboxylate of Gly76 of ubiquitin to a protein substrate. Ubiquitin can be attached to proteins by an isopeptide bond to the ε-amino group of substrate lysine residues, a thioester bond to substrate cysteine residues, an oxy-ester bond to substrate serine or threonine residues, or peptide bond to the N^{α}-amino group of a protein substrate [113, 114]. The attachment of a single moiety of ubiquitin to a protein (mono-ubiquitylation) can alter the activity and/or location of the protein and can even signal the protein for destruction in lysosomes [115–117]. Chains of ubiquitin can also form through isopeptide bonds between the C-terminal carboxyl group of an incoming ubiquitin to one of the seven lysine residues of a ubiquitin (K6, K11, K27, K29, K33, K48 and K63) moiety covalently attached to a protein substrate [118]. Of these chains, the K63- and K48-linked chains of ubiquitin are best characterized. The K63-linked ubiquitin chains have non-proteolytic roles such as signalling for sorting into the multivesicular body pathway [119], while the K48-linked ubiquitin chains typically target a protein for destruction by 26S proteasomes [120].

Ubiquitylation involves a series of enzyme catalyzed reactions with the formation of adenylated ubiquitin and thioester intermediates that result in the covalent attachment of ubiquitin to proteins (Fig. 11.7a) [121–123]. In the first step, a ubiquitin-activating enzyme (E1) adenylates the C-terminal carboxylate of Gly76 of ubiquitin (part of the a conserved diglycine motif, Gly75Gly76) in an ATP-dependent reaction that releases PPi. The adenylated form of ubiquitin is attacked by a conserved cysteine residue of E1 to form a thioester intermediate between the C-terminal Gly76 of ubiquitin and the conserved catalytic cysteine of the E1. In the next step, a conserved cysteine residue of an ubiquitin-conjugating enzyme (E2) attacks the E1-ubiquitin thioester intermediate. This E2-mediated reaction results in the transfer of ubiquitin to the E2 enzyme and formation of a thioester bond between the E2 and the C-terminal Gly76 of ubiquitin. Finally, ubiquitin is transferred to the substrate protein with assistance from an ubiquitin ligase (E3). The E3 enzyme either directly transfers ubiquitin from E2 to the protein substrate or forms a thioester intermediate between a conserved cysteine of E3 and the C-terminal Gly76

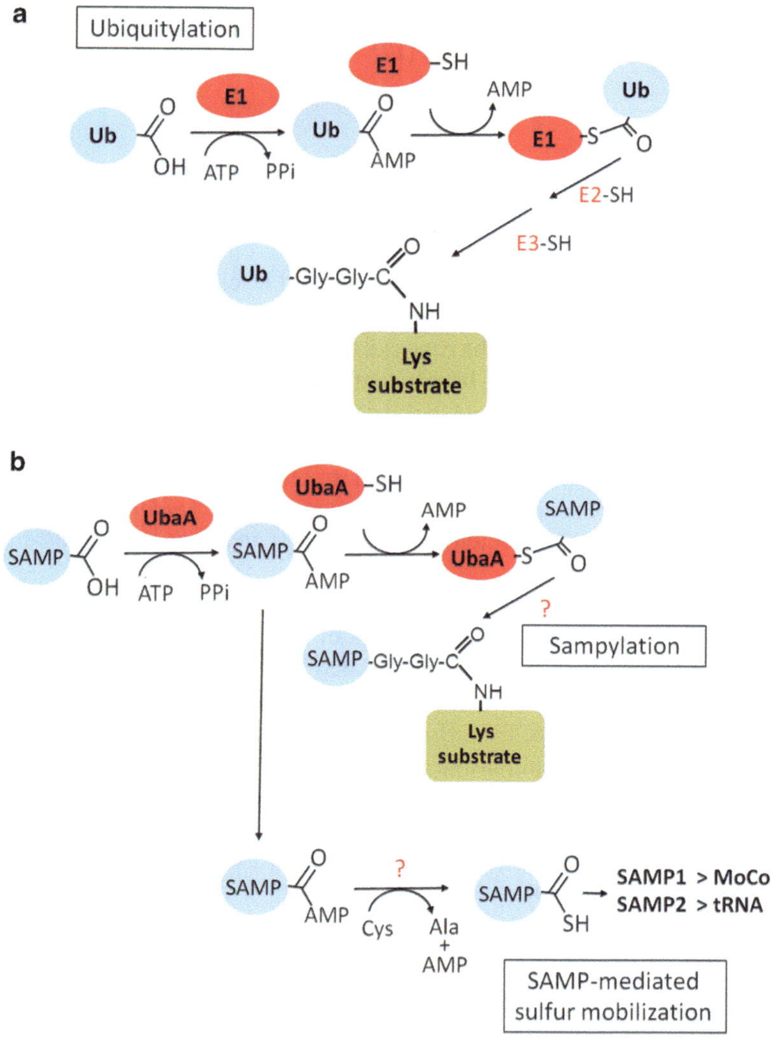

Fig. 11.7 Ubiquitylation and sampylation. (**a**) Eukaryotes use an elaborate E1-E2-E3 mediated mechanism for the attachment of ubiquitin to protein substrates. (**b**) Similarly to eukaryotes, small archaeal ubiquitin-like modifier proteins (termed *SAMPs*) can form protein conjugates in the archaeon *H. volcanii* by a pathway that is dependent upon the synthesis of an E1 homolog (termed *UbaA*). Based on genome sequence, E2 and E3 homologs are not predicted for this pathway. UbaA and SAMP proteins appear to also be linked to sulfur incorporation pathways (such as the biosynthesis of MoCo and thiolated tRNA) that require an E1-type adenylation reaction for the formation of a thiocarboxylated sulfur carrier protein with a ubiquitin-type β-grasp fold

of ubiquitin prior to transfer. In the formation of ubiquitin chains, the concerted action of the E1-E2-E3 enzymes is repeated with the transfer of ubiquitin to one of the seven lysine residues of ubiquitin that is attached to the modified protein [124]. In some cases, the ubiquitylation process is more elaborate with the use of a ubiquitin chain elongation factor termed an E4 [122], while other ubiquitylation events occur independent of an E3-ubiquitin ligase [125–127].

Multifunctional Roles of Proteins Related to Ubiquitin and E1 in Bacteria

While ubiquitin is not conserved, at the amino acid level, among either Archaea or bacteria, proteins with a predicted β-grasp ubiquitin-like fold are common to all organisms [128]. For example, the bacterial proteins ThiS and MoaD both exhibit a ubiquitin-like structure and have been extensively studied for their role in the incorporation of sulfur into thiamine and molybdenum cofactors, respectively [129, 130]. Additional biochemical roles for bacterial proteins with a ubiquitin-like structure have been identified that are independent of protein modification including QbsE-mediated thioquinolobactin siderophore biosynthesis in *Pseudomonas fluorescens* [131], CysO-mediated cysteine biosynthesis in *Mycobacterium tuberculosis* [132, 133] and others [134].

Similar to eukaryotic ubiquitin, bacterial proteins with a ubiquitin-like fold can be activated at their C-terminus by enzymes of the E1-like superfamily [135]. For example, the bacterial E1 homologs, MoeB and ThiF, associate with and adenylate the C-terminal carboxylate of their cognate ubiquitin-like partner protein, MoaD and ThiS, respectively [136, 137]. However, in contrast to ubiquitin, the adenylated form of the bacterial ubiquitin-like proteins is typically modified by an E1-like enzyme to accept sulfur from either a cysteine desulfurase or rhodanese [138, 139]. This transfer of sulfur results in the formation of a C-terminal thiocarboxylated form of the ubiquitin-like protein that can be used as a source of activated-sulfur for biosynthetic reactions [140].

E1 and Ubiquitin-Like Protein Homologs of Archaea

While all Archaea encode homologs of E1 and proteins with a ubiquitin-like protein structure based on genome sequence, the role of these proteins in archaeal cell function has only recently been examined. Using the halophilic archaeon *H. volcanii* as a model system, two ubiquitin-like proteins, termed SAMP1 and SAMP2 (small archaeal modifier protein 1 and 2) and an E1 homolog (UbaA, ubiquitin-like activating protein of Archaea) were examined for their role in protein modification and/or sulfur incorporation into biomolecules such as molybdenum cofactor and tRNA (2-thiouracil) [141, 142]. Similar to most archaeal species, halophilic Archaea are predicted to synthesize biomolecules that require sulfur such as molybdenum/tungsten cofactors [143]. Furthermore, like most Archaea, *H. volcanii* is predicted to encode multiple ubiquitin-like proteins and a single E1 homolog, but not E2 or E3 homologs [144–146]. Thus, Archaea were anticipated to use ubiquitin-like proteins for sulfur incorporation into biomolecules and not for protein conjugation [128].

Sampylation – An Archaeal Form of Ubiquitylation

Recently it was shown that ubiquitin-like proteins can be conjugated to proteins in Archaea, by a process termed sampylation that is dependent upon the synthesis of an E1 homolog (Fig. 11.7b). These surprising findings are based on the following experimental evidence. Using *H. volcanii* as a model system, the ubiquitin-like proteins SAMP1 (87 a.a.) and SAMP2 (66 a.a.) were expressed with an N-terminal FLAG-tag and the formation of SAMP1/2 protein conjugates was analyzed by α-FLAG immunoblot [141]. Under certain growth conditions, an array of large α-FLAG specific protein bands was detected for both FLAG-SAMP1 and FLAG-SAMP2 expression strains [141]. Deletion of the conserved C-terminal diglycine motif of the ubiquitin-like protein or deletion of the *ubaA* gene (encoding the *H. volcanii* E1 homolog UbaA) abolished the ability to detect SAMP protein conjugates in *H. volcanii* cells [141, 142]. This effect could be complemented *in trans* by the wild type *ubaA* (but not by *ubaA* with a mutation in the conserved active site cysteine) suggesting that SAMP1 and SAMP2 are activated by an E1-like mechanism [142].

To determine the type of peptide bond formed between archaeal ubiquitin-like proteins and their protein substrates, SAMP2 was selected for further analysis [141], as it contains a lysine residue immediately preceding its C-terminal diglycine motif (i.e. KGG). The location of this lysine residue enabled the use of tryptic digestion and tandem mass spectrometry (MS/MS) to analyze the SAMP2 protein conjugates, in a similar to approach to that used to analyze ubiquitylated proteomes of eukaryotes [147]. If an isopeptide bond was present, between the C-terminal carboxylate of Gly66 in SAMP2 and an internal lysine residue on the protein substrate, the modified lysine residue of the protein substrate would be resistant to cleavage by trypsin and hence retain the GG footprint derived from the SAMP2 C-terminus. With this approach, SAMP2-protein conjugates were isolated by α-FLAG affinity chromatography from an *H. volcanii* FLAG-SAMP2 expression strain (compared to the control strain) and isopeptide bonds were detected between Gly66 (of SAMP2) and the lysine residues of eight different proteins predicted to mediate a variety of functions, from metabolism to transcription [141]. Similar to ubiquitylation [148], SAMP2 modification (samp2ylation) was also detected on multiple sites within a single target protein (i.e. a TATA-box binding protein and tandem rhodanese domain protein) [141]. Furthermore, Gly66 of SAMP2 also formed an isopeptide bond with one of the two SAMP2 lysine residues (i.e. Lys58) revealing the presence of K58-linked SAMP2 chains [141]. Whether the SAMP2 chains are associated with a protein substrate remains to be determined. In eukaryotes, ubiquitin genes are translated into polypeptide chains of repeating units of ubiquitin fused head-to-tail [149–151] or ubiquitin fused to unrelated amino acid sequences [152]. These ubiquitin-protein fusions are cleaved by deubiquitylases to generate free pools of mature ubiquitin [153]. Unlike ubiquitin, SAMP2 is translated as a single protein. While speculative, it is possible that archaeal cells might regulate the pools of free SAMP2 by the synthesis and cleavage of unanchored SAMP2 chains. Alternatively, if SAMP2 chains are anchored, this might enhance the diversity of SAMP modifications available for protein targeting.

Using various strains of *H. volcanii* expressing either FLAG-SAMP1 or FLAG-SAMP2, together with α-FLAG immunoprecipitation, several interacting proteins were identified by tandem MS, following "in-gel" digested with trypsin [141]. Cells used for this analysis were grown in different culture conditions to enhance coverage of proteins modified by SAMP1 and/or SAMP2 (termed the "sampylome"). While SAMP modification sites were not mapped using this second approach, 32 proteins specific for the FLAG-SAMP expression strains were identified by MS/MS. These included the 8 proteins described above (with their SAMP2 modification sites mapped) as well as 24 additional homologs of proteins involved in ubiquitylation, sulfur incorporation, stress responses, metabolism and information processing such as transcription and translation [141].

Regulation of Sampylation

Environmental conditions can signal changes in the level of SAMP protein conjugates that are formed in *H. volcanii* cells. For example, growth in the presence of dimethyl sulfoxide and other culture conditions can increase the levels of SAMP protein conjugates [141, 142]. Thus, sampylation is regulated, most likely at the post-transcriptional level, based on the use of a constitutive promoter for expression of the genes encoding the SAMP proteins throughout these experiments [141, 142].

Functional Role of Sampylation

Sampylation (the formation of SAMP protein conjugates) appears to mark some proteins for degradation by proteasomes in Archaea. However, the evidence supporting a biological connection between sampylation and proteasomes remains indirect. The first piece of evidence in support of this association is that SAMP1-modified proteins accumulate in *H. volcanii* cells that contain chromosomal deletions of the genes encoding the proteasomal CP α1-subunit (*psmA*) or the Rpt-like ATPase PAN-A (*panA*) [141]. Thus, SAMP1-modified proteins may not be efficiently degraded by proteasomes in these mutant cells. In addition to this finding, many of the proteins that are modified by SAMP1 and/or SAMP2 or associated with these SAMP proteins [141] were also found to accumulate in *H. volcanii* cells that are disrupted in proteasome function either by deletion of *panA* [154] or treatment with the proteasome-specific inhibitor *clasto*-Lactacystin β-lactone [155]. However, currently there is no direct evidence that sampylation of a protein leads to its degradation by the proteasome.

To further understand the role of sampylation in archaeal cell function, the genes encoding the ubiquitin-like protein modifiers (SAMP1 and SAMP2) and the E1 homolog (UbaA) were deleted from the genome of *H. volcanii* [142]. The resulting mutant strains were examined for biochemical and phenotypic differences compared to wild type cells. With this approach, UbaA was found to be required for the formation (or stabilization) of SAMP1 and SAMP2 protein conjugates in *H. volcanii* [142]. Thus, an E1-type mechanism of SAMP protein activation has been proposed (Fig. 11.7b).

In addition to protein conjugation, UbaA and SAMP2 appear essential for the thiolation of tRNA (most likely 2-thiouridine) based on comparison of tRNA from wild type and mutant cells, using [(*N*-acryloylamino)phenyl]mercuric chloride (APM) gel electrophoresis followed by hybridization to a specific probe for lysine tRNAs with anticodon UUU ($tRNA^{Lys}{}_{UUU}$) [142]. Because E1- and ubiquitin-like proteins can catalyze sulfur incorporation into tRNA in bacteria and eukaryotes [156, 157], the biochemical differences in the level of thiolated tRNA in *H. volcanii ubaA* and *samp2* mutants (devoid of slow migrating "thiolated" $tRNA^{Lys}{}_{UUU}$) compared to wild type are thought to be independent of protein conjugation (Fig. 11.7b).

UbaA and SAMP1 also seem important for sulfur incorporation into the pterin-based molybdenum cofactor (MoCo) (Fig. 11.7b). This is based on the finding that, in comparison to wild type cells, *ubaA* and *samp1* mutant cells are unable to grow under anaerobic conditions with dimethyl sulfoxide (DMSO) as a terminal electron acceptor or produce DMSO reductase activity when switched to anaerobic conditions with DMSO [142]. DMSO reductase requires incorporation of MoCo into the catalytic subunit DmsA for activity. Similarly to wild type, the *ubaA* and *samp1* mutant strains were still able to produce transcript specific for *dmsA* (encoding the catalytic subunit of DMSO reductase).

Together, these results suggest UbaA and SAMP proteins mediate not only sampylation but also sulfur incorporation into tRNA and MoCo (Fig. 11.7b). The E1 homolog, UbaA, appears at the crossroads of activating the SAMP proteins for formation of an isopeptide bond between the C-terminal carboxylate of SAMP and the target protein and for the formation of a putative C-terminal thiocarboxylate group on the SAMP1 and SAMP2 proteins for sulfur transfer to MoCo and tRNA, respectively. Thus, any phenotypic differences that are detected for archaeal mutants of E1 and ubiquitin-like homologs (in comparison to wild type cells) will need to be further examined for whether these changes are due to perturbations in sampylation (protein conjugation) or sulfur transfer to biomolecules.

Summary

While Archaea are often thought of as prokaryotes, due to the absence of a membrane-bound nucleus, these incredible microbes have self-compartmentalized proteases and protein conjugation systems that are closely related to the ubiquitin-proteasome systems of eukaryotes. Unlike bacteria, which have multiple energy-dependent proteases in their cytosol [i.e. Lon, Clp and HslUV (or proteasomes in actinobacteria)], Archaea appear highly dependent on the proteasome system (with an unknown contribution by the membrane-bound Lon protease) as their major energy-dependent proteases. Based on numerous studies that have investigated how archaeal PAN and CP complexes function at the atomic level, an elaborate mechanism of how proteasomes degrade unfolded and/or folded proteins is rapidly emerging. Surprisingly, Archaea also contain a system in which ubiquitin-like proteins are conjugated to substrate proteins, through an apparent E1-like mechanism termed sampylation. Sampylation is predicted to exist in all Archaea and occurs independent of E2 or E3 homologs. Whether or not sampylation marks proteins for degradation by the proteasome remains to be determined; however, indirect evidence suggests these two pathways are functionally connected. Improving our understanding of how proteins are targeted for degradation in archaea promises to be an exciting area of research. Sampylation is likely a simplified system for the covalent marking proteins, which has close evolutionary roots to ubiquitylation and other related pathways that play an important role in eukaryotic cell function. Interestingly, more elaborate protein conjugation mechanisms than sampylation may exist in a select group of bacteria and archaea that harbour E1, E2 and E3 homologs, based on recent metagenomic sequencing and comparative genomics [144, 158].

Acknowledgments This work was funded in part by grants from the National Institutes of Health (GM57498) and the Department of Energy Office of Basic Energy Sciences (DE-FG02-05ER15650).

References

1. Gottesman S (1999) Regulation by proteolysis: developmental switches. Curr Opin Microbiol 2(2):142–147
2. Wolf DH, Hilt W (2004) The proteasome: a proteolytic nanomachine of cell regulation and waste disposal. Biochim Biophys Acta 1695(1–3):19–31
3. Lupas A, Flanagan JM, Tamura T, Baumeister W (1997) Self-compartmentalizing proteases. Trends Biochem Sci 22(10):399–404
4. Cha SS, An YJ, Lee CR, Lee HS et al (2010) Crystal structure of Lon protease: molecular architecture of gated entry to a sequestered degradation chamber. EMBO J 29(20):3520–3530
5. Effantin G, Maurizi MR, Steven AC (2010) Binding of the ClpA unfoldase opens the axial gate of ClpP peptidase. J Biol Chem 285(19):14834–14840
6. Lee ME, Baker TA, Sauer RT (2010) Control of substrate gating and translocation into ClpP by channel residues and ClpX binding. J Mol Biol 399(5):707–718
7. Smith DM, Benaroudj N, Goldberg A (2006) Proteasomes and their associated ATPases: a destructive combination. J Struct Biol 156(1):72–83
8. Ammelburg M, Frickey T, Lupas AN (2006) Classification of AAA+proteins. J Struct Biol 156(1):2–11
9. Neuwald AF, Aravind L, Spouge JL, Koonin EV (1999) AAA+: a class of chaperone-like ATPases associated with the assembly, operation, and disassembly of protein complexes. Genome Res 9(1):27–43
10. Sauer RT, Baker TA (2011) AAA+ proteases: ATP-fueled machines of protein destruction. Annu Rev Biochem 80:587–612

11. Gur E, Ottofuelling R, Dougan DA (2013) Machines of destruction – AAA+proteases and the adaptors that control them. In: Dougan DA (ed) Regulated proteolysis in microorganisms. Springer, Subcell Biochem 66:3–33
12. Samanovic M, Li H, Darwin KH (2013) The Pup-Proteasome system of Mycobacterium tuberculosis. In: Dougan DA (ed) Regulated proteolysis in microorganisms. Springer, Subcell Biochem 66:267–295
13. Maupin-Furlow JA, Humbard MA, Kirkland PA, Li W et al (2006) Proteasomes from structure to function: perspectives from Archaea. Curr Top Dev Biol 75:125–169
14. Bar-Nun S, Glickman MH (2012) Proteasomal AAA-ATPases: structure and function. Biochim Biophys Acta 1823(1):67–82
15. Besche H, Tamura N, Tamura T, Zwickl P (2004) Mutational analysis of conserved AAA+ residues in the archaeal Lon protease from Thermoplasma acidophilum. FEBS Lett 574(1–3):161–166
16. Fukui T, Eguchi T, Atomi H, Imanaka T (2002) A membrane-bound archaeal Lon protease displays ATP-independent proteolytic activity towards unfolded proteins and ATP-dependent activity for folded proteins. J Bacteriol 184(13):3689–3698
17. De Mot R, Nagy I, Walz J, Baumeister W (1999) Proteasomes and other self-compartmentalizing proteases in prokaryotes. Trends Microbiol 7(2):88–92
18. Lupas A, Zuhl F, Tamura T, Wolf S et al (1997) Eubacterial proteasomes. Mol Biol Rep 24(1–2):125–131
19. Lupas A, Zwickl P, Baumeister W (1994) Proteasome sequences in eubacteria. Trends Biochem Sci 19(12):533–534
20. Tamura T, Nagy I, Lupas A, Lottspeich F et al (1995) The first characterization of a eubacterial proteasome: the 20S complex of Rhodococcus. Curr Biol 5(7):766–774
21. Bochtler M, Ditzel L, Groll M, Hartmann C et al (1999) The proteasome. Annu Rev Biophys Biomol Struct 28:295–317
22. Voges D, Zwickl P, Baumeister W (1999) The 26S proteasome: a molecular machine designed for controlled proteolysis. Annu Rev Biochem 68:1015–1068
23. Larsen CN, Finley D (1997) Protein translocation channels in the proteasome and other proteases. Cell 91(4):431–434
24. Brannigan JA, Dodson G, Duggleby HJ, Moody PC et al (1995) A protein catalytic framework with an N-terminal nucleophile is capable of self-activation. Nature 378(6555):416–419
25. Ekici OD, Paetzel M, Dalbey RE (2008) Unconventional serine proteases: variations on the catalytic Ser/His/Asp triad configuration. Protein Sci 17(12):2023–2037
26. Oinonen C, Rouvinen J (2000) Structural comparison of Ntn-hydrolases. Protein Sci 9(12):2329–2337
27. Seemuller E, Lupas A, Stock D, Lowe J et al (1995) Proteasome from Thermoplasma acidophilum: a threonine protease. Science 268(5210):579–582
28. Dodson G, Wlodawer A (1998) Catalytic triads and their relatives. Trends Biochem Sci 23(9):347–352
29. Kisselev AF, Songyang Z, Goldberg AL (2000) Why does threonine, and not serine, function as the active site nucleophile in proteasomes? J Biol Chem 275(20):14831–14837
30. Maupin-Furlow JA, Aldrich HC, Ferry JG (1998) Biochemical characterization of the 20S proteasome from the methanoarchaeon Methanosarcina thermophila. J Bacteriol 180(6):1480–1487
31. Zwickl P, Kleinz J, Baumeister W (1994) Critical elements in proteasome assembly. Nat Struct Biol 1(11):765–770
32. Groll M, Heinemeyer W, Jager S, Ullrich T et al (1999) The catalytic sites of 20S proteasomes and their role in subunit maturation: a mutational and crystallographic study. Proc Natl Acad Sci U S A 96(20):10976–10983
33. Jager S, Groll M, Huber R, Wolf DH et al (1999) Proteasome beta-type subunits: unequal roles of propeptides in core particle maturation and a hierarchy of active site function. J Mol Biol 291(4):997–1013
34. Bauer MW, Bauer SH, Kelly RM (1997) Purification and characterization of a proteasome from the hyperthermophilic archaeon Pyrococcus furiosus. Appl Environ Microbiol 63(3):1160–1164

35. Dahlmann B, Kopp F, Kuehn L, Niedel B et al (1989) The multicatalytic proteinase (prosome) is ubiquitous from eukaryotes to archaebacteria. FEBS Lett 251(1–2):125–131
36. Maupin-Furlow JA, Ferry JG (1995) A proteasome from the methanogenic archaeon Methanosarcina thermophila. J Biol Chem 270(48):28617–28622
37. Wilson HL, Aldrich HC, Maupin-Furlow J (1999) Halophilic 20S proteasomes of the archaeon Haloferax volcanii: purification, characterization, and gene sequence analysis. J Bacteriol 181(18):5814–5824
38. Arendt CS, Hochstrasser M (1999) Eukaryotic 20S proteasome catalytic subunit propeptides prevent active site inactivation by N-terminal acetylation and promote particle assembly. EMBO J 18(13):3575–3585
39. Kwon YD, Nagy I, Adams PD, Baumeister W et al (2004) Crystal structures of the Rhodococcus proteasome with and without its pro-peptides: implications for the role of the pro-peptide in proteasome assembly. J Mol Biol 335(1):233–245
40. Zuhl F, Seemuller E, Golbik R, Baumeister W (1997) Dissecting the assembly pathway of the 20S proteasome. FEBS Lett 418(1–2):189–194
41. Li D, Li H, Wang T, Pan H et al (2010) Structural basis for the assembly and gate closure mechanisms of the Mycobacterium tuberculosis 20S proteasome. EMBO J 29(12): 2037–2047
42. Lin G, Hu G, Tsu C, Kunes YZ et al (2006) Mycobacterium tuberculosis prcBA genes encode a gated proteasome with broad oligopeptide specificity. Mol Microbiol 59(5):1405–1416
43. Wilson HL, Ou MS, Aldrich HC, Maupin-Furlow J (2000) Biochemical and physical properties of the Methanococcus jannaschii 20S proteasome and PAN, a homolog of the ATPase (Rpt) subunits of the eucaryal 26S proteasome. J Bacteriol 182(6):1680–1692
44. Kaczowka SJ, Maupin-Furlow JA (2003) Subunit topology of two 20S proteasomes from Haloferax volcanii. J Bacteriol 185(1):165–174
45. Ginzburg M, Sachs L, Ginzburg BZ (1970) Ion metabolism in a Halobacterium. I. Influence of age of culture on intracellular concentrations. J Gen Physiol 55(2):187–207
46. Mevarech M, Frolow F, Gloss LM (2000) Halophilic enzymes: proteins with a grain of salt. Biophys Chem 86(2–3):155–164
47. Falb M, Aivaliotis M, Garcia-Rizo C, Bisle B et al (2006) Archaeal N-terminal protein maturation commonly involves N-terminal acetylation: a large-scale proteomics survey. J Mol Biol 362(5):915–924
48. Kirkland PA, Humbard MA, Daniels CJ, Maupin-Furlow JA (2008) Shotgun proteomics of the haloarchaeon Haloferax volcanii. J Proteome Res 7(11):5033–5039
49. Soppa J (2010) Protein acetylation in archaea, bacteria, and eukaryotes. Archaea 2010 (Article ID 820681): 9. doi:10.1155/2010/820681
50. Kisselev AF, Akopian TN, Goldberg AL (1998) Range of sizes of peptide products generated during degradation of different proteins by archaeal proteasomes. J Biol Chem 273(4):1982–1989
51. Kisselev AF, Akopian TN, Woo KM, Goldberg AL (1999) The sizes of peptides generated from protein by mammalian 26 and 20S proteasomes. Implications for understanding the degradative mechanism and antigen presentation. J Biol Chem 274(6):3363–3371
52. Gallastegui N, Groll M (2010) The 26S proteasome: assembly and function of a destructive machine. Trends Biochem Sci 35(11):634–642
53. Vigneron N, Van den Eynde BJ (2012) Proteasome subtypes and the processing of tumor antigens: increasing antigenic diversity. Curr Opin Immunol 24(1):84–91
54. Zwickl P, Grziwa A, Puhler G, Dahlmann B et al (1992) Primary structure of the Thermoplasma proteasome and its implications for the structure, function, and evolution of the multicatalytic proteinase. Biochemistry 31(4):964–972
55. Zwickl P, Lottspeich F, Baumeister W (1992) Expression of functional Thermoplasma acidophilum proteasomes in Escherichia coli. FEBS Lett 312(2–3):157–160
56. Kocabiyik S, Ozdemir I, Zwickl P, Ozdogan S (2010) Molecular cloning and co-expression of Thermoplasma volcanium proteasome subunit genes. Protein Expr Purif 73(2):223–230
57. Medalia N, Sharon M, Martinez-Arias R, Mihalache O et al (2006) Functional and structural characterization of the Methanosarcina mazei proteasome and PAN complexes. J Struct Biol 156(1):84–92

58. Groll M, Brandstetter H, Bartunik H, Bourenkow G et al (2003) Investigations on the maturation and regulation of archaebacterial proteasomes. J Mol Biol 327(1):75–83
59. Robb FT, Maeder DL, Brown JR, DiRuggiero J et al (2001) Genomic sequence of hyperthermophile, Pyrococcus furiosus: implications for physiology and enzymology. Methods Enzymol 330:134–157
60. Madding LS, Michel JK, Shockley KR, Conners SB et al (2007) Role of the beta1 subunit in the function and stability of the 20S proteasome in the hyperthermophilic archaeon Pyrococcus furiosus. J Bacteriol 189(2):583–590
61. Zhou G, Kowalczyk D, Humbard MA, Rohatgi S et al (2008) Proteasomal components required for cell growth and stress responses in the haloarchaeon Haloferax volcanii. J Bacteriol 190(24):8096–8105
62. Karadzic I, Maupin-Furlow J, Humbard M, Prunetti L, Singh P, Goodlett DR (2012) Chemical cross-linking, mass spectrometry, and in silico modeling of proteasomal 20S core particles of the haloarchaeon Haloferax volcanii. Proteomics 12(11):1806–1814
63. Reuter CJ, Kaczowka SJ, Maupin-Furlow JA (2004) Differential regulation of the PanA and PanB proteasome-activating nucleotidase and 20S proteasomal proteins of the haloarchaeon Haloferax volcanii. J Bacteriol 186(22):7763–7772
64. Humbard MA, Reuter CJ, Zuobi-Hasona K, Zhou G et al (2010) Phosphorylation and methylation of proteasomal proteins of the haloarcheon Haloferax volcanii. Archaea 2010:481725
65. Humbard MA, Stevens SM Jr, Maupin-Furlow JA (2006) Posttranslational modification of the 20S proteasomal proteins of the archaeon Haloferax volcanii. J Bacteriol 188(21):7521–7530
66. Humbard MA, Zhou G, Maupin-Furlow JA (2009) The N-terminal penultimate residue of 20S proteasome alpha1 influences its N(alpha) acetylation and protein levels as well as growth rate and stress responses of Haloferax volcanii. J Bacteriol 191(12):3794–3803
67. Stadtmueller BM, Hill CP (2011) Proteasome activators. Mol Cell 41(1):8–19
68. Finley D, Tanaka K, Mann C, Feldmann H et al (1998) Unified nomenclature for subunits of the Saccharomyces cerevisiae proteasome regulatory particle. Trends Biochem Sci 23(7):244–245
69. Glickman MH, Rubin DM, Coux O, Wefes I et al (1998) A subcomplex of the proteasome regulatory particle required for ubiquitin-conjugate degradation and related to the COP9-signalosome and eIF3. Cell 94(5):615–623
70. Bult CJ, White O, Olsen GJ, Zhou L et al (1996) Complete genome sequence of the methanogenic archaeon, Methanococcus jannaschii. Science 273(5278):1058–1073
71. Zwickl P, Ng D, Woo KM, Klenk HP et al (1999) An archaebacterial ATPase, homologous to ATPases in the eukaryotic 26S proteasome, activates protein breakdown by 20S proteasomes. J Biol Chem 274(37):26008–26014
72. Stolz A, Hilt W, Buchberger A, Wolf DH (2011) Cdc48: a power machine in protein degradation. Trends Biochem Sci 36(10):515–523
73. Buchberger A (2013) Roles of Cdc48 in regulated protein degradation in yeast. In: Dougan DA (ed) Regulated proteolysis in microorganisms. Springer, Subcell Biochem 66:3–195–222
74. Gerega A, Rockel B, Peters J, Tamura T et al (2005) VAT, the thermoplasma homolog of mammalian p97/VCP, is an N domain-regulated protein unfoldase. J Biol Chem 280(52):42856–42862
75. Barthelme D, Sauer RT (2012) Identification of the Cdc48•20S proteasome as an ancient AAA+ proteolytic machine. Science 337(6096):843–846
76. Lowe J, Stock D, Jap B, Zwickl P et al (1995) Crystal structure of the 20S proteasome from the archaeon T. acidophilum at 3.4 A resolution. Science 268(5210):533–539
77. Groll M, Ditzel L, Lowe J, Stock D et al (1997) Structure of 20S proteasome from yeast at 2.4 A resolution. Nature 386(6624):463–471
78. Unno M, Mizushima T, Morimoto Y, Tomisugi Y et al (2002) Structure determination of the constitutive 20S proteasome from bovine liver at 2.75 A resolution. J Biochem 131(2):171–173
79. Groll M, Bajorek M, Kohler A, Moroder L et al (2000) A gated channel into the proteasome core particle. Nat Struct Biol 7(11):1062–1067
80. Kohler A, Cascio P, Leggett DS, Woo KM et al (2001) The axial channel of the proteasome core particle is gated by the Rpt2 ATPase and controls both substrate entry and product release. Mol Cell 7(6):1143–1152

81. Whitby FG, Masters EI, Kramer L, Knowlton JR et al (2000) Structural basis for the activation of 20S proteasomes by 11S regulators. Nature 408(6808):115–120
82. Benaroudj N, Zwickl P, Seemuller E, Baumeister W et al (2003) ATP hydrolysis by the proteasome regulatory complex PAN serves multiple functions in protein degradation. Mol Cell 11(1):69–78
83. Smith DM, Kafri G, Cheng Y, Ng D et al (2005) ATP binding to PAN or the 26S ATPases causes association with the 20S proteasome, gate opening, and translocation of unfolded proteins. Mol Cell 20(5):687–698
84. Zhang F, Hu M, Tian G, Zhang P et al (2009) Structural insights into the regulatory particle of the proteasome from Methanocaldococcus jannaschii. Mol Cell 34(4):473–484
85. Rabl J, Smith DM, Yu Y, Chang SC et al (2008) Mechanism of gate opening in the 20S proteasome by the proteasomal ATPases. Mol Cell 30(3):360–368
86. Yu Y, Smith DM, Kim HM, Rodriguez V et al (2010) Interactions of PAN's C-termini with archaeal 20S proteasome and implications for the eukaryotic proteasome-ATPase interactions. EMBO J 29(3):692–702
87. Religa TL, Sprangers R, Kay LE (2010) Dynamic regulation of archaeal proteasome gate opening as studied by TROSY NMR. Science 328(5974):98–102
88. Bochtler M, Hartmann C, Song HK, Bourenkov GP et al (2000) The structures of HslU and the ATP-dependent protease HslU-HslV. Nature 403(6771):800–805
89. Rohrwild M, Pfeifer G, Santarius U, Muller SA et al (1997) The ATP-dependent HslVU protease from Escherichia coli is a four-ring structure resembling the proteasome. Nat Struct Biol 4(2):133–139
90. Sousa MC, Trame CB, Tsuruta H, Wilbanks SM et al (2000) Crystal and solution structures of an HslUV protease-chaperone complex. Cell 103(4):633–643
91. Smith DM, Chang SC, Park S, Finley D et al (2007) Docking of the proteasomal ATPases' carboxyl termini in the 20S proteasome's alpha ring opens the gate for substrate entry. Mol Cell 27(5):731–744
92. Ramachandran R, Hartmann C, Song HK, Huber R et al (2002) Functional interactions of HslV (ClpQ) with the ATPase HslU (ClpY). Proc Natl Acad Sci U S A 99(11):7396–7401
93. Ma CP, Willy PJ, Slaughter CA, DeMartino GN (1993) PA28, an activator of the 20S proteasome, is inactivated by proteolytic modification at its carboxyl terminus. J Biol Chem 268(30):22514–22519
94. Forster A, Masters EI, Whitby FG, Robinson H et al (2005) The 1.9 A structure of a proteasome-11S activator complex and implications for proteasome-PAN/PA700 interactions. Mol Cell 18(5):589–599
95. Forster A, Whitby FG, Hill CP (2003) The pore of activated 20S proteasomes has an ordered 7-fold symmetric conformation. EMBO J 22(17):4356–4364
96. Zhang F, Wu Z, Zhang P, Tian G et al (2009) Mechanism of substrate unfolding and translocation by the regulatory particle of the proteasome from Methanocaldococcus jannaschii. Mol Cell 34(4):485–496
97. Djuranovic S, Hartmann MD, Habeck M, Ursinus A et al (2009) Structure and activity of the N-terminal substrate recognition domains in proteasomal ATPases. Mol Cell 34(5):580–590
98. Medalia N, Beer A, Zwickl P, Mihalache O et al (2009) Architecture and molecular mechanism of PAN, the archaeal proteasome regulatory ATPase. J Biol Chem 284(34):22952–22960
99. Navon A, Goldberg AL (2001) Proteins are unfolded on the surface of the ATPase ring before transport into the proteasome. Mol Cell 8(6):1339–1349
100. Matouschek A (2003) Protein unfolding – an important process in vivo? Curr Opin Struct Biol 13(1):98–109
101. Hinnerwisch J, Fenton WA, Furtak KJ, Farr GW et al (2005) Loops in the central channel of ClpA chaperone mediate protein binding, unfolding, and translocation. Cell 121(7):1029–1041
102. Inobe T, Kraut DA, Matouschek A (2008) How to pick a protein and pull at it. Nat Struct Mol Biol 15(11):1135–1136
103. Martin A, Baker TA, Sauer RT (2008) Pore loops of the AAA+ClpX machine grip substrates to drive translocation and unfolding. Nat Struct Mol Biol 15(11):1147–1151

104. Martin A, Baker TA, Sauer RT (2008) Diverse pore loops of the AAA+ClpX machine mediate unassisted and adaptor-dependent recognition of ssrA-tagged substrates. Mol Cell 29(4):441–450
105. Wang J, Song JJ, Seong IS, Franklin MC et al (2001) Nucleotide-dependent conformational changes in a protease-associated ATPase HslU. Structure 9(11):1107–1116
106. Smith DM, Fraga H, Reis C, Kafri G et al (2011) ATP binds to proteasomal ATPases in pairs with distinct functional effects, implying an ordered reaction cycle. Cell 144(4):526–538
107. Ehlers C, Kopp F, Dahlmann B (1997) Screening for molecules interacting with proteasomes in Thermoplasma acidophilum. Biol Chem 378(3–4):249–253
108. Kusmierczyk AR, Kunjappu MJ, Kim RY, Hochstrasser M (2011) A conserved 20S proteasome assembly factor requires a C-terminal HbYX motif for proteasomal precursor binding. Nat Struct Mol Biol 18(5):622–629
109. Hirano Y, Hendil KB, Yashiroda H, Iemura S et al (2005) A heterodimeric complex that promotes the assembly of mammalian 20S proteasomes. Nature 437(7063):1381–1385
110. Le Tallec B, Barrault MB, Courbeyrette R, Guerois R et al (2007) 20S proteasome assembly is orchestrated by two distinct pairs of chaperones in yeast and in mammals. Mol Cell 27(4):660–674
111. Li X, Kusmierczyk AR, Wong P, Emili A et al (2007) beta-Subunit appendages promote 20S proteasome assembly by overcoming an Ump1-dependent checkpoint. EMBO J 26(9):2339–2349
112. Hochstrasser M (2009) Origin and function of ubiquitin-like proteins. Nature 458(7237):422–429
113. Iwai K, Tokunaga F (2009) Linear polyubiquitination: a new regulator of NF-kappaB activation. EMBO Rep 10(7):706–713
114. Wang X, Herr RA, Hansen TH (2012) Ubiquitination of substrates by esterification. Traffic 13(1):19–24
115. Frappier L, Verrijzer CP (2011) Gene expression control by protein deubiquitinases. Curr Opin Genet Dev 21(2):207–213
116. Hicke L, Dunn R (2003) Regulation of membrane protein transport by ubiquitin and ubiquitin-binding proteins. Annu Rev Cell Dev Biol 19:141–172
117. Ramaekers CH, Wouters BG (2011) Regulatory functions of ubiquitin in diverse DNA damage responses. Curr Mol Med 11(2):152–169
118. Xu P, Duong DM, Seyfried NT, Cheng D et al (2009) Quantitative proteomics reveals the function of unconventional ubiquitin chains in proteasomal degradation. Cell 137(1):133–145
119. Lauwers E, Erpapazoglou Z, Haguenauer-Tsapis R, Andre B (2010) The ubiquitin code of yeast permease trafficking. Trends Cell Biol 20(4):196–204
120. Pickart CM (1997) Targeting of substrates to the 26S proteasome. FASEB J 11(13):1055–1066
121. Metzger MB, Weissman AM (2010) Working on a chain: E3s ganging up for ubiquitylation. Nat Cell Biol 12(12):1124–1126
122. Pastore A (2010) Further insights into the ubiquitin pathway: understanding the scarlet letter code. Structure 18(8):891–892
123. Scheffner M, Nuber U, Huibregtse JM (1995) Protein ubiquitination involving an E1-E2-E3 enzyme ubiquitin thioester cascade. Nature 373(6509):81–83
124. Ye Y, Rape M (2009) Building ubiquitin chains: E2 enzymes at work. Nat Rev Mol Cell Biol 10(11):755–764
125. Hochstrasser M (2007) Ubiquitin ligation without a ligase. Dev Cell 13(1):4–6
126. Hoeller D, Hecker CM, Wagner S, Rogov V et al (2007) E3-independent monoubiquitination of ubiquitin-binding proteins. Mol Cell 26(6):891–898
127. Sorkin A (2007) Ubiquitination without E3. Mol Cell 26(6):771–773
128. Iyer LM, Burroughs AM, Aravind L (2006) The prokaryotic antecedents of the ubiquitin-signaling system and the early evolution of ubiquitin-like beta-grasp domains. Genome Biol 7(7):R60
129. Jurgenson CT, Begley TP, Ealick SE (2009) The structural and biochemical foundations of thiamin biosynthesis. Annu Rev Biochem 78:569–603
130. Schwarz G, Mendel RR, Ribbe MW (2009) Molybdenum cofactors, enzymes and pathways. Nature 460(7257):839–847

131. Godert AM, Jin M, McLafferty FW, Begley TP (2007) Biosynthesis of the thioquinolobactin siderophore: an interesting variation on sulfur transfer. J Bacteriol 189(7):2941–2944
132. Jurgenson CT, Burns KE, Begley TP, Ealick SE (2008) Crystal structure of a sulfur carrier protein complex found in the cysteine biosynthetic pathway of Mycobacterium tuberculosis. Biochemistry 47(39):10354–10364
133. O'Leary SE, Jurgenson CT, Ealick SE, Begley TP (2008) O-phospho-L-serine and the thiocarboxylated sulfur carrier protein CysO-COSH are substrates for CysM, a cysteine synthase from Mycobacterium tuberculosis. Biochemistry 47(44):11606–11615
134. Burroughs AM, Balaji S, Iyer LM, Aravind L (2007) Small but versatile: the extraordinary functional and structural diversity of the beta-grasp fold. Biol Direct 2:18
135. Burroughs AM, Iyer LM, Aravind L (2009) Natural history of the E1-like superfamily: implication for adenylation, sulfur transfer, and ubiquitin conjugation. Proteins 75(4):895–910
136. Lehmann C, Begley TP, Ealick SE (2006) Structure of the Escherichia coli ThiS-ThiF complex, a key component of the sulfur transfer system in thiamin biosynthesis. Biochemistry 45(1):11–19
137. Schmitz J, Wuebbens MM, Rajagopalan KV, Leimkuhler S (2007) Role of the C-terminal Gly-Gly motif of Escherichia coli MoaD, a molybdenum cofactor biosynthesis protein with a ubiquitin fold. Biochemistry 46(3):909–916
138. Dahl JU, Urban A, Bolte A, Sriyabhaya P et al (2011) The identification of a novel protein involved in molybdenum cofactor biosynthesis in Escherichia coli. J Biol Chem 286(41):35801–35812
139. Zhang W, Urban A, Mihara H, Leimkuhler S et al (2010) IscS functions as a primary sulfur-donating enzyme by interacting specifically with MoeB and MoaD in the biosynthesis of molybdopterin in Escherichia coli. J Biol Chem 285(4):2302–2308
140. Mueller EG (2006) Trafficking in persulfides: delivering sulfur in biosynthetic pathways. Nat Chem Biol 2(4):185–194
141. Humbard MA, Miranda HV, Lim JM, Krause DJ et al (2010) Ubiquitin-like small archaeal modifier proteins (SAMPs) in Haloferax volcanii. Nature 463(7277):54–60
142. Miranda HV, Nembhard N, Su D, Hepowit N et al (2011) E1- and ubiquitin-like proteins provide a direct link between protein conjugation and sulfur transfer in archaea. Proc Natl Acad Sci U S A 108(11):4417–4422
143. Zhang Y, Gladyshev VN (2008) Molybdoproteomes and evolution of molybdenum utilization. J Mol Biol 379(4):881–899
144. Burroughs AM, Iyer LM, Aravind L (2011) Functional diversification of the RING finger and other binuclear treble clef domains in prokaryotes and the early evolution of the ubiquitin system. Mol Biosyst 7(7):2261–2277
145. Hartman AL, Norais C, Badger JH, Delmas S et al (2010) The complete genome sequence of Haloferax volcanii DS2, a model archaeon. PLoS One 5(3):e9605
146. Makarova KS, Koonin EV (2010) Archaeal ubiquitin-like proteins: functional versatility and putative ancestral involvement in tRNA modification revealed by comparative genomic analysis. Archaea 2010. doi: pii: 710303
147. Wagner SA, Beli P, Weinert BT, Nielsen ML et al (2011) A proteome-wide, quantitative survey of in vivo ubiquitylation sites reveals widespread regulatory roles. Mol Cell Proteomics 10(10):M111 013284
148. Yang XJ (2005) Multisite protein modification and intramolecular signaling. Oncogene 24(10):1653–1662
149. Baker RT, Board PG (1987) The human ubiquitin gene family: structure of a gene and pseudogenes from the Ub B subfamily. Nucleic Acids Res 15(2):443–463
150. Lund PK, Moats-Staats BM, Simmons JG, Hoyt E et al (1985) Nucleotide sequence analysis of a cDNA encoding human ubiquitin reveals that ubiquitin is synthesized as a precursor. J Biol Chem 260(12):7609–7613

151. Wiborg O, Pedersen MS, Wind A, Berglund LE et al (1985) The human ubiquitin multigene family: some genes contain multiple directly repeated ubiquitin coding sequences. EMBO J 4(3):755–759
152. Ozkaynak E, Finley D, Solomon MJ, Varshavsky A (1987) The yeast ubiquitin genes: a family of natural gene fusions. EMBO J 6(5):1429–1439
153. Katz EJ, Isasa M, Crosas B (2010) A new map to understand deubiquitination. Biochem Soc Trans 38(Pt 1):21–28
154. Kirkland PA, Gil MA, Karadzic IM, Maupin-Furlow JA (2008) Genetic and proteomic analyses of a proteasome-activating nucleotidase A mutant of the haloarchaeon Haloferax volcanii. J Bacteriol 190(1):193–205
155. Kirkland PA, Reuter CJ, Maupin-Furlow JA (2007) Effect of proteasome inhibitor clasto-lactacystin-beta-lactone on the proteome of the haloarchaeon Haloferax volcanii. Microbiology 153(Pt 7):2271–2280
156. Leidel S, Pedrioli PG, Bucher T, Brost R et al (2009) Ubiquitin-related modifier Urm1 acts as a sulphur carrier in thiolation of eukaryotic transfer RNA. Nature 458(7235):228–232
157. Shigi N, Sakaguchi Y, Asai S, Suzuki T et al (2008) Common thiolation mechanism in the biosynthesis of tRNA thiouridine and sulphur-containing cofactors. EMBO J 27(24):3267–3278
158. Nunoura T, Takaki Y, Kakuta J, Nishi S et al (2011) Insights into the evolution of Archaea and eukaryotic protein modifier systems revealed by the genome of a novel archaeal group. Nucleic Acids Res 39(8):3204–3223

Index

D.A. Dougan (ed.), *Regulated Proteolysis in Microorganisms*, Subcellular Biochemistry 66,
DOI 10.1007/978-94-007-5940-4,

MIX
Papier aus verantwortungsvollen Quellen
Paper from responsible sources
FSC® C105338

If you have any concerns about our products,
you can contact us on
ProductSafety@springernature.com

In case Publisher is established outside the EU,
the EU authorized representative is:
Springer Nature Customer Service Center GmbH
Europaplatz 3, 69115 Heidelberg, Germany

Printed by Libri Plureos GmbH
in Hamburg, Germany